Lecture Notes in Computer Science 3875

Commenced Publication in 1973
Founding and Former Series Editors:
Gerhard Goos, Juris Hartmanis, and Jan van Leeuwen

Shmuel Ur   Eyal Bin
Yaron Wolfsthal (Eds.)

Hardware and Software Verification and Testing

First International Haifa Verification Conference
Haifa, Israel, November 13-16, 2005
Revised Selected Papers

 Springer

Volume Editors

Shmuel Ur
Eyal Bin
Yaron Wolfsthal
IBM Labs, Haifa University, Mount Carmel, Haifa 31905, Israel
E-mail: {ur,bin,wolfstal}@il.ibm.com

Library of Congress Control Number: Applied for

CR Subject Classification (1998): D.2.4-5, D.2, D.3, F.3

LNCS Sublibrary: SL 2 – Programming and Software Engineering

ISSN   0302-9743
ISBN   3-540-32604-9 Springer Berlin Heidelberg New York
ISBN   978-3-540-32604-5 Springer Berlin Heidelberg New York

Springer is a part of Springer Science+Business Media

springer.com

© Springer-Verlag Berlin Heidelberg 2006

Typesetting: Camera-ready by author, data conversion by Scientific Publishing Services, Chennai, India
Printed on acid-free paper   SPIN: 11678779   06/3142   5 4 3 2 1 0

Preface

The First Haifa Verification Conference was held at the IBM Haifa Research Lab and at the Haifa University in Israel from November 13 to16, 2005. The conference incorporated three different workshops that took place separately in previous years. The IBM Verification Workshop is now its sixth year, the IBM Software Testing Workshop is now in its fourth year, and the PADTAD Workshop on testing and debugging multi-threaded and parallel software was held for the third time. The Verification Conference was a three-day, single-track conference followed by a one-day tutorial on the testing and review of multi-threaded code.

The conference presented a unique combination of fields that brought together the hardware and software testing communities. Merging the different communities under a single roof gave the conference a distinctive flavor and provided the participants with added benefits. While the applications in these separate fields are different, the techniques used are often very similar. By offering lectures in these disparate but related disciplines, the conference engendered an environment of collaboration and discovery.

The conference emphasized applicability to real world challenges, which was very important to the many attendees coming from industry. A relatively large number of invited speakers presented topics of great interest to the audience. These outstanding speakers included Sharad Malik, Amir Pnueli, Cindy Eisner, Yoav Katz, Yoav Hollander, and Michael Rosenfield on hardware verification, Thomas Wolf, Scott Stoller, Dan Quinlan, Yael Dubinsky, Orit Hazzan, Bernd Finkbeiner, and Thomas Ball on software verification, and Scott Stoller and Dan Quinlan for the PADTAD track.

Thirty-one papers, from 11 countries, were submitted and thoroughly reviewed by the Program Committee and additional referees, with an average of 3.5 reviews per paper. Of the papers submitted, fourteen were accepted. The acceptance was based on the score received and in the case of an equal score, industry papers were given preference. Some of the invited speakers were also encouraged to submit papers. This volume is composed of the 14 papers accepted by the committee and three invited papers.

Attendance at the conference was very high, with more than 170 participants (225 registered) from 12 countries (Israel, USA, Germany, Switzerland, Brazil, Finland, India, Canada, Denmark, Bulgaria, Italy, Portugal). The facilities provided by the IBM Haifa Research Labs and the Caesarea Edmond Benjamin de Rothschild Foundation Institute for Interdisciplinary Applications of Computer Science (C.R.I.) were remarked upon very favorably by the attendees, as was the proficiency of the administrative assistants.

We would like to thank our sponsors, IBM and CRI, the Organizing Committee, and the Program Committee. Our appreciation goes out to the administrative assistants, especially Vered Aharon and Efrat Shalom-Hillman from IBM and Ornit Bar-Or from CRI, and to Yair Harry our web master. We would also like to thank all the authors who contributed their work. Special thanks go to Yaniv Eytani, the Local Arrangement Chair, for his boundless energy in working with the invited speakers.

We hope this conference is the first of many and that these communities continue to work together and learn from each other. There is a lot of knowledge and insight that can be shared by the hardware verification and testing communities, and we hope this conference represents a step in this important direction.

November 2005   Shmuel Ur
                                                                    Program Chair

Organization

The Haifa Verification Conference was organized by:

Program Chair
Shmuel Ur, IBM Haifa Labs, Israel (ur@il.ibm.com)

Verification Conference Organizing Committee
Shmuel Ur (ur@il.ibm.com)
Eyal Bin (bin@il.ibm.com)
Eitan Farchi (farchi@il.ibm.com)
Yaron Wolfsthal (wolfstal@il.ibm.com)
Avi Ziv (aziv@il.ibm.com)

Verification Track Chair
Avi Ziv, IBM Haifa Labs, Israel (aziv@il.ibm.com)

Software Testing Track Chair
Eitan Farchi, IBM Haifa Labs, Israel (farchi@il.ibm.com)

PADTAD Chair
Shmuel Ur, IBM Haifa Labs, Israel (ur@il.ibm.com)

Proceedings Chair
Tsvi Kuflik, University of Haifa, Israel (tsvikak@mis.hevra.haifa.ac.il)

Local Arrangements Chair
Yaniv Eytani, University of Haifa, Israel (ieytani@cslx.haifa.ac.il)

Program Committee
Jose Nelson Amaral, University of Alberta, Canada (amaral@cs.ualberta.ca)
Yosi Ben-Asher, University of Haifa, Israel (yosi@cs.haifa.ac.il)
Valeria Bertachio, Department of Electrical Engineering and Computer Science, University of Michigan, USA (valeria@unich.edu)
Angelos Bilas, University of Toronto, Canada (bilas@eecg.toronto.edu)
Eyal Bin, IBM Haifa Labs, Israel (bin@il.ibm.com)
Roderick Bloem, Graz University of Technology, Austria (Roderick.Bloem@ist.TUGratz.at)

Table of Contents

Hardware Verification

Software Testing

PADTAD

Path-Based System Level Stimuli Generation

Shady Copty, Itai Jaeger, and Yoav Katz

IBM Research Lab in Haifa
{shady, itaij, katz}@il.ibm.com

Abstract. Over the last few years, there has been increasing emphasis on integrating ready-made components (IP, cores) into complex System on a Chip
(SoC) designs. The verification of such designs poses new challenges. At the
heart of these challenges lies the requirement to verify the integration of several
previously designed components in a relatively short time. Simulation-based
methods are the main verification vehicle used for system-level functional verification of SoC designs; therefore, stimuli generation plays an important role in
this field.

Our work offers a solution for efficiently dealing with the verification of
systems with multiple configurations and derivative systems, a common challenge in the context of system verification. We present a generation scheme in
which the system behavior is defined using a combination of transaction-based
modeling, local component behavior, and the topology of the system. We show
how this approach allows the implementation of the verification plan using
high level constructs and promotes the reuse of verification IP between
systems.

The ideas described below were implemented as part of X-Gen, a system-
level test-case generator developed and used in IBM.

1 Introduction

Verification consumes up to 70% of the design effort in today's multi-million gate
ASICs, reusable Intellectual Property (IP), and System-on-a-Chip (SoC) designs
[1]. A variety of approaches are used to ensure that all bugs are caught before the
design is cast in silicon. Verification is usually performed from the bottom up, starting with unit verification and chip verification and following through to system level
verification.

The objective of system level verification is to verify the integration of previously
verified components [2]. In system verification, the verification challenges faced in
unit and chip verification are intensified, due to the increasing complexity of the system. As system verification is done at a later stage in the project, tight schedules and
resource limitations are common. Careful formulation and efficient execution of the
verification plan are crucial for success.

Simulation-based verification methods verify the system behavior by simulating
the HDL code and driving stimuli into the design. They use either a reference model
or alternative checking mechanisms to verify that the design is operating as required.
The verification plan must identify the major risk areas of integration and devise scenarios to identify errors in these areas. Capturing the essence of the system, these
scenarios typically include interactions between IPs [3].

S. Ur, E. Bin, and Y. Wolfsthal (Eds.): Haifa Verification Conf. 2005, LNCS 3875, pp. 1–13, 2006.
© Springer-Verlag Berlin Heidelberg 2006

Subsequently, the verification engineer must map the verification plan scenarios into stimuli that will cause the system to reach a potentially illegal state and expose a bug in the design. This mapping is a challenge because the verification plan is formulated in natural language with a high level of abstraction, while the stimuli must be precise and detailed enough to be executed in simulation.

To assist the verification engineer, stimuli generation tools attempt to simplify the process by supplying a higher level language for describing the required stimuli. The stimuli generator's input language must be expressive enough to allow the specification of the required scenario, while the generation capabilities of the tool must be sufficiently elaborate to generate executable stimuli for these scenarios. Transaction-based verification methods improve the efficiency of the verification process by allowing the verification engineer to develop stimuli from a system level perspective. They separate the stimuli generation into two layers. The top layer is the stimuli specification, which orchestrates transaction level activity in the system without regard to the specific details of signal-level protocols on the design's interfaces. The bottom layer provides the mapping between transaction level requests and the detailed signal-level protocols on the design's interfaces [4].

The elusive nature of hardware bugs and the amount of stimuli required to cover the system behaviors has caused directed random stimuli generation to become the preeminent verification approach. In directed random stimuli generation, the user specifies partial requirements on the stimuli and the stimuli are randomly augmented with additional stimuli, which may either be required for simulation or may improve the likelihood of hitting interesting events.

Over the last decade, constraint satisfaction problem (CSP) techniques have been used successfully in the domain of random stimuli generation [5][6][7]. Given the user scenario, generated stimuli must be architecturally legal. The combination of the user scenario requirements and the architectural restrictions is represented as a CSP, where the variables are different stimuli attributes and the constraints are formed from an architectural model of the system and from the user scenario. Once the CSP is solved, the solved values for the variables are either used to drive the stimuli into the design or to create a test case when offline generation is used. However, finding the appropriate CSP formulation for system level generation requires an extension of existing solution schemes.

We describe a method of system level stimuli generation that extends the transaction-based verification approach and provides better support for multiple system configurations. This method was implemented in X-Gen [8], a state-of-the-art system level test case generator, which is used in IBM for the system level verification of high end servers and SoCs.

This paper is organized as follows: Section 2 presents the main challenges faced during system level stimuli generation. Section 3 describes the current approaches to system level stimuli generation and discusses their strengths and limitations. Section 4 describes the proposed solution scheme, which aims to overcome these limitations. Section 5 analyzes the advantages and disadvantages of the presented approach. We conclude with a summary.

2 System Level Stimuli Generation Challenges

2.1 Bridging the Abstraction Gap Between Verification Plan and Stimuli

The system level verification process is initiated by the verification engineer formulating a system level verification plan. The plan must identify the integration risks and describe a set of scenarios that are the focus of the verification. These are typically scenarios that involve intricate interactions of several previously verified cores or other complex scenarios that were not verified in lower levels. As an example, consider an SMP server, as depicted in Figure 1. The system is composed of several processor cores connected to memory via a central processor local bus (PLB). I/O devices connect to a peripheral bus (PHB) and the PHB is connected to the PLB by an I/O bridge. If the architecture defines a coherent memory model, the caches of the CPU and IO devices must implement a complex coherency protocol. This protocol is a prime candidate for system verification.

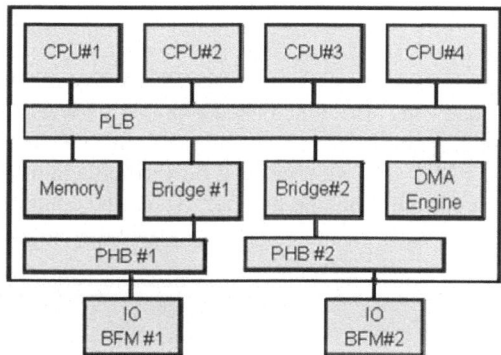

Fig. 1. High end server configuration

The first challenge we face is bridging the gap between the high level abstract scenarios in the verification plan and the low level details needed to generate the actual stimuli. For example, a verification plan scenario may involve generating several DMA transfers to the same address. This one line description needs to be converted to multiple register and memory initializations, processor instructions, and BFM commands.

Selecting the correct level of abstraction for the scenario specification language is crucial to the success of the stimuli generator. If the level of abstraction of the stimuli specifications is too high, it will not be possible to formulate the scenarios required for verification. If the level of abstraction is too low, the verification process will be too time consuming and it will not be possible to effectively cover all the scenarios. High level specification languages also make the verification environment more adaptive to changes in the design and enable the reuse of stimuli specification between designs [9].

2.2 Coordinated Stimuli Between Cores

System level scenarios typically require coordinated stimuli between cores. One example of a verification task, used to verify the correctness of a cache coherency protocol, would be to generate stimuli that cause multiple processors and IO devices to access the same cache lines at the same time. This requires the stimuli generator to coordinate stimuli from multiple devices.

Paths also play a significant role in the terminology used to describe scenarios in the verification plan. Many hardware problems, such as buffer overruns, are exposed when stress testing a particular component. A common verification task would be to select a component and generate stimuli that will cause data traffic to be transferred through the component.

2.3 Reusing of Test Specification and Verification Environment Code

Because the design IP is reused, we also require the verification IP to be reused; otherwise verification will become a gating factor. Reuse of verification IP is important between the unit, core, and system verification environments, as well as between similar or follow-on designs.

2.4 Adapting to Design Changes, Multiple Configurations, and Derivative Systems

As a recurring theme in SoC reuse, systems tend be designed with support for multiple configurations, targeting different balance points between performance, power, and cost. For example, a family of servers can range from a single processor machine to a multi-processor SMP. The system must be verified against several possible configurations. In each configuration, the number, types, and connectivity of components may change. Also, in follow-on systems, some previously used components may be

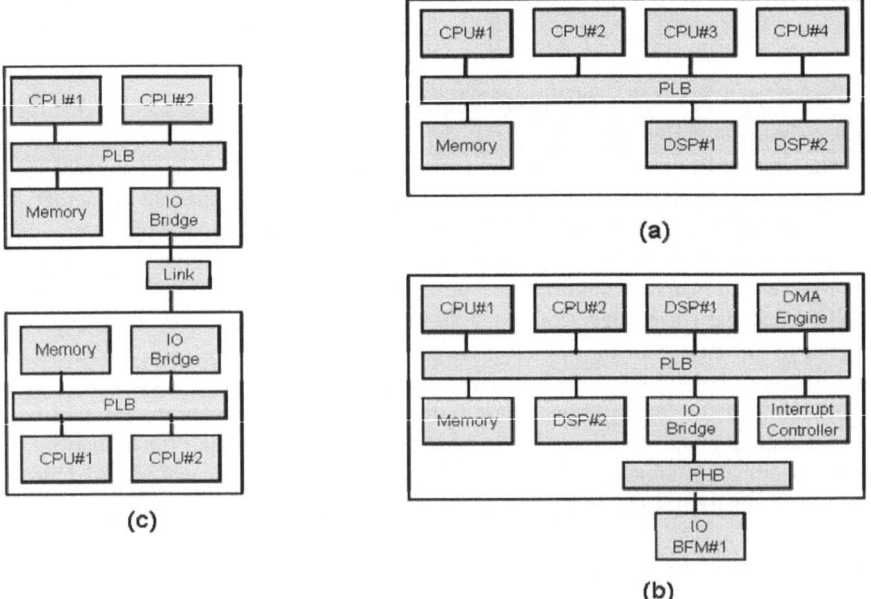

Fig. 2. Multiple configurations in a family of systems

replaced with newer components that have new implementation and enhanced specifications. The new systems must be verified both for backward compatibility and for
new capabilities. The ability to quickly adapt to such changes is crucial for the successful and timely execution of the verification plan. Figure 2 shows several configurations of an SMP server.

3 Current Approaches to System Level Stimuli Generation

3.1 Combining Core Level Stimuli Generators

The simplest approach to generate stimuli for the system is to combine the existing
lower level stimuli generators. In this approach, we connect the cores, remove any
stimuli generation on the interfaces connecting the cores, and use the external interface drivers to create stimuli for the system. This approach is very simple and costs
very little to implement because it reuses the existing verification IP.

The major drawback of this approach is that it is not suitable for system level
functional verification. There is no way to express scenarios that involve multiple
components. Additionally, while the drivers generate simultaneous stimuli over the
interfaces, there is no way to coordinate the stimuli.

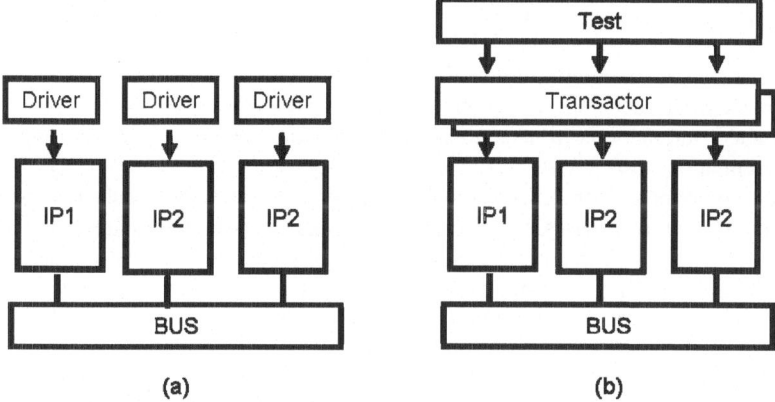

Fig. 3. Traditional approaches to system level stimuli generation: on the left, combining the
core level stimuli generator; on the right, using Transaction Based Verification (TBV)

3.2 Transaction-Based Verification

Transaction-based verification (TBV) approach is a common methodology [1][4][10],
which does allow system level verification. The test specification is expressed as a
series of abstract system level transactions. Examples of such transactions are CPU to
memory data transfers, BFM initiated interrupts, and DMA transfers. Each transaction has a set of parameters that can be specified, for example, the source and target
address of a DMA transaction. A transaction is converted to the actual stimuli by a
transactor code, which maps the abstract transaction into the low level signal or resource initialization.

This approach has several advantages. First, since tasks in the verification plan are expressed as scenarios, they are naturally mapped to transactions. Transactions allow coordinate stimuli to be generated by driving the interfaces of multiple cores. Transactions define a consistent interface to drive the system behavior, regardless of the actual implementation details. This enables the same test specification to be reused when the system configuration changes and components are replaced or added.

Although this approach decouples the interface of the transaction from its implementation, it does not as support reuse as may be expected. Each transaction is implemented by a transactor, which is a monolithic piece of code. Thus, any addition of a new core in the system or a change to the behavior of a core requires multiple transactor code segments to be modified. This results in high development costs and lower productivity.

4   A Combined System Modeling and Stimuli Generation Scheme

This section presents a scheme of stimuli generation that extends the transaction-based approach. In this approach, the generator receives as input the user specification describing the scenario to be generated and an abstract system model. The generator then solves a series of CSPs and uses their solution to generate the actual stimuli.

4.1   Abstract System Model

The abstract system model describes the system under verification and is based on three basic constructs: the components (cores) from which the system under test is constructed, interactions (transactions) between components, and the configuration of the system under test.

Each component represents a physical entity in the system, such as a processor core, I/O bridge, or a DMA. The component contains its current state, usually in the form of registers and memory. A component interfaces with other components through its interface ports. On each port, we model logical properties that represent the data transferred between the component and its neighboring component. The component also models its behavior by specifying constraints between the properties of the different ports and its internal state. For example, consider the I/O bridge in Figure 4. The bridge has two ports, one connected to the PLB and one connected to the PHB. The PLB port can contain properties such as *data (D)*, *system address (SA)*, and *operation (O)*. The PHB port can contain the properties *data (D)*, *I/O address (IA)*, and *operation (O)*. One of the roles of the I/O bridge in the system is to translate between the system address on the PLB and the I/O address on the PHB bus. This is achieved by a constraint between the two properties. This constraint may also use the values of the component's registers and memory.

The modeling of interactions serves two purposes. First, as in classical transaction-based modeling, interactions are the building blocks of the language used by the verification engineer writing the test specifications. For example, a test specification would include statements such as "generate 10 DMAs and 20 processor load/store interactions that access the same cache line". Second, interactions describe behavior of the system that involves several components. An interaction is modeled as a set of

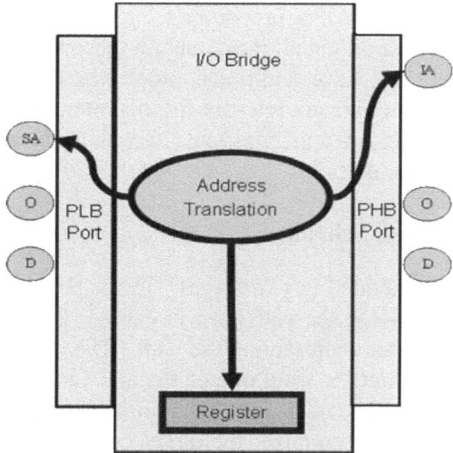

Fig. 4. I/O Bridge model

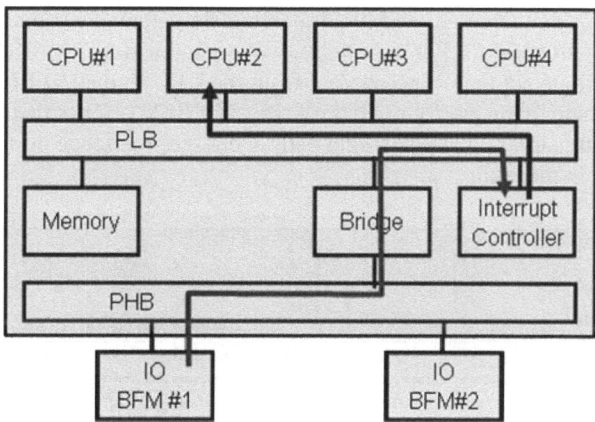

Fig. 5. I/O device initiated interrupt interaction model, modeled as two acts

separate *acts*. Each act represents a single data transfer across a path in the system. The initiator and target components on the path are called *actors*. Each act specifies the components that are allowed to be its initiator and target actors, and describes restrictions on the paths allowed between the initiator and the target. Additional constraints can be placed on the components participating in different acts.

Figure 5 depicts an I/O device-initiated interrupt interaction. It is composed of two acts. In the first act, the I/O device initiates an interrupt request on the PHB bus and this request is forwarded to the interrupt controller. In the second act, the interrupt controller signals the interrupt line of the target CPU.

Each actor (initiator and target) specifies a set of properties that are relevant for the interaction. These properties define the interface of the interaction in the test specification. In Figure 5, for example, the initiator I/O device interface can contain the *interrupt priority* and *interrupt line* properties. Different types of I/O devices may take the role of the act initiator, but all must contain these two properties on their port. These components typically have additional properties on their port (such as *I/O address*), but these properties are not relevant for this interaction.

The configuration of the system represents the physical and logical connection between the core's interface ports.

4.2 Two Stage Generation Scheme

A single interaction is generated in two stages. In the first stage, called *path selection*, we form a CSP to choose the path used in each of the different acts. For each act, we create a CSP variable representing the path. The domain of the CSP variable is the set of all paths that initiate from one of the allowed initiators and terminate at one of the allowed targets. The additional constraints modeled on the act are used to enforce path restrictions. For example, the target component of one act may be required to be the same as the initiator component of another act.

Consider, for example, the system described in Figure 6. Suppose a verification engineer writes a test template with the following statement: "Generate 10 CPU initiated MMIOs to I/O BFM". For simplicity, we assume a single MMIO interaction is modeled as a single act that transfers data from the CPU to the I/O BFM. In the system presented here, there are four CPUs and two I/O BFMs. Thus there are eight possible path combinations. Once a single path is chosen, we focus on the components that participate in the path.

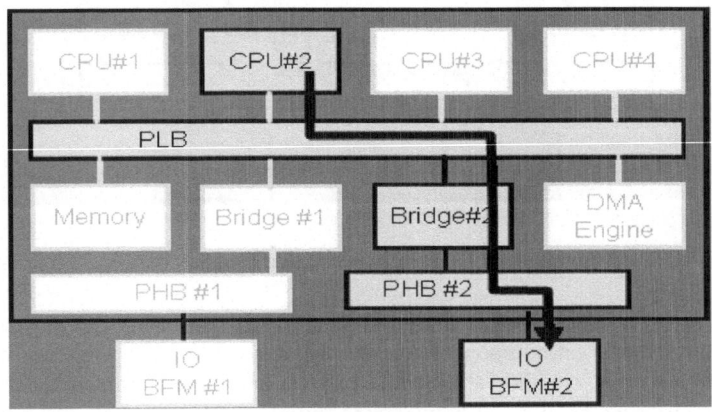

Fig. 6. A selected path in an MMIO interaction.There are five components on the path: *CPU #2, PLB, Bridge#2, PHB#2 and IO BFM#2.*

Once the paths are chosen, a second stage, referred to as *property selection*, constructs a second CSP network. The potential variables in the CSP are the properties of the participating components in the selected path. We first calculate which of these properties are relevant for this interaction, because not all properties are relevant for all interactions.For example, an *interrupt priority* property is not relevant for an MMIO interaction.

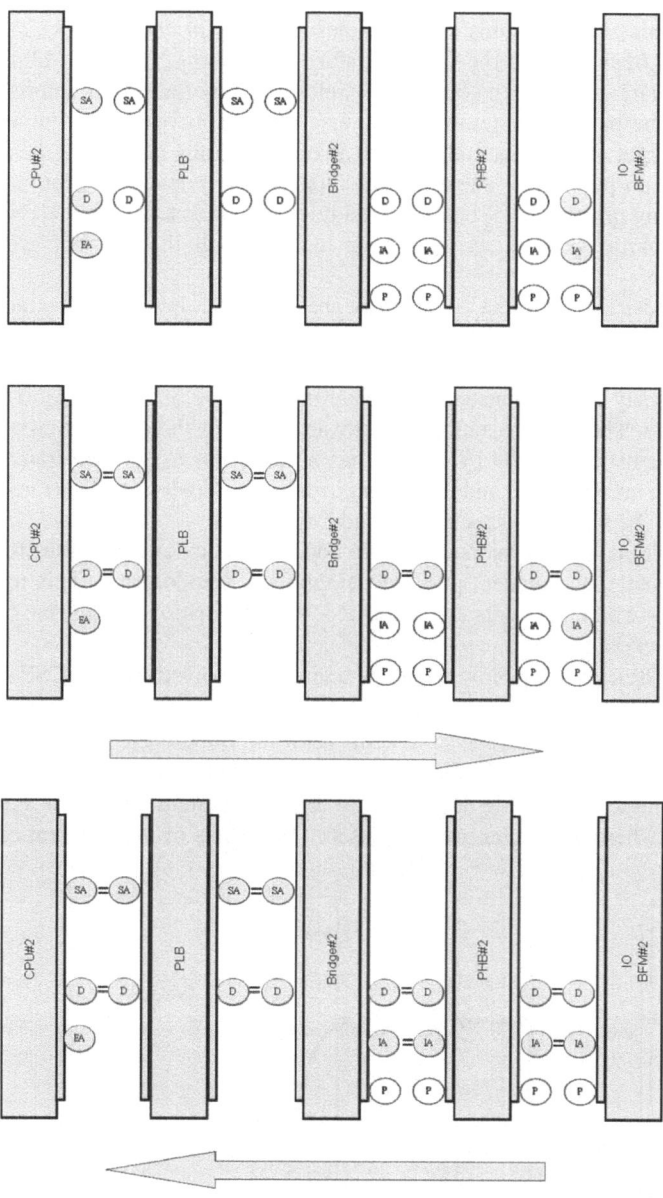

Fig. 7. The relevant properties for the interaction are computed in an iterative process. Initially, all properties modeled on the initiator or target actors are considered relevant. Consequently, any property on an intermediate component port that is relevant on a neighboring component, is also considered relevant and is added to the CSP.

The calculation proceeds as follows: First, all the properties appearing on the initiator and target actors of the interaction are added to the CSP network. As explained above, these properties were identified as relevant during the modeling of the interaction's actors. In this example, these are *system address (SA)*, *data (D)* and *effective address (EA)* on the CPU initiator, and the *data (D)* and *I/O address (IA)* on the BFM target. We then compute the relevant properties of the intermediate components in the path. We iteratively traverse the path between the initiator component and the target component and back. If two neighboring ports have the same property, and one property has already been considered relevant in the interaction, the neighboring property is also considered relevant and is added to the CSP network. At the end of the process we will have identified all the relevant properties of the interaction.

The constraints in this CSP network are the constraints of the interaction and some of the constraints of the components that participate in the selected path. For each component, we add all constraints that relate to the relevant properties. For example, the CPU may have a constraint to relate the *effective address (EA)* to the *system address (SA)*. The buses usually just forward data, so their properties are linked with equality constraints. The I/O bridge has a constraint to perform translation between the *system address (SA)* and the *I/O address (IA)*. Finally, the BFM may limit the *I/O address* to be within its specified I/O address region.

In addition, equality constraints are added between any variables representing the same property on adjacent ports of neighboring components. This reflects the idea that when two components are connected, the information is the same on both ends of the connection.

The CSP network is now complete and can be solved using a CSP solver. Despite the local approach to constructing the CSP, global architectural restrictions are enforced. For example, the system address on the initiating CPU is synchronized with the I/O address of the BFM device according the I/O bridge translation rules.

The solution values are used either to drive the stimuli into the design or to create a test case when offline generation is used. The states of the different components are

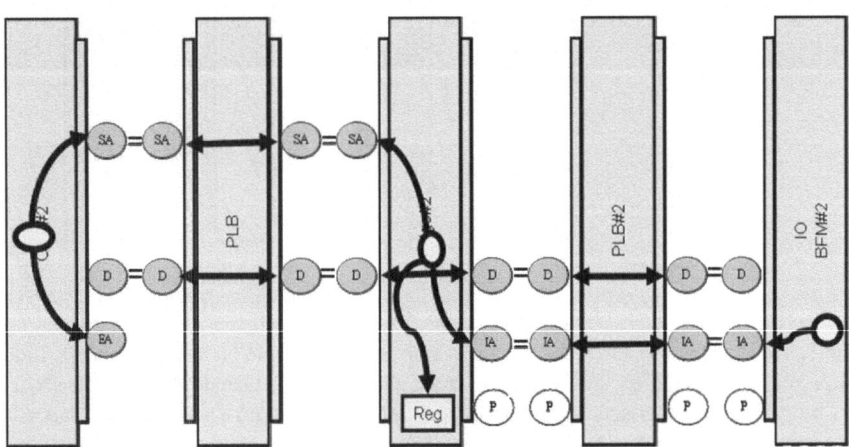

Fig. 8. The constraints in the property selection stage are composed from the interaction constraints, component constraints that relate to relevant properties, and equality constraints between the same properties on adjacent ports

also updated by the result of the interaction (e.g., the BFM's memory content is updated with the results of the transfer).

5 Discussion

This section discusses the advantages and disadvantages of the generation approach presented in the previous section, and compares it to the traditional transaction-based verification approach.

One can easily notice how the suggested approach provides better modularity and improved opportunities for reuse. If a new component of an existing type is added to the system (for example, an additional processor), the only change required is to the system configuration. In traditional transaction-based verification, different transactors must be modified to be made aware of the new component.

In case where an existing component behavior changes, for example, the translation mechanism of the I/O bridge changes, any transactors that utilize that component need to be changed. On the other hand, in the suggested approach, only the modified component needs to be changed. The only modification required is to the modeling of the translation constraint on the new bridge type and to the configuration used to generate the stimuli. Similarly, when a new I/O bridge is introduced into the system, the only change is the definition of the new component. When a new interaction is added to the system, a new transactor needs to be implemented in the traditional approach, and similarly, a new interaction is required in the suggested approach. Additional changes may be necessary in existing components to support the new interaction.

Table 1. Effects of design changes on stimuli generation code

DUT change	Pure transaction based changes	Suggested approach changes
New component of existing type	Modify multiple transactors	Change configuration
Changes to existing component	Modify multiple transactors	Modify component
New component with same interactions	Modify multiple transactors	New component
New system level interaction	New transactor	New interaction + Modify components

The decoupling of the components and their connection topology from the interactions yields additional benefits. The generation scheme, in which the participating components and the paths between them are selected, and subsequently the additional properties are solved, is scalable with respect to system size. The path selection CSP consists of a small number of variables, while the property selection CSP problem formed for each interaction, consists of variables and constraints of only the relevant components on the path.

The generation scheme enables the generation of coordinated stimuli. There is explicit control over the participating components in each interaction. Furthermore, the current internal state of the participating components is accessible during interaction generation. The state may be used to generate stimuli that cause contention on previously used hardware resources. For example, each memory component can hold a list of the last accessed addresses, and new interactions can use this list to potentially collide with previously generated interactions. In addition, it is possible to bias the random selection of paths toward interesting scenarios. For example, errors can be exposed by selecting multiple interactions that stress the same intermediate bus.

Notwithstanding, the presented approach suffers from some disadvantages. Identifying the optimal way to partition the behavior of the system between components and interactions requires expertise. One must decide on the relevant level of detail for each interaction determining which properties and constraints are part of the interaction interface and which are low level details to be modeled on the components. These decisions usually have a high impact on the quality of the model and the ability to support effective reuse.

While the two stage generation scheme allows the tool to scale to large configurations, it suffers from the inherent drawback of CSP network partitioning techniques. The solution for the first stage may not be extendable to a solution for the second stage. Consider a system with several memory components, each implementing a part of the global memory map. If the test specification contains a DMA interaction to a specific address, then a legal solution for the path selection might select a target memory whose address range is not inclusive of the required address. Subsequently, this will cause the second stage to fail and will require the solver to backtrack to the first stage and attempt to find a new set of participating components and paths. Although these backtracks can cause severe performance degradation, some look-ahead techniques can be applied to reduce their relative incidence.

6   Summary

We examined several key challenges faced in system level stimuli generation and the ways the current generation approaches address these challenges. We proposed a novel integrated method for system modeling and random stimuli generation. The method extends the traditional transaction-based verification approach by modeling the system behavior as interactions, components and a topology. We described how this kind of decoupling of the system behavior provides improved modularity, which, among other things, enables better support for verifying multiple system configurations.The method described in this paper was fully implemented in X-Gen [8], IBM's system-level test case generator, and was successfully used in the verification of IBM's most complex designs in recent years. X-Gen was the sole system-level generator used for verification of the Power5 family of servers [11], as well as of the Cell [12].

References

[1]   Rashinkar P., Paterson, P., Singh L., System-on-a-chip Verification – Methodology and Techniques, Kluwer Academic Publishers, 2001.

[2]   Albin K., Nuts and Bolts of Core and SoC Verification, 38th Design Automation Conference (DAC '01), 2001

[3]   Berry G., Bouali A., Dormoy J. , Blanc L., Top-level Validation of System-on-Chip in Esterel Studio, Proceedings of the IEEE International High Level Design Validation and Test Workshop, 2001.

[4]   Brahme, et al. The Transaction-Based Verification Methodology, Cadence Berkeley Labs, Technical Report # CDNL-TR-2000-0825, August 2000.

[5]   Aharon A., Goodman D., Levinger M., Lichtenstein Y., Malka Y., Metzger C., Molcho M., and Shurek G., Test Program Generation for Functional Verification of PowerPC Processors in IBM, Proceedings of the Design Automation Conference, 1995.

[6]   Bin E., Emek R., Shurek G., and Ziv A., Using a constraint satisfaction formulation and solution techniques for random test program generation, IBM Systems Journal, Vol 41, No 3, 2002.

[7]   Allon A., Almog E., Fournier L., Marcus E., Rimon M., Vinov M., and Ziv A., Genesys-Pro: Innovations in Test Program Generation for Functional Processor Verification, IEEE Design & Test of Computers, 2004.

[8]   Emek R., Jaeger I., Naveh Y., Bergman G., Aloni G., Katz Y., Farkash M., Dozoretz I., and Goldin A., X-Gen: A Random Test-Case Generator for Systems and SoCs, Proceedings of the IEEE International High Level Design Validation and Test Workshop, 2002.

[9]   Practical Approaches to SOC Verification, Mosensoson G.http://www.verisity.com/download/practical_soc_verification.pdf

[10] Besyakov V., Shleifman D., Constrained Random Test Environment for SoC Verification using VERA, SNUG Boston, 2002

[11] http://www-03.ibm.com/systems/power/

[12] D. Pham, S.Asano, M. Bolliger, M. Day, H. Hofstee, C. Johns, J. Kahle, Kameyama, J. Keaty,Y. Masubuchi, M. Riley, D. Shippy, D. Stasiak, M.Wang, J.Warnock, S.Weitzel, D.Wendel, T.Yamazaki, K.Yazawa, The Design and Implementation of a First-Generation CELL Processor, Proceedings of the IEEE International Solid-State Circuits Conference (ISSCC 2005).

The Safety Simple Subset*

Shoham Ben-David[1], Dana Fisman[2,3,**], and Sitvanit Ruah[3]

[1] University of Waterloo
[2] Weizmann Institute of Science
[3] IBM Haifa Research Lab

Abstract. Regular-LTL (RLTL), extends LTL with regular expressions, and it is the core of the IEEE standard temporal logic PSL. Safety formulas of RLTL, as well as of other temporal logics, are easier to verify than other formulas. This is because verification of safety formulas can be reduced to invariance checking using an auxiliary automaton recognizing violating prefixes.

In this paper we define a special subset of safety RLTL formulas, called RLTL^{LV}, for which the automaton built is *linear* in the size of the formula. We then give two procedures for constructing such an automaton, the first provides a translation into a regular expression of linear size, while the second constructs the automaton directly from the given formula. We have derived the definition of RLTL^{LV} by combining several results in the literature, and we devote a major part of the paper to reviewing these results and exploring the involved relationships.

1 Introduction

The specification language PSL [11] is an IEEE standard temporal logic which is widely used in industry, both for simulation and for model checking. While PSL is a rich and expressive language, some of its formulas are hard to implement and expensive to verify. For this reason, an effort has been made in the PSL language reference manual (LRM), to define the *simple subset*, which should consist of formulas that are easy to implement and to verify. The LRM, however, provides no proof for the simplicity of the subset.

In this paper we deal with the simple subset of the LRM. We note that this subset, as defined in the LRM, is geared towards simulation tools. In order to accommodate model checking as well, we focus on the subset of the LRM simple subset, that consists of safety formulas only. A formula φ is a safety formula if every violation of φ occurs after a finite execution of the system. Verification of safety formulas is easier for model checking than verification of general formulas.

The core of PSL is the language Regular-LTL (RLTL), which extends linear temporal logic LTL [18] with regular expressions. The verification of a general RLTL formula typically involves the construction of an automaton on *infinite* words with size exponential in the size of the formula [4]. The restriction to safety formulas allows the use of automata on *finite* words. The use of a finite-word automata is important. It allows

---

* This work was partially supported by the European Community project FP6-IST-507219 (PROSYD).

** The work of this author was carried out at the John von-Neumann Minerva Center for the Verification of Reactive Systems.

S. Ur, E. Bin, and Y. Wolfsthal (Eds.): Haifa Verification Conf. 2005, LNCS 3875, pp. 14–29, 2006.

verification of safety formulas to be reduced to invariance checking, by stating the invariance "The violation automaton is not in an accepting state". Invariance verification is, for most of the verification methods, easier to perform than verification of a general RLTL formula, making safety formulas easier to verify than others.

Kupferman and Vardi in [15], classify safety formulas according to efficiency of verification. They define pathologically safe formulas, and show that non-pathologically safe formulas are easier to verify than pathologically safe ones. In particular, they show that violation of non-pathologically safe formulas can be detected by an automaton on finite words with size exponential in the size of the formula.

In this document we define a subset of safety RLTL formulas, for which violation can be detected by an automaton on finite words with size *linear* in the size of the formula. This subset corresponds to the definition of the safety simple subset in PSL's LRM. We term this subset RLTL^{LV}, where LV stands for "linear violation". The definition of RLTL^{LV} restricts the syntax of the formulas to be specified. However, experience shows that the vast majority of the safety formulas written in practice, are expressible in RLTL^{LV}. Thus this subset is both easier to verify and very useful in practice.

We provide two procedures for the construction of a linear-sized automaton (NFA) for an RLTL^{LV} formula. The first goes through building a linear-sized regular expression, and the other directly constructs the automaton from the given formula. These procedure can serve PSL tool implementors.

We have derived the definition of RLTL^{LV} by combining several results in the literature. We devote a major part of the paper to reviewing these results and exploring the involved relationships.

Maidl [16] has studied the common subset of the temporal logics LTL and ACTL. She defined a syntactic subset of LTL, LTL^{DET}, such that every formula in the common fragment of LTL and ACTL has an equivalent in LTL^{DET}. By this she strengthened the result of Clarke and Draghicescu [5] who gave a characterization of the CTL formulas that can be expressed in LTL and the result of Kupferman and Vardi [14] who solved the opposite problem of deciding whether an LTL formula can be specified in the alternation free μ-calculus. She further showed that for formulas in LTL^{DET} there exists a 1-weak Büchi automaton, linear in the size of the formula, recognizing the negation of the formula. The subset RLTL^{LV} can be seen as the safety subset of LTL^{DET}, augmented with regular expressions. The augmentation with regular expressions is important as it increases the expressive power. While Maidl was engaged with the expressiveness of LTL and ACTL, it is interesting to observe that the safety subset of LTL^{DET} is associated with efficient verification.

In [1] Beer et al. have defined the logic RCTL, an extension of CTL with regular expressions via a suffix implication operator. They were interested in "on-the-fly model checking", which in our terminology, is called invariance verification. We therefore term their subset RCTL^{OTF}, where OTF stands for "on-the-fly". Beer et al. gave a procedure to translate an RCTL^{OTF} formula into a regular expression. Our results extend theirs, in the sense that the subset RCTL^{LV} which corresponds to RLTL^{LV} subsumes RCTL^{OTF}. We elaborate more on this in Section 4.

The remainder of the paper is organized as follows. In Section 2 we provide some preliminaries. In Section 3 we define RLTL^{LV} and provide a succinct construction yielding a regular expression of linear size recognizing violation. In Section 4 we discuss the relation of this subset with the results of Kupferman and Vardi [15], Maidl [16] and Beer et al. [1]. In Section 5 we provide a direct construction for an NFA of linear size, and in Section 6 we conclude.

2 Preliminaries

2.1 Notations

Given a set of atomic propositions $\mathbb{P}$, we use $\Sigma_{\mathbb{P}}$ to denote the alphabet $2^{\mathbb{P}} \cup \{\top, \bot\}$. That is, the set of interpretations of atomic propositions extended with two special symbols $\top$ and $\bot$.[1] We use $\mathbb{B}_{\mathbb{P}}$ to denote the set of boolean expressions over $\mathbb{P}$, which we identify with $2^{2^{\mathbb{P}}}$. That is, every boolean expression is associated with the set of interpretations satisfying it. The boolean expressions *true* and *false* denote the elements $2^{\mathbb{P}}$ and $\emptyset$ of $\mathbb{B}_{\mathbb{P}}$, respectively. When the set of atomic propositions is assumed to be known, we often omit the subscript $\mathbb{P}$ from $\Sigma_{\mathbb{P}}$ and $\mathbb{B}_{\mathbb{P}}$.

We denote a boolean expression by b, c or d, a letter from Σ by ℓ (possibly with subscripts) and a word from Σ by u, v, or w. The *concatenation* of u and v is denoted by uv. If u is infinite, then $uv = u$. The empty word is denoted by ϵ, so that $w\epsilon = \epsilon w = w$. If $w = uv$, we define $w/u = v$ and we say that u is a *prefix* of w, denoted $u \preceq w$, that v is a *suffix* of w, and that w is an *extension* of u, denoted $w \succeq u$.

We denote the length of a word w by $|w|$. The empty word $w = \epsilon$ has length 0, a finite non-empty word $w = (s_0 s_1 s_2 \cdots s_n)$ has length $n + 1$, and an infinite word has length ∞. We use i, j, and k to denote non-negative integers. For $i < |w|$, we use w^i to denote the $(i + 1)^{th}$ letter of w (since counting of letters starts at zero). We denote by $\overline{w}$ the word obtained by replacing every $\top$ in w with a $\bot$ and vice versa. We refer to $\overline{w}$ as the *dual* of w.

We denote a set of finite/infinite words by U, V or W and refer to them as *properties*. The *concatenation* of U and V, denoted UV, is the set $\{uv \mid u \in U, v \in V\}$. Define $V^0 = \{\epsilon\}$ and $V^k = VV^{k-1}$ for $k \geq 1$. The *Kleene closure* of V, denoted V^*, is the set $V^* = \bigcup_{k<\omega} V^k$. The infinite concatenation of V to itself is denoted V^ω. The union $V^* \cup V^\omega$ is denoted V^∞. For a letter ℓ we use ℓ^* and ℓ^ω as abbreviations of $\{\ell\}^*$ and $\{\ell\}^\omega$, respectively.

2.2 Regular Expressions and Finite Automata

Traditional regular expressions are defined using the operators of concatenation ($\cdot$), union ($\cup$) and Kleene closure ($*$). We define regular expressions using also the operator *fusion* ($\circ$), also known as *overlapping concatenation*. The fusion operator does not add expressive power to regular expressions (see e.g. [1]).

Definition 1 (Regular Expressions (REs)). *Let b be a boolean expression. The set of REs is recursively defined as follows:*

$$r ::= b \mid r \cdot r \mid r \cup r \mid r^* \mid r \circ r$$

---

[1] The role of $\top$ and $\bot$ in the alphabet is explained in Section 2.3.

The semantics of regular expressions is defined inductively, using the semantics of boolean expressions in the base case. For a boolean expression $b \in \mathbb{B}$ and a letter $\ell \in \Sigma$, we define the boolean satisfaction relation $\Vdash$ as follows. For $\ell \in 2^{\mathbb{P}}$, we define $\ell \Vdash b \Longleftrightarrow \ell \in b$. We define $\top \Vdash b$ and $\bot \not\Vdash b$. Note that in particular $\top \Vdash \textit{false}$ and $\bot \not\Vdash \textit{true}$.

Definition 2 (Tight Satisfaction). *Let v denote a finite (possibly empty) word over Σ; b denote a boolean expression; and r, r_1, and r_2 denote REs. The notation $v \models r$ means that v tightly satisfies r. The relation $\models$ is defined as follows:*

- $v \models b \Longleftrightarrow |v| = 1$ *and* $v^0 \Vdash b$
- $v \models r_1 \cdot r_2 \Longleftrightarrow \exists v_1, v_2$ *s.t.* $v = v_1 v_2$ *and* $v_1 \models r_1$ *and* $v_2 \models r_2$
- $v \models r_1 \cup r_2 \Longleftrightarrow v \models r_1$ *or* $v \models r_2$
- $v \models r^* \Longleftrightarrow$ *either* $v = \epsilon$ *or* $\exists v_1, v_2$ *s.t.* $v_1 \neq \epsilon$, $v = v_1 v_2$, $v_1 \models r$ *and* $v_2 \models r^*$
- $v \models r_1 \circ r_2 \Longleftrightarrow \exists v_1, v_2$ *and* ℓ *s.t.* $v = v_1 \ell v_2$ *and* $v_1 \ell \models r_1$ *and* $\ell v_2 \models r_2$

We use $\mathcal{L}(r)$ to denote the set of finite words tightly satisfying r. That is $\mathcal{L}(r) = \{w \in \Sigma^* \mid w \models r\}$.

The literature usually provides different definitions for automata over finite vs. infinite words. When infinite words are considered by *finite acceptance* [20] (i.e. an automaton accepts an infinite word if it accepts some prefix of the word by the standard notion of acceptance for finite words) the same automaton can serve for both finite and infinite words. Automata are usually defined over a syntactic alphabet. We define them over a semantic alphabet, using a given set of atomic propositions as follows.

Definition 3 (NFA, CO-UFA). *An automaton $\mathcal{A}$ is a tuple $\mathcal{A} = \langle P, Q, Q_0, \delta, F \rangle$ consisting of the following components:*

- $P = \{p_1, \ldots, p_n\}$: *A finite set of* atomic propositions
- Q: *A finite set of* automata *locations*
- $Q_0 \subseteq Q$: *A set of* initial *locations*
- $\delta \subseteq Q \times \mathbb{B}_P \times Q$: *A* transition relation
- $F \subseteq Q$: *A set of* final *locations*

Let $\mathcal{A}$ be an automaton for which the above components have been defined. The input to $\mathcal{A}$ is a finite/infinite word over Σ_P. We define a run of $\mathcal{A}$ over a word $w = w^0 w^1 \ldots$ to be a finite or infinite non-empty sequence $\sigma : q_0 q_1 \ldots$ of locations in Q satisfying the requirements of initiality i.e. that $q_0 \in Q_0$; and of consecution i.e. that $|\sigma| \leq |w| + 1$ and for each $0 \leq j < |w|$ there exists $b \in \mathbb{B}_P$ such that $(q_j, b, q_{j+1}) \in \delta$ and $w^j \Vdash b$. A run satisfying the requirement of maximality i.e. that it is either infinite, or terminates at a location q_k which has no successors w.r.t. δ is termed a maximal run. Let $\sigma : q_0 q_1 q_2 \ldots$ be a run of $\mathcal{A}$ over a word w. The run σ accepts w iff w is finite and $q_{|w|} \in F$ or w is infinite and there exists $0 \leq i < |w|$ such that $q_i \in F$. The run σ co-accepts w iff $q_i \notin F$ for all i such that $0 \leq i < |w|$. If acceptance is determined by the existence of at least one accepting run then we refer to $\mathcal{A}$ as an NFA (non-deterministic or existential finite automaton). If acceptance is determined by the fact that all runs are co-accepting then we refer to $\mathcal{A}$ as a CO-UFA (co-universal finite automaton).

The expressive power of regular expressions is the same as that of automata over finite words (both NFAs and CO-UFAs) [10]. Moreover, as stated formally below, given a regular expression r (that may be composed using fusion as well) it is possible to generate an equivalent NFA which is *linear* in the size of the RE.

Fact 4 (see e.g. [12]). *Let r be an RE. There exists an NFA N_r with size linear in $|r|$ such that for every word v, N_r accepts v iff $v \models r$.*

2.3   The Logic RLTL

The logic Regular-LTL (RLTL) extends LTL [18] in two ways. First it interprets the formulas over finite (possibly truncated and possibly empty) as well as infinite words. Second it adds regular expressions to the logic.

In order to incorporate finite words, the syntax of LTL is extended to include two *next-time* operators, weak (X) and strong (X!) [17, pp. 272-273]. The semantics distinguish between the weak and strong versions only on finite words and only on the last letter of a finite word: $X \varphi$ holds on the last letter of any word for any φ, and $X! \varphi$ does not. Similarly, since the logic is interpreted over the empty word as well, there are two versions of a boolean expression: weak and strong [7]. The strong boolean expression is satisfied over a word if the first letter satisfies the boolean expression, and the weak boolean expression is satisfied also if there is no first letter, i.e. if the word is empty.

Regular expressions are added to the logic via the operator $\longmapsto$ or its dual $\diamondsuit\!\!\!\!\rightarrow$. We refer to the $\longmapsto$ operator as the *suffix implication operator* $\longmapsto$, since $r \longmapsto \varphi$ (read r "suffix-implies" φ) requires that *if* there exists a non-empty prefix of the path tightly satisfying r *then* the suffix starting at the last letter of the prefix should satisfy φ. We refer to the $\diamondsuit\!\!\!\!\rightarrow$ operator as the *suffix conjunction operator*, since $r \diamondsuit\!\!\!\!\rightarrow \varphi$ (read r "suffix-and" φ) demands the existence of a non-empty prefix of the path tightly satisfying r *and* that the suffix starting at the last letter of the prefix satisfy φ. These operators are essentially the "diamond" and "box" modalities of propositional dynamic logic (PDL) [9], resp.

The syntax of RLTL is formally defined as follows.

Definition 5 (RLTL). *Let b a boolean expression and r a regular expression (RE). The set of RLTL formulas is recursively defined as follows:*

$$\varphi ::= b! \mid \neg\varphi \mid \varphi \wedge \varphi \mid X!\varphi \mid \varphi U \varphi \mid r \longmapsto \varphi$$

Additional operators are defined as syntactic sugaring of the above operators:

- $\varphi \vee \psi \overset{\text{def}}{=} \neg(\neg\varphi \wedge \neg\psi)$
- $\varphi W \psi \overset{\text{def}}{=} \neg(\neg\psi U (\neg\varphi \wedge \neg\psi))$
- $X \varphi \overset{\text{def}}{=} \neg(X! \neg\varphi)$
- $\varphi \to \psi \overset{\text{def}}{=} \neg\varphi \vee \psi$
- $F \varphi \overset{\text{def}}{=} true\, U \varphi$
- $G \varphi \overset{\text{def}}{=} \varphi W false$
- $b \overset{\text{def}}{=} \neg(\neg b!)$
- $r \diamondsuit\!\!\!\!\rightarrow \varphi \overset{\text{def}}{=} \neg(r \longmapsto \neg\varphi)$
- $r! \overset{\text{def}}{=} r \diamondsuit\!\!\!\!\rightarrow true$

The formula $r!$ holds on a word w iff there exists a prefix u of w which satisfies r tightly.

The semantics of RLTL formulas is defined inductively, using the semantics of boolean and regular expressions in the base case, as follows.

Definition 6 (Formula Satisfaction). *Let v denote a word over Σ; b a boolean expression; r an* RE; *and φ and ψ* RLTL *formulas. The notation $v \models \varphi$ means that v satisfies φ. The relation $\models$ is defined as follows:*

1. *$v \models b! \Longleftrightarrow |v| > 0$ and $v^0 \Vdash b$*
2. *$v \models \neg\varphi \Longleftrightarrow \bar{v} \not\models \varphi$*
3. *$v \models \varphi \wedge \psi \Longleftrightarrow v \models \varphi$ and $v \models \psi$*
4. *$v \models X! \varphi \Longleftrightarrow |v| > 1$ and $v^{1\cdots} \models \varphi$*
5. *$v \models \varphi U \psi \Longleftrightarrow \exists\, 0 \le k < |v|$ s.t. $v^{k\cdots} \models \psi$ and $\forall\, 0 \le j < k,\ v^{j\cdots} \models \varphi$*
6. *$v \models r \mapsto \psi \Longleftrightarrow \forall\, 0 \le j < |v|$ s.t. $v^{0\cdots j} \equiv r,\ v^{j\cdots} \models \psi$*

We use $[\![\varphi]\!]$ to denote the set of finite/infinite words satisfying φ. That is $[\![\varphi]\!] = \{w \in \Sigma^\infty \mid w \models \varphi\}$.

Note that, as suggested in [7], the semantics is defined with respect to finite (possibly empty) as well as infinite words over the alphabet $\Sigma = 2^{\mathbb{P}} \cup \{\top, \bot\}$. That is, the alphabet consists of the set of interpretations of atomic propositions extended with two special symbols $\top$ and $\bot$. Below we mention a few facts about the role of $\top$ and $\bot$ in the semantics, for a deep understanding please refer to [7].

By definition we have that $\top$ satisfies every boolean expression including *false* while $\bot$ satisfies no boolean expression and in particular, not even the boolean expression *true*. By the inductive definition of the semantics, we get that $\top^\omega$ satisfies every formula while $\bot^\omega$ satisfies none. Using the notions of [8] if $v\top^\omega \models \varphi$ we say that v *satisfies φ weakly* and denote it $v \models^{-} \varphi$. If $v\bot^\omega \models \varphi$ we say that v *satisfies φ strongly* and denote it $v \models^{+} \varphi$. If $v \models \varphi$ we often say that v *satisfies φ neutrally*. The *strength relation theorem* [8] gives us that if v satisfies φ strongly then it also satisfies it neutrally, and if v satisfies φ neutrally then it also satisfies it weakly. The *prefix/extension theorem* [8] gives us that if v satisfies φ strongly then any extension w of v also satisfies φ strongly. And, if v satisfies φ weakly then any prefix u of v also satisfies φ weakly.

Safety and Liveness. Intuitively, a formula is said to be safety iff it characterizes that "something bad" will never happen. A formula is said to be liveness iff it characterizes that "something good" will eventually happen. More formally, a formula φ defines a *safety* property iff any word violating φ contains a finite prefix all of whose extensions violate φ. A formula φ defines a *liveness* property iff any arbitrary finite word has an extension satisfying φ. We use the definitions of safety and liveness as suggested in [7]. These definitions modify those of [17] to reason about finite words as well.

Definition 7 (Safety, Liveness). *Let $W \subseteq \Sigma^\infty$.*

- *W is a* safety *property if for all $w \in \Sigma^\infty - W$ there exists a finite prefix $u \preceq w$ such that for all $v \succeq u$, $v \in \Sigma^\infty - W$.*
- *W is a* liveness *property if for all finite u, there exists $v \succeq u$ such that $v \in W$.*

A formula φ is said to be a *safety (liveness)* formula iff $[\![\varphi]\!]$ is a safety (liveness) property. Some formulas are neither safety nor liveness. For instance, $G\, p$ is a safety formula, $F\, q$ is a liveness formula, and $p \cup q$, equivalent to $(p \, W \, q) \wedge F\, q$, is neither.

3   The Subset of Linear Violation

Recall that a formula is said to be safety iff it characterizes that "something bad" will never happen. Thus a safety formula is violated iff something bad has happened. Since the "bad thing" happens after a finite execution, for any safety formula it is possible to characterize the set of violating prefixes by an automaton on finite words. In this document we focus on formulas for which this violation can be characterized by an automaton on finite words and by a regular expression, both of *linear* size.

Intuitively, a word v violates a formula φ iff v carries enough evidence to conclude that φ does not hold on v and any of its extensions. Using the terminology of [8] we get that a word v violates φ iff $v \models^+ \neg \varphi$. Thus, we define:

Definition 8. *Let φ be a safety formula. We say that a set of words L recognizes violation of φ iff $L = \{v \mid v \models^+ \neg \varphi\}$.*

The following definition captures the set of RLTL formulas whose violation can be recognized by a automaton or a regular expression, both of linear size. We denote this set by RLTL^{LV} (where LV stands for "linear violation").

Definition 9 (RLTL^{LV}). *If b is a boolean expression, r is an RE and φ, φ_1 and φ_2 are RLTL^{LV} then the following are in RLTL^{LV}:*

1. b
2. $\varphi_1 \wedge \varphi_2$
3. $X\varphi$
4. $(b \wedge \varphi_1) \vee (\neg b \wedge \varphi_2)$
5. $(b \wedge \varphi_1) \, W \, (\neg b \wedge \varphi_2)$
6. $r \mapsto \varphi$

Construction Via Violating RE

In the following we define inductively the RE recognizing the violation of RLTL^{LV}. By Fact 4 this RE can be translated into an automaton on finite words linear in the size of the RE. Since the RE itself is linear in the size of the input formula, this gives a procedure for generating an automaton of linear size for every formula in RLTL^{LV}. In Section 5 we provide a direct procedure for generating an automaton of linear size. The benefits of the regular expression are in its succinctness and in the ability to use existing code for translating regular expressions into finite automata.

Definition 10 (Violating RE). *Let b be a boolean expression, r an RE, and $\varphi, \varphi_1, \varphi_2$ RLTL^{LV} formulas. The violating RE of an RLTL^{LV} formula φ, denoted $\mathcal{V}(\varphi)$ is defined as follows:*

1. $\mathcal{V}(b) = \neg b$
2. $\mathcal{V}(\varphi_1 \wedge \varphi_2) = \mathcal{V}(\varphi_1) \cup \mathcal{V}(\varphi_2)$
3. $\mathcal{V}(X\varphi) = true \cdot \mathcal{V}(\varphi)$
4. $\mathcal{V}((b \wedge \varphi_1) \vee (\neg b \wedge \varphi_2)) = (b \circ \mathcal{V}(\varphi_1)) \cup (\neg b \circ \mathcal{V}(\varphi_2))$
5. $\mathcal{V}((b \wedge \varphi_1) \, W \, (\neg b \wedge \varphi_2)) = b^* \cdot ((b \circ \mathcal{V}(\varphi_1)) \cup (\neg b \circ \mathcal{V}(\varphi_2)))$
6. $\mathcal{V}(r \mapsto \varphi) = r \circ \mathcal{V}(\varphi)$

That is, a boolean expression is violated iff its negation holds. The conjunction of two formulas is violated iff either one of them is violated. The formula $X\varphi$ is violated iff

φ is violated at the next letter. The formula $(b \wedge \varphi_1) \vee (\neg b \wedge \varphi_2)$ is violated iff either b holds and φ_1 is violated or b does not hold and φ_2 is violated. The formula $(b \wedge \varphi_1) \mathsf{W} (\neg b \wedge \varphi_2)$ is violated iff after a finite number of letters satisfying b either b holds and φ_1 is violated or b does not hold and φ_2 is violated. The formula $r \mapsto \varphi$ is violated iff φ is violated after a prefix tightly satisfying r.

The following theorem, which is proven in the appendix, states that $\llbracket \mathcal{V}(\varphi)! \rrbracket$ recognizes violation of the formula φ. That is, that a word v has a prefix u tightly satisfying $\mathcal{V}(\varphi)$ if and only if $v \models^+ \neg\varphi$.

Theorem 11. *Let φ be an* RLTL^{LV} *formula over* $\mathbb{P}$ *and let v be a word over* $\Sigma_{\mathbb{P}}$. *Then,* $|\mathcal{V}(\varphi)| = O(|\varphi|)$ *and* $\llbracket \mathcal{V}(\varphi)! \rrbracket$ *recognizes violation of φ.*

Thus, as stated by the following corollary, for any formula φ of RLTL^{LV} one can construct a monitor recognizing the violation of φ, with size linear in φ.

Corollary 12. *Let φ be an* RLTL^{LV} *formula. Then there exists an* NFA *of linear size recognizing the violation of φ.*

Proof Sketch. Direct corollary of Theorem 11 and Fact 4. $\square$

In Section 5 we give a direct construction for such an automaton.

4 Discussion

As mentioned in the introduction, we have derived the definition of RLTL^{LV} by combining several results in the literature. In this section we discuss the relation of RLTL^{LV} to other logics in the literature associated with efficient verification, thus clarifying the nature of this subset.

4.1 Relation to Classification of Safety Formulas

In [15] Kupferman and Vardi classify safety formulas according to whether all, some, or none of their bad prefixes are informative, and name them *intentionally* safe, *accidentally* safe and *pathologically* safe, respectively. A finite word v is a *bad prefix* for φ iff for any extension w of v, $w \not\models \varphi$. A finite word v is an *informative bad prefix* for φ iff $v \models^+ \neg\varphi$.[2] It follows that the violating regular expression $\mathcal{V}(\varphi)$ characterizes exactly the set of informative bad prefixes of φ.

Theorem 13. *Let φ be an* RLTL^{LV} *formula. Then* $\llbracket \mathcal{V}(\varphi)! \rrbracket$ *defines the set of informative bad prefixes for φ.*

Proof Sketch. Follows immediately from Theorem 11 and Definition 8. $\square$

The subset RLTL^{LV} is syntactically weak by definition (since all formulas are in positive normal form and use only weak operators). Thus, by [15] all RLTL^{LV} formulas

---

[2] The concept of an informative prefix was defined syntactically in [15]. We have used the semantic equivalent definition provided by [8].

are non-pathologically safe (i.e. they are either intentionally safe or accidentally safe). Therefore, every computation that violates an RLTL^{LV} formula has at least one informative bad prefix.

Theorem 14. *All formulas of* RLTL^{LV} *are non-pathologically safe.*

Proof Sketch. The formulas of RLTL^{LV} are syntactically weak by definition. They are non-pathologically safe by [15, Theorem 5.3]. □

Kupferman and Vardi in [15], characterize an automaton as *fine* for φ iff for every word violating φ it recognizes at least one informative bad prefix for φ. They show that violation of non-pathologically safe formulas can be detected by a fine alternating automaton on infinite words of linear size or by a fine non-deterministic automaton on finite words of *exponential* size. Corollary 12 gives us that for RLTL^{LV} violation can be detected by a fine *linear*-sized finite automaton on finite words.

4.2   Expressibility and Relation to the Common Fragment of LTL and ACTL

In [16] Maidl studied the subset of ACTL formulas which have an equivalent in LTL. She defined the fragments ACTL^{DET} and LTL^{DET} of ACTL and LTL, respectively, and proved that any formula of LTL that has an equivalent in ACTL is expressible in LTL^{DET}. And vice versa, any formula of ACTL that has an equivalent in LTL is expressible in ACTL^{DET}.

  Below we repeat the definition of LTL^{DET}, and give its restriction to safety formulas which we denote LTL^{LV}.[3]

Definition 15 (LTL^{DET},LTL^{LV}). *Let b be a boolean expression.*

  – *The set of* LTL^{DET} *formulas is recursively defined as follows:*

$$\varphi ::= b \mid \varphi \wedge \varphi \mid (b \wedge \varphi) \vee (\neg b \wedge \varphi) \mid X\varphi \mid (b \wedge \varphi)\, W\, (\neg b \wedge \varphi) \mid (b \wedge \varphi)\, U\, (\neg b \wedge \varphi)$$

  – *The set of* LTL^{LV} *formulas is recursively defined as follows:*

$$\varphi ::= b \mid \varphi \wedge \varphi \mid (b \wedge \varphi) \vee (\neg b \wedge \varphi) \mid X\varphi \mid (b \wedge \varphi)\, W\, (\neg b \wedge \varphi)$$

Theorem 16. LTL^{LV} *is a strict subset of* RLTL^{LV}.

Proof Sketch. Follows from Definitions 9 and 15. □

Maidl additionally showed that the negation of a formula φ in LTL^{DET} is recognizable by a 1-weak Büchi automaton with size linear in φ. Thus intuitively, the safety formulas of LTL^{DET} should have an NFA of linear size. Theorem 16 together with Corollary 12 state that indeed the restriction of LTL^{DET} to safety lies in the subset of formulas whose violation can be detected by an NFA of linear size.

  Note that LTL^{LV} is not only syntactically weaker than RLTL^{LV} it is also semantically weaker than RLTL^{LV} (i.e. it has less expressive power). This can be seen by taking the

---

[3] The set ACTL^{DET} and its restriction to safety formulas, which we denote ACTL^{LV}, are obtained by replacing the operators X with AX, W with AW, and U with AU in the definitions of LTL^{DET} and LTL^{LV}, respectively.

same example as Wolper's in showing that LTL is as expressive as star free omega regular languages rather than entire omega regular languages [21]. That is, by showing that the property "b holds at every even position (yet b may hold on some odd positions as well)" is not expressible in LTL^{LV} (since it is not expressible in LTL) while it is expressible in RLTL^{LV} by the formula: $true \cdot (true \cdot true)^* \mapsto b$.

4.3   Relation to on-the-fly Verification of RCTL

In [1] Beer et al. defined the logic RCTL which extends CTL with regular expressions via a suffix implication operator. In addition they defined a subset of this logic which can be verified by on-the-fly model checking (we term this subset RCTL^{OTF}, where OTF stands for "on-the-fly"). A formula can be verified by on-the-fly model checking iff it can be verified by model checking an invariant property over a parallel composition of the design model with a finite automaton. Their definitions can be rephrased as follows.

Definition 17 (RCTL,RCTL^{OTF}). *Let b be a boolean expression, and r a regular expression.*

– *The set of* RCTL *formulas is recursively defined as follows:*

$$\varphi ::= b \mid \neg\varphi \mid \varphi \wedge \varphi \mid \textbf{\textit{EX}}\,\varphi \mid \textbf{\textit{EG}}\,\varphi \mid \varphi\textbf{\textit{EU}}\,\varphi \mid r \mapsto \varphi$$

Additional operators are defined as syntactic sugaring of the above operators:

- $\varphi \vee \psi \stackrel{\text{def}}{=} \neg(\neg\varphi \wedge \neg\psi)$
- $\varphi\,\textbf{\textit{EW}}\,\psi \stackrel{\text{def}}{=} \textbf{\textit{EG}}\,\varphi \vee (\varphi\,\textbf{\textit{EU}}\,\psi)$
- $\textbf{\textit{AF}}\,\varphi \stackrel{\text{def}}{=} \neg(\textbf{\textit{EG}}\,\neg\varphi)$
- $\textbf{\textit{EF}}\,\varphi \stackrel{\text{def}}{=} true\,\textbf{\textit{EU}}\,\varphi$
- $\varphi\,\textbf{\textit{AU}}\,\psi \stackrel{\text{def}}{=} \textbf{\textit{AF}}\,\psi \wedge (\varphi\,\textbf{\textit{AW}}\,\psi)$
- $\textbf{\textit{AX}}\,\varphi \stackrel{\text{def}}{=} \neg(\textbf{\textit{EX}}\,\neg\varphi)$
- $\varphi \to \psi \stackrel{\text{def}}{=} \neg\varphi \vee \psi$
- $\varphi\,\textbf{\textit{AW}}\,\psi \stackrel{\text{def}}{=} \neg(\neg\psi\,\textbf{\textit{EU}}\,\neg(\varphi \vee \psi))$
- $\textbf{\textit{AG}}\,\varphi \stackrel{\text{def}}{=} \neg(\textbf{\textit{EF}}\,\neg\varphi)$

– *The set of* RCTL^{OTF} *formulas is recursively defined as follows:*

$$\varphi ::= b \mid \varphi \wedge \varphi \mid b \to \varphi \mid \textbf{\textit{AX}}\varphi \mid \textbf{\textit{AG}}\varphi \mid r \mapsto \varphi$$

It can easily be seen that the subset obtained from RCTL^{OTF} by omitting the path quantifiers (replacing the operators AX with X and AG with G) is a strict subset of RLTL^{LV}. And as stated by the following theorem, the subset of RCTL formulas that can be verified on-the-fly can be extended to include all formulas in the subset obtained from RLTL^{LV} by adding the universal path quantifier.

Theorem 18. *Let* RCTL^{LV} *be the subset of* RCTL *formulas obtained by adding the universal path quantifier to the operators* X *and* W *in the definition of* RLTL^{LV}. *Then*

1. RCTL^{OTF} *is a strict subset of* RCTL^{LV}, *and*
2. *for every formula φ in* RCTL^{LV} *there exists an* NFA *of linear size recognizing the informative bad prefixes of φ.*

Proof Sketch. The first item is correct by definition. The second item follows by applying the construction of the violating regular expression (Definition 10) to the formula obtained by removing the path quantifiers, using Maidl's result (that an ACTL formula has an equivalent in LTL iff it has an equivalent in ACTL^{DET} [16, Theorem 2]) together with Clarke and Draghicescu's result [5] (that a CTL formula φ has an equivalent in LTL iff it is equivalent to the formula φ' obtained by removing the path quantifiers from φ) and Corollary 12.   □

4.4   Relation to PSL

The language reference manual (LRM) of PSL [11] defines *the simple subset*, which intuitively conforms to the notion of monotonic advancement of time, left to right through the property, as in a timing diagram. For example, the property $G(a \rightarrow Xb)$ is in the simple subset whereas the property $G((a \wedge Xb) \rightarrow c)$ is not. This characteristic should in turn ensure that properties within the subset can be verified easily by simulation tools, though no construction or proof for this is given. The exact definition is given in terms of restriction on operators and their operands as can be depicted in [11, subsection 4.4.4, p. 29].

   As explained in the introduction since we are interested in model checking as well, we focus on the safety formulas of the simple subset. There are two differences between the definition of the safety simple subset in the LRM and the definition of RLTL^{LV}.

1. The definition in the LRM considers weak regular expressions while RLTL does not. A construction for the safety simple subset of the LRM, including weak regular expressions, is available in [19]. The reason we exclude weak regular expressions is that their verification involves determinism which results in an automaton exponential rather than linear in the size of the formula.
2. The restrictions in the LRM are a bit more restrictive than those in RLTL^{LV}. To be exact, for the operators $\vee$ and W the simple subset allows only one operand to be non-boolean, whereas RLTL^{LV} allows both to be non-boolean, conditioned they can be conjuncted with some Boolean and its negation. It is quite clear that the motivation for over restricting these operators in the definition of the simple subset in the LRM is to simplify the restricting rule.

5   Direct Construction

In the following we give a direct construction for an NFA recognizing the violation of a formula. This automaton recognize a bad informative prefix for any computation violating the given formula: if a final location is reached on some run, then the observed prefix is a bad informative prefix. The advantage of this construction over the previous one is in it being a direct one, thus enabling various optimizations.

Theorem 19 . *Let φ be a RLTL^{LV} formula over $\mathbb{P}$ and let w be a word over $\Sigma_{\mathbb{P}}$. Then there exists an NFA $\mathcal{N}_\varphi$ linear in the size of φ recognizing violation of φ.*

In the following subsection we provide the construction, we give the proof of its correctness in the appendix.

As $\mathcal{N}_\varphi$ recognizes bad prefixes we can reduce the verification of φ to verification of the invariant property AG ($\neg$"The automaton $\mathcal{N}_\varphi$ is in a final state") on a parallel composition of the design model with $\mathcal{N}_\varphi$. In order to state this formally one needs to define a symbolic transition system which can model both a given design and an NFA, and define the result of a parallel composition of two such systems. An appropriate mathematical model for this is a discrete transition system DTS (see e.g. [3]), inspired from the fair transition system of [13]. The following corollary states the result formally using the notion of a DTS.

Corollary 20. *Let φ be an* RLTL^{LV} *formula. There exists an* NFA *$\mathcal{N}_\varphi = (P, S, S_0, \delta, F)$ linear in the size of φ such that for any model M*

$$\mathcal{D}(M) \models \varphi \quad iff \quad \mathcal{D}(M) \,\|\, \mathcal{D}(\mathcal{N}_\varphi) \models \mathsf{AG}(\neg \mathsf{at}(F))$$

where $\mathcal{D}(M)$ and $\mathcal{D}(\mathcal{N}_\varphi)$ are the DTSs *for M and $\mathcal{N}_\varphi$ respectively, and $\mathsf{at}(F)$ is a boolean expression stating that $\mathcal{N}_\varphi$ is in a final location.*

Note that if we define a CO-UFA $\mathcal{C}_\varphi$ with the same components of N_φ (i.e. $\mathcal{C}_\varphi = \langle P, S, S_0, \delta, F \rangle$) then $\mathcal{C}_\varphi$ accepts a word w iff $w \models \varphi$.

Constructing an NFA for a Given RLTL^{LV} Formula

Let r be an RE such that $\epsilon \notin \mathcal{L}(r)$, b a boolean expression, $\varphi, \varphi_1, \varphi_2$ RLTL^{LV} formulas. For the induction construction, let $\mathcal{N}_\varphi = \langle P, Q, Q_0, \delta, F \rangle$, $\mathcal{N}_{\varphi_1} = \langle P^1, Q^1, Q_0^1, \delta^1, F^1 \rangle$ and $\mathcal{N}_{\varphi_2} = \langle P^2, Q^2, Q_0^2, \delta^2, F^2 \rangle$.

1. Case b.

$$\mathcal{N}_b = \langle P, \{q_0, q_1, q_2\}, \{q_1\}, \delta, \{q_0\} \rangle$$

   where P is the set of state variables in b and

$$\delta = \{(q_1, b, q_2), (q_1, \neg b, q_0), (q_0, \mathit{true}, q_0), (q_2, \mathit{true}, q_2)\}.$$

2. Case $\varphi_1 \wedge \varphi_2$.

$$\mathcal{N}_{\varphi_1 \wedge \varphi_2} = \langle P^1 \cup P^2, Q^1 \cup Q^2, Q_0^1 \cup Q_0^2, \delta^1 \cup \delta^2, F^1 \cup F^2 \rangle$$

   A run of $\mathcal{N}_{\varphi_1 \wedge \varphi_2}$ has a non-deterministic choice between a run of $\mathcal{N}_{\varphi_1}$ and a run of $\mathcal{N}_{\varphi_2}$. In any choice it should not reach a state in either F_1 or F_2. The resulting NFA is described in Figure 1.

3. Case $\mathsf{X}\varphi$.

$$\mathcal{N}_{\mathsf{X}\varphi} = \langle P, Q \cup \{s_0\}, \{s_0\}, \delta', F \rangle$$

   where s_0 is a new state and $\delta' = \delta \cup \bigcup_{q \in Q_0}(s_0, \mathit{true}, q)$. The resulting NFA is described in Figure 2.

4. Case $(b \wedge \varphi_1) \vee (\neg b \wedge \varphi_2)$.

$$\mathcal{N}_{(b \wedge \varphi_1) \vee (\neg b \wedge \varphi_2)} = \langle P^1 \cup P^2, \{s_0\} \cup Q^1 \cup Q^2, \{s_0\}, \delta', F^1 \cup F^2 \rangle$$

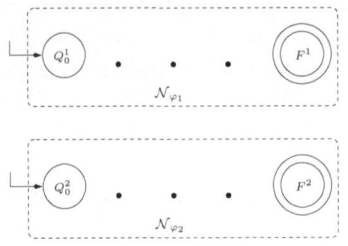

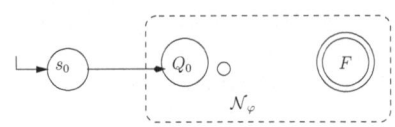

Fig. 1. An NFA for $\varphi_1 \wedge \varphi_2$ **Fig. 2.** An NFA for $\mathsf{X}\varphi$

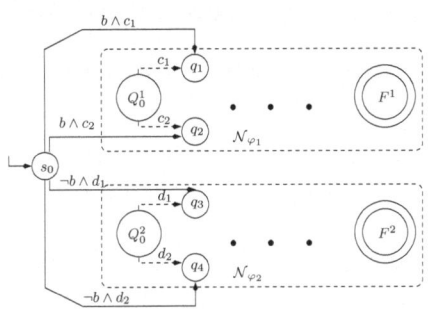

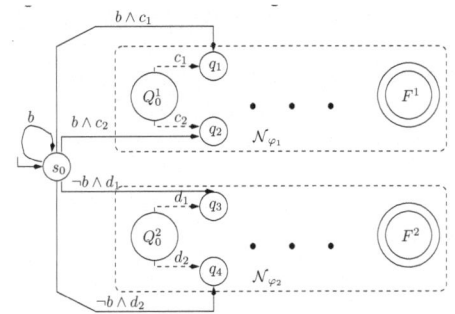

Fig. 3. An NFA for $(b \wedge \varphi_1) \vee (\neg b \wedge \varphi_2)$ **Fig. 4.** An NFA for $(b \wedge \varphi_1) \mathsf{W} (\neg b \wedge \varphi_2)$

where

$$\delta' = \delta^1 \cup \delta^2 \cup \\
\bigcup_{q_1 \in Q_0^1} \bigcup_{(q_1,c,q_2) \in \delta^1} (s_0, b \wedge c, q_2) \\
\bigcup_{q_1 \in Q_0^2} \bigcup_{(q_1,c,q_2) \in \delta^2} (s_0, \neg b \wedge c, q_2)$$

A run of $\mathcal{N}_{(b\wedge\varphi_1)\vee(\neg b\wedge\varphi_2)}$ starts in a new state s_0, if b holds it continues on $\mathcal{N}_{\varphi_1}$ otherwise it continues on $\mathcal{N}_{\varphi_2}$. The resulting NFA is described in Figure 3.

5. Case $(b \wedge \varphi_1) \mathsf{W} (\neg b \wedge \varphi_2)$.

$$\mathcal{N}_{(b\wedge\varphi_1)\mathsf{W}(\neg b\wedge \varphi_2)} = \langle P^1 \cup P^2, \{s_0\} \cup Q^1 \cup Q^2, \{s_0\}, \delta', F^1 \cup F^2\rangle$$

where

$$\delta' = (s_0, b, s_0) \cup \delta^1 \cup \delta^2 \cup \\
\bigcup_{q_1 \in Q_0^1} \bigcup_{(q_1,c,q_2) \in \delta^1} (s_0, b \wedge c, q_2) \\
\bigcup_{q_1 \in Q_0^2} \bigcup_{(q_1,c,q_2) \in \delta^2} (s_0, \neg b \wedge c, q_2)$$

The resulting NFA is described in Figure 4.

6. Case $r \mapsto \varphi_2$

Let $N = \langle \mathbb{B}_P, Q, Q_0, \delta, F\rangle$ be a non-deterministic automata on finite words accepting $\mathcal{L}(r)$ (as in [2] for example). Convert it to an NFA $\mathcal{N}_1$ recognizing violation

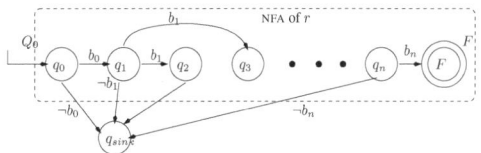

Fig. 5. An NFA for $r \mapsto \mathit{false}$

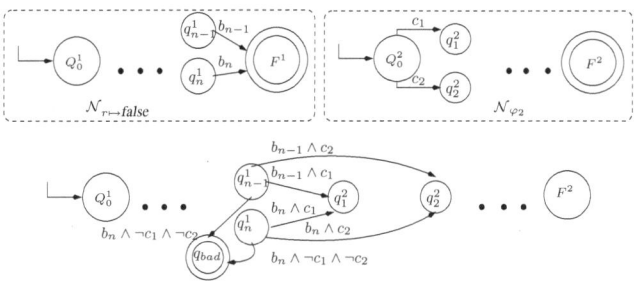

Fig. 6. An NFA for $r \mapsto \varphi_2$

of $[\![r \mapsto \mathit{false}]\!]$ by sending every "missing edge" to a new trapping state and making the set of final states of N the set of bad states of $\mathcal{N}_1$. The resulting NFA, $\mathcal{N}_1 = \langle P^1, Q^1, Q_0^1, \delta^1, F^1 \rangle$ is defined as follows.

$$\mathcal{N}_1 = \langle P, Q \cup \{q_{sink}\}, Q_0, \delta^1, F \rangle$$

where P is the set of atomic propositions in r and

$$\delta^1 = \bigcup_{(q_1, c, q_2) \in \delta} \{(q_1, c, q_2).(q_1, \neg c, q_{sink})\}.$$

The resulting NFA (recognizing violation of $[\![r \mapsto \mathit{false}]\!]$) is described in Figure 5. Let $\mathcal{N}_2 = \mathcal{N}_{\varphi_2} = \langle P^2, Q^2, Q_0^2, \delta^2, F^2 \rangle$ be the NFA, as constructed by induction for φ_2. Then $\mathcal{N}_{r \mapsto \varphi_2}$ is constructed by concatenation of $\mathcal{N}_1$ and $\mathcal{N}_2$ as follows:

$$\mathcal{N}_{r \mapsto \varphi_2} = \langle P^1 \cup P^2, Q^1 \cup Q^2 \cup \{q_{bad}\}, Q_0^1, \delta', F^2 \cup \{q_{bad}\} \rangle$$

where q_{bad} is a new state, and

$$\delta' = \begin{aligned} &\delta^1 \cup \delta^2 \cup \\ &\{(q_1, c_1 \wedge c_2, q_4) \mid \exists q_2 \in F^1, q_3 \in Q_0^2 \text{ s.t.} \\ &\qquad\qquad (q_1, c_1, q_2) \in \delta^1, (q_3, c_2, q_4) \in \delta^2\} \cup \\ &\{(q_1, c_1 \wedge \bigwedge_{\{c' \mid \exists q_3 \in Q_0^2, q_4 \in Q^2 \text{ s.t. } (q_3, c', q_4) \in \delta^2\}} \neg c', q_{bad}) \mid \exists q_2 \in F^1 \text{ s.t. } (q_1, c_1, q_2) \in \delta^1\} \end{aligned}$$

The resulting NFA is described in Figure 6.

6 Conclusions

We have defined a subset of safety RLTL formulas. This subset, while consisting of most formulas run in hardware verification, enjoys efficient verification algorithms. Verification of general RLTL formulas requires an exponential Büchi automaton. Verification of RLTL^{LV} formulas can be reduced to invariance checking using an auxiliary automaton on finite words (NFA). Moreover, the size of the generated NFA is linear in the size of the formula.

We have presented two procedures for the construction of a linear-sized automaton for an RLTL^{LV} formula. The first goes through building an equivalent linear-sized regular expression (RE), and the other directly constructs the automaton from the given formula. Both methods provide algorithms that can be easily implemented by tool developers. We note that for traditional regular expressions, the existence of an NFA of linear size does not imply the existence of a regular expression of linear size, as REs are exponentially less succinct than NFAs [6]. Since our translation to regular expression involves the *fusion* operator, it would be interesting to find whether the result of [6] holds for this type of regular expressions as well.

The PSL language reference manual (LRM) [11], defines a subset of the language, called *the simple subset*, which intuitively should consist of the formulas which are easy to verify by state of the art verification methods. However, there is no justification for choosing the defined subset, and in particular no proof that it meets the intuition it should. The restrictions made on RLTL in the definition of the simple subset in the LRM reflect the restrictions in RLTL^{LV}. Having proved that RLTL^{LV} formulas are easy to verify by both model checking and simulation, we view this paper as providing the missing proof.

Acknowledgments

We would like to thank Cindy Eisner and Orna Lichtenstein for their helpful comments on an early draft of the paper.

References

1. I. Beer, S. Ben-David, and A. Landver. On-the-fly model checking of RCTL formulas. In *Proc. 10th International Conference on Computer Aided Verification (CAV'98)*, LNCS 1427, pages 184–194. Springer-Verlag, 1998.
2. S. Ben-David, D. Fisman, and S. Ruah. Automata construction for regular expressions in model checking, June 2004. IBM research report H-0229.
3. S. Ben-David, D. Fisman, and S. Ruah. Embedding finite automata within regular expressions. In 1st International Symposium on Leveraging Applications of Formal Methods. Springer-Verlag, November 2004.
4. D. Bustan, D. Fisman, and J. Havlicek. Automata construction for PSL. Technical Report MCS05-04, The Weizmann Institute of Science, May 2005.
5. E.M. Clarke and I.A. Draghicescu. Expressibility results for linear-time and branching-time logics. In *Proc. Workshop on Linear Time, Branching Time, and Partial Order in Logics and Models for Concurrency*, LNCS 354, pages 428–437. Springer-Verlag, 1988.

6. Andrzej Ehrenfeucht and Paul Zeiger. Complexity measures for regular expressions. In *STOC '74: Proceedings of the sixth annual ACM symposium on Theory of computing*, pages 75–79, New York, NY, USA, 1974. ACM Press.

7. C. Eisner, D. Fisman, and J. Havlicek. A topological characterization of weakness. In *PODC '05: Proceedings of the twenty-fourth annual ACM SIGACT-SIGOPS symposium on Principles of distributed computing*, pages 1–8, New York, NY, USA, 2005. ACM Press.

8. C. Eisner, D. Fisman, J. Havlicek, Y. Lustig, A. McIsaac, and D. Van Campenhout. Reasoning with temporal logic on truncated paths. In *The 15th International Conference on Computer Aided Verification (CAV)*, LNCS 2725, pages 27–40. Springer-Verlag, July 2003.

9. M. J. Fischer and R. E. Ladner. Propositional dynamic logic of regular programs. In *J. Comput. Syst. Sci.*, pages 18(2), 194–211, 1979.

10. J.E. Hopcroft and J.D. Ullman. *Introduction to Automata Theory, Languages, and Computation*. Addison-Wesley, 1979.

11. IEEE. IEEE standard for property specification language (PSL), October 2005.

12. Christian Josef Kargl. A Sugar translator. Master's thesis, Institut für Softwaretechnologie, Technische Universität Graz, Graz, Austria, December 2003.

13. Y. Kesten, A. Pnueli, and L. Raviv. Algorithmic verification of linear temporal logic specifications. volume 1443, pages 1–16, 1998.

14. O. Kupferman and M.Y. Vardi. Freedom, weakness, and determinism: From linear-time to branching-time. In *Proc. 13th IEEE Symposium on Logic in Computer Science*, June 1995.

15. Orna Kupferman and Moshe Y. Vardi. Model checking of safety properties. In *Proc. 11th International Conference on Computer Aided Verification (CAV)*, LNCS 1633, pages 172–183. Springer-Verlag, 1999.

16. Monika Maidl. The common fragment of CTL and LTL. In *IEEE Symposium on Foundations of Computer Science*, pages 643–652, 2000.

17. Z. Manna and A. Pnueli. *The Temporal Logic of Reactive and Concurrent Systems: Specification*. Springer-Verlag, New York, 1992.

18. A. Pnueli. A temporal logic of concurrent programs. *Theoretical Computer Science*, 13:45–60, 1981.

19. S. Ruah, D. Fisman, and S. Ben-David. Automata construction for on-the-fly model checking PSL safety simple subset, June 2005. Research Report H-0234.

20. Moshe Y. Vardi and Pierre Wolper. Reasoning about infinite computations. *Information and Computation*, 115(1):1–37, 15 1994.

21. P. Wolper. Temporal logic can be more expressive. *Information and Control*, 56(1/2):72–99, 1983.

A Case for Runtime Validation of Hardware

Sharad Malik

Dept. of Electrical Engineering,
Princeton University
sharad@princeton.edu

Abstract. Increasing hardware design complexity has resulted in significant challenges for hardware design verification. The growing "verification gap" between the complexity of what we can verify and what we can fabricate/design is indicative of a crisis that is likely to get only worse with increasing complexity. A variety of methodology and tool solutions have been proposed to deal with this crisis, but there is little optimism that a single solution or even a set of cooperative solutions will be scalable to enable future design verification to be cost effective. It is time we reconcile ourselves to the fact that hardware, like software, will be shipped with bugs in it. One possible solution to deal with this inevitable scenario is to provide support for runtime validation that detects functional failures at runtime and then recovers from such failures. Such runtime validation hardware will increasingly be used to handle dynamic operational failures caused by reduced reliability of devices due to large process variations as well as increasing soft errors. Expanding the use of such hardware to deal with functional design failures provides for an on-chip insurance policy when design errors inevitably slip through the verification process. This paper will discuss the strengths and weaknesses of this form of design validation, some possible forms this may take, and implications on design methodology.

1 The Verification Gap

Unlike area, performance or power, hardware design complexity cannot be quantified precisely. Probably the closest acceptable measure for design complexity is the number of states in the system. While the value of this measure may be debatable due to the uncertain role played by memory in increasing design complexity, it is perhaps the best measure we have. The number of states is exponential in the number of state bits; which in turn tends to be proportional to the number of transistors in the design. This, thanks to Moore's Law has roughly doubled every few years or so. Thus, the exponential growth rate due to Moore's Law combined with the exponential dependence of the number of states on the number of state bits provides for a first order estimate of the growth rate of design complexity that is doubly exponential over time. However, at the same time, the exponential growth rate due to Moore's Law has led to faster computation that we can bring to bear in terms of simulation cycles and speed of formal verification algorithms. This still leaves us with a complexity growth rate that is increasing exponentially compared to our ability to deal with it. This has

S. Ur, E. Bin, and Y. Wolfsthal (Eds.): Haifa Verification Conf. 2005, LNCS 3875, pp. 30–42, 2006.
© Springer-Verlag Berlin Heidelberg 2006

led to the so-called verification gap between designs that we can verify and designs that we can fabricate at reasonable cost [1].

This exponential gap between the growing complexity and our ability to deal with it is somewhat corroborated by some field data in the context of microprocessors that shows an increase of design bugs that is linear in the number of transistors and thus exponential over time [2]. This increase in complexity and the corresponding bug rate have led to an exponential increase in the verification effort, as measured by both the number of simulation vectors, as well as the number of engineer years [3]. This effort and cost continues to be justified given the high cost of post-manufacture bugs. A post-manufacture, but pre-deployment bug may result in one or more respins of silicon and each respin is estimated to run at several millions of dollars in mask and other costs. Recent articles point to a study by Collet International showing that the number of respins is increasing in recent years. Further, it is claimed that 71% of respins for Systems-on-a-Chip are due to logic bugs [3]. A further cost of delays due to logic bugs discovered late in the design and manufacturing process is the lost market opportunity. Significant delays in product deployment may wipe out large profit margins or even lead to cancellation of projects. Post-deployment bugs may be even more costly. The Pentium FDIV bug reportedly cost Intel several hundred million dollars.

Given the large costs of and increasing number of bugs, verification efforts on design projects will continue to grow to cope with this situation. However, an exponential growth in verification costs is untenable even in the near future. In fact, the limited data on increasing number of respins due to logic bugs is indicative of the growing number of bugs that continue to stay in designs even late in the process, despite large verification efforts. Such occurrences will only increase with time. I claim that soon we will need to reconcile ourselves to the fact that hardware, like software will be shipped with bugs. This may seem inconceivable for hardware for at least a couple of reasons. Software bugs result in failures relatively sporadically, given that software tends to be extensively tested at speed. In comparison, hardware validation is predominantly through simulation, which is done at a tiny fraction of the speed of eventual deployment. Thus, bugs in hardware may manifest themselves relatively often when run at speed. This is unacceptable. Second, software bugs can generally be dealt with by resetting the system to some safe state. It is unclear how to do robust, practical and reliable recovery for hardware. Clearly, some creative ideas are needed to deal with this scenario.

2 A Possible Alternative

This paper proposes the general use of runtime validation as a mechanism to deal with hardware bugs that are triggered post-deployment. This involves the integration of two specific tasks in hardware along with the original design. The first is checking for functional failures and the second is recovering from them. Runtime validation is not intended to replace existing pre-fabrication verification, but rather complement it. It should be viewed as an insurance policy that is exercised in unforeseen circumstances. This mechanism is intended to improve designs and the design process in two distinct ways. The first is to provide robustness against the design errors that are increasingly likely to escape through the verification process. The second, and equally important, is to rein in verification costs by scaling back verification efforts given the availability of runtime validation to catch escaped bugs. Currently chips are verified to death; runtime validation will help verify them to life.

Runtime validation has some favorable characteristics. First, it does not need to deal with all possible behaviors of the design, but rather only the actual behavior. Thus, in some sense it is insensitive to the "state space explosion problem", as it never actually visits the complete design space. Instead, it just needs to monitor the current system state, which is linear in the size of the design. Second, the checking circuits do not need to understand the cause of the bug, but rather just recognize the symptom when the bug is exercised. For example, for a division operation it is sufficient to determine that the result is incorrect using a relatively simpler multiplication, rather than determine the cause for the incorrect result. In another example, if a bus is deadlocked, it is sufficient for the checker to recognize the deadlock, rather than determine the cause of the deadlock. These examples also help illustrate the different forms the checkers may take – checking the results of computations as well as checking temporal properties.

The recovery process is not likely to be easy though. Guaranteeing forward progress in the face of a functional failure has two requirements. First, there needs to be an alternate computation path that is completely different from the current design, and yet easily verifiable pre-fabrication so as to be trusted at runtime. Second, corrupt computation needs to be avoided using either backward (rollback), or forward (commit results only after validation) recovery mechanisms. However, as with integrity checking there is one favorable aspect to recovery. The recovery circuits do not need to fix the bug, but rather just ensure correct forward progress. For the divider, recovery can fix the result of the particular buggy case by using a simpler albeit slower circuit. In the case of the deadlocked bus, the recovery could release all resources and sequentially process all outstanding bus requests in a "safe" albeit slow mode. Both these examples illustrate the trade-off between design speed and complexity. Generally, we trade-off design complexity in favor of speed; going with complex but high performance designs. The general philosophy for recovery circuits in this setting is to make this trade-off in the opposite direction, going with simplicity at the cost of speed. If recovery is triggered rarely, its amortized cost will be negligible.

Besides the above rationale, there is another important reason to consider the use of runtime validation for functional failures. There is increasing concern about dynamic design failures due to a variety of sources. Increasing process variations for succeeding technology generations will make it less likely for worst case design to result in expected design improvements. Various possibilities for "Better than worst-case design" are being explored which must allow for some form of occasional failure that must be recovered from [4]. Further, operational failures may be triggered by thermal conditions, aging, or energized particle hits (soft-errors), which are all expected to have greater impact with finer device geometries. Given the dynamic nature of these failures, there is no alternative but to deal with them at runtime. Thus, runtime checking and recovery is likely to be used in some form in emerging designs anyway. It will be helpful to examine if their scope can be expanded to consider functional failures as well.

While runtime validation for functional, transient and other failure modes all require some form of checking to detect the failure, the recovery process may be very different for different failure modes. Fig. 1 provides a classification of error sources along two axes. The x-axis refers to the temporal nature of the failure – a failure may be transient or permanent. The y-axis specifies the spatial nature of the failure. It is deterministic if all instantiations (in space) of the design will fail and probabilistic if an instantiation will fail only probabilistically. The figure also shows how various failures are classified along these two axes. The simplest case of transient and probabilistic failures covers many different failures. For this point in the failure space, each instance of the design/component need not fail, nor each usage of such an instance fail. Soft-errors as well as failures due to variations in processing and operating conditions fall in this category. Also in this category are failures due to aggressive deployment of designs, for example by aggressive clocking for high performance and/or lowering of supply voltage for low power as has been suggested for "Better than Worst-Case Design" scenarios [4]. More complex is the case of manufacturing defects. In this case the failure is temporally permanent, every invocation of the design/component will fail. However, it is spatially probabilistic as not all instantiations need fail. Finally, functional failures, or failures due to design bugs, occupy the most difficult part of this classification space. Here, the failure is permanent in time, and all instances of the design will fail in the same way. This classification also provides some insight into possible runtime solutions for different failure modes. Replication in space, for example by using triple-modular redundancy (TMR), is helpful with spatially probabilistic errors since different instances fail only probabilistically. Replication in time, for example using identical redundant threads of computation on the same processor, is helpful with temporally transient errors. However, neither of these simple forms of failure tolerance works with functional bugs due to their permanent, deterministic nature. A growing body of work in the architectural community that is geared towards transient failures is thus not applicable here and other forms of solutions are needed. However, given that solutions for

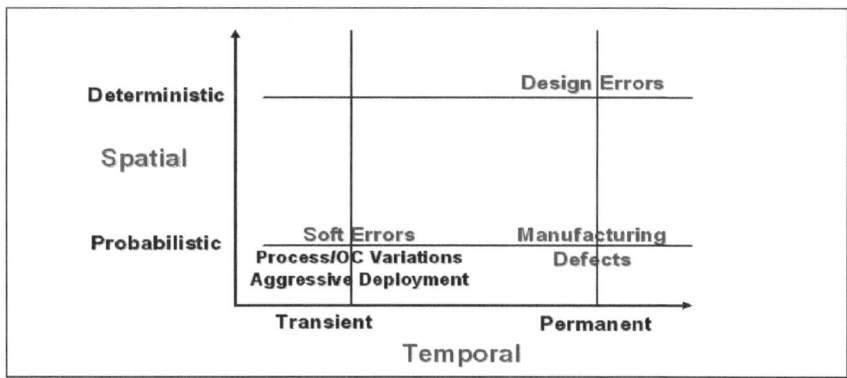

Fig. 1. Classification of Failure Modes

functional failures must in some sense be stronger than spatial or temporal replication, they will dominate and thus suffice for all other forms of failure. This is a very attractive attribute and makes it interesting to consider dealing with these various forms of failure in a unified way.

3   Runtime Validation: Specific Forms

This section will examine some instances of current and recent work on runtime validation capable of dealing with functional failures.

3.1   Instruction Set Processor Validation

The DIVA project [5] provides the first complete solution for runtime validation for an instruction set processor. In this solution a complex core processor is validated using a much simpler, and thus slower, checker processor. The core processor is similar to a modern superscalar processor with significant micro-architectural support for high performance. However, before the results of execution are committed (register and memory updates), they are validated by the simpler checker processor. The checker processor is provided the instructions in program order, along with the instructions' input operands/output result values as computed in the core processor. It recomputes the results of the execution, compares them with those of the core processor, and if no mismatch is detected, these are committed. In case of an error in the core processor, its results are flushed, the results of the checker used for this instruction and computation restarts in the core processor after the incorrect instruction.

Note that as discussed before, the core processor design trades-off complexity for speed, while the checker processor does the reverse. However, as the checker processor in this case performs both the checking and recovery functions, it needs to keep up with the core processor. A particularly elegant solution of letting the checker processor use the performance enhancing features of the core processor enables this. (This is enabled by a particularly elegant solution of using the pre-execution in the core processor to streamline the re-execution in the checker processor.) The core processor takes care of all the performance inefficiencies such as cache misses, branch mis-predictions and data dependencies, providing the checker with the final results of the performance enhancing features. The checker in turn validates them and the results of the computation. Any error in the performance aspects will only result in performance penalties and not functional errors. A net consequence is that this enables the much simpler and thus slower checker processor to keep up in speed with the core processor. The checker processor is made very simple so that it is easily amenable to formal verification techniques [6]. Further, it is made electrically more robust by using larger transistors. Finally, additional support is provided to make it resilient to soft errors. Note that as discussed earlier, the checker processor is only checking for errors, both functional and operational, and not their cause. Further, the recovery mechanism is only responsible for making forward progress and not fixing the cause of the errors.

3.2 Concurrent Processing

The DIVA solution for validating a complex processor exploits the fact that the committed instruction stream provides for a natural checkpointing of total computation at the end of each instruction. However, in many other cases it is less clear what needs to be checked to ensure correct computation. For example, a large number of errors in hardware are due to complex interactions of concurrent components. These are hard to capture with appropriate simulation traces and are also the bane of formal verification tools which need to deal with the large product state spaces of the concurrent components. In most cases the local sequential behavior of each component may be correct, but there may be an error in their concurrent interaction. Consequently some component uses an incorrect shared value for its computation. How do we check for correctness of concurrent interactions using runtime validation?

A specific effort directed towards examining this issue in the context of processors with simultaneous multi-threading (SMT) considers the various interactions between threads and checks that their shared data is used correctly [7]. The intra-thread correctness for each thread is handled using a DIVA style checker. The instructions in individual threads may be processed out of order in the core SMT processor; however each checker, as in DIVA, processes instructions in order. For the inter-thread correctness it points out three different interaction mechanisms between threads and provides for their correctness checkers.

3.2.1 Memory Consistency Between Threads and Checkers
The threads may share data through memory locations. Thus, the values read from shared memory can depend on the order in which reads and writes to these shared locations are processed. Further, the checkers must process these in an order consistent with the race resolution in the core processor; otherwise the results of the core processor and its checker will start to diverge. We denote the execution of the threads in the core processor as "main threads", and the execution in the checkers as "checker threads". The checker threads are not in lock step with the main threads, and in fact may have a lag of many cycles. Thus, it is entirely possible that the checker threads may proceed in a consistent but different order from the main threads resulting in computational divergence and a "false error", just because the main threads selected a different, but also consistent, processing order. For example, Fig. 2 shows the execution order in the main threads and the replay order in the checkers for a critical section mutual-exclusion code, which should run correctly under the sequential consistency memory model. The expected load/store values given the observed order are enclosed in the parenthesis. In this scenario, both loads in the checker processors will find a result mismatch and declare an error, which will nullify the subsequent execution and trigger recovery. However, this is actually unnecessary, since the main processor execution already maintains the mutual exclusion property required by the program. For applications with more than two threads, in the worst case, a false-negative error in one thread can invalidate useful work in other dependent threads in a domino effect, resulting in increased performance penalty and wasted energy.

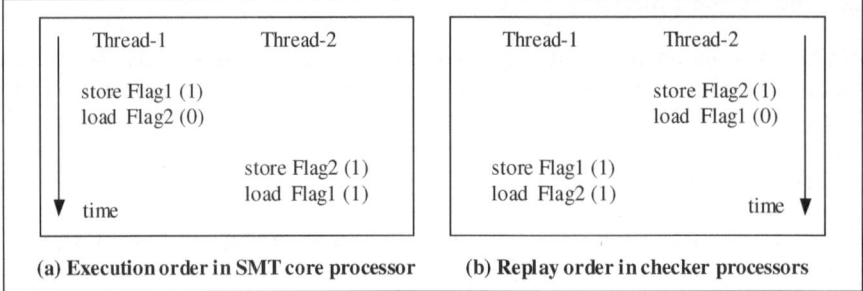

Fig. 2. Example of False Errors

The proposed solution exploits the fact that in the SMT processor, the load/store unit is thread-aware and responsible for correctly handling aliasing memory instructions. Before a memory instruction issues, it must search load/store queue entries from all threads to determine if a load needs to stall because of a prior conflicting store with unresolved value, or a store needs to forward the value to some later load. For example, the notion of "prior" and "later" can be determined by the order in which the instructions are dispatched into the pipeline. The correct memory ordering that the core processor is supposed to respect can be captured with simple runtime monitoring hardware support. The checker processors use this to ensure that the order in which they process the loads and stores are consistent with the main threads they are checking, while conforming to the memory consistency model.

3.2.2   Correctness of Synchronization Instructions

SMT architectures may support explicit synchronization instructions such as test-and-set, locks or semaphores. They create additional verification challenges. First, tailored hardware primitives can be used on SMT architecture to enable more efficient synchronization operation. In this case the synchronization may involve changing states within the processor, such as the shared variables in a lock box or semaphore table. Accordingly, the resilience against operational and functional failures is accomplished by providing a shadow version of these states for use by the checker processors. Again, as with shared memory accesses, the checker processors need to follow consistent replay order of accessing these locations to avoid false errors. Second, synchronization instructions have semantics that go beyond the normal computation in the thread. For example, a thread may be blocked or unblock other threads by executing a synchronization instruction. These specific semantics are enforced by the checker processors with dedicated synchronization hardware. The design is typically simpler than required in the main processor, because the checker re-execution does not involve complicated techniques such as speculation or out-of-order execution.

3.2.3   Forwarding of Erroneous Results

The results of computation of one thread may be used by another thread. In aggressive SMT, these results may be forwarded from the first to the second thread even before they have been committed to memory. Thus, if there is an error in these results, it may not have been caught by the checker of the first thread before it is forwarded. Thus, once the error is detected for the producing thread, not only does that stop and recover, but so does the consuming thread, since its computation result is also erroneous at that point. This is accomplished by preventing speculative inter-thread state change from being committed until the related traces of both the producing thread and consuming thread have been validated to be correct.

3.3 General Property Checking

Recent years have seen the emergence of design specification accompanying designs in the form of properties that must hold at various points in the design. This has even been formalized through the use of property specification languages [8] Currently these are used in two different ways. During simulation/emulation, these are used in the form of assertions that are triggered when a specific trace violates that property. During formal verification, these are used as temporal properties that need to be checked using formal verification techniques such as model checking. An easy extension of this to runtime validation is to synthesize critical property checkers as part of the final design. In fact, this is already done for use of these checkers in emulation, where they are implemented using FPGAs. Their synthesis as simple finite state machine implementations of the property specified in temporal logic is relatively straightforward [9]. In simulation or emulation when a checker detects failure – computation can stop and the failure diagnosed as part of design debugging. However in runtime validation, detection of failure must be followed by recovery. This may be non-trivial. As described earlier, this must ensure forward progress, typically by redoing computation through an alternative simpler implementation. This may not always be straightforward.

3.4 Interactions with Model Checking

As pointed out earlier, unlike model checking, runtime checkers do not face the state explosion problem. A runtime checker needs to only monitor the current state of the system (and possibly some local state of the checker). The only behavior of interest is what the system experiences. This is a major strength. However, in comparison with model checking, there are also a few weaknesses. It may be difficult to monitor distributed state in the system. For state that may be widely distributed across the chip or across chips, this may be a problem. The spatial distribution of state is not a problem for model checking, since this just ends up as variables in the software model. The second limitation of runtime checkers is that unlike model checking, they cannot be used to prove liveness properties, i.e. those properties where some outcome must eventually happen. The uncertainly of the period associated with "eventually" makes this impossible for a checker that does not get to see the entire state space. However, this is less of a problem in practice. Most liveness checks tend to be bounded in time in practice, i.e. the desired outcome must happen in a bounded number of steps. This bound can easily be built in a checker. In fact the use of "eventually" in temporal logic may be seen as a simplifying abstraction of real time that makes model checking easier. For runtime checkers, this "simplifying" assumption can be removed. Finally, while fairness constraints can be implemented in runtime checkers, this would need significant state in the checker and is not practical for most cases. In practice however, fairness properties tend not to be common for hardware designs, and this limitation should not be a major one.

Thus, overall we see complementary strengths of model checking and runtime checkers. The state explosion problem for model checkers does not exist for runtime checking and the distributed monitoring problem for runtime checkers is not a problem for model checking. This makes it tempting to consider the complementary integration of these two techniques. The following overall framework using compositional reasoning is one possible way of doing this.

Compositional reasoning has been used in various ways in formal verification to manage state space complexity [10]. The basic idea is illustrated using the following meta-example. Fig. 3 (a) shows a simple Design Under Verification (DUV), where components A and B interact with each other. Fig. 3 (b) shows how one can use compositional reasoning to verify each component separately. Here, A is verified under some assumptions regarding B. Similarly, B needs to be verified with some assumptions about A. In both cases, the assumptions need to be verified too. This simple method can be used for theorem proving, model checking or any kind of verification method including simulation and runtime validation. For example, runtime checkers can validate the assumptions, and model checking completes the verification. In using assumptions in compositional reasoning, in general, you need to be careful about cyclic dependencies, i.e. proving assumptions about A may depend on assumptions about B and vice-versa. However, with runtime checking of assumptions this is not a problem, as the assumptions are independently validated at runtime for the actual circuit behavior. When verification of A requires more detailed assumptions about B, it may be easier to use B', a highly abstracted version of B, to verify A as in Fig. 3 (c). In that case, runtime validation can help validate that B actually behaves like B', from A's point of view. Finally, it is also possible to use the runtime validation assumptions regarding the interaction of A and B as in Fig. 3 (d), and verify the model all together. All of the above cases exploit the fact that the runtime checkers need to monitor only local information and then their guarantees can be used to speed up offline model checking. Note also in the above that in addition to

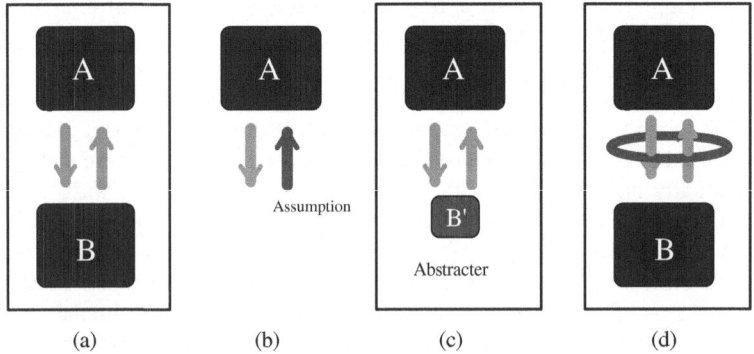

(a)   (b)   (c)   (d)

Fig. 3. Using Runtime Checking for Assumptions and Abstractions

the runtime checkers, it will be necessary to specify how these guarantees are maintained in case of functional failures, i.e. recovery circuits need to be specified for each of the properties being checked by the runtime checkers. This overall framework has been used to some benefit with some case-studies in recent work [11].

4 A General RTL Methodology

The previous section described some specific instances of using runtime validation. It is instructive to examine if some of the ideas in those solutions can be generalized to a general hardware design methodology. Specifically it would be helpful if runtime validation could be encapsulated in the RTL design process, from design specification through synthesis, with design tools taking up as much of the burden for this as possible. This section proposes the basics of such a methodology.

The essence of runtime validation is the ability to specify checking and recovery functions for any computation. Thus, at the RT level, it would be helpful if the design specification language would provide a construct of the form:

```
design(D){

        ...

}while checker(C){
}else recovery(R)
```

Informally, the above construct provides the ability to specify in addition to the design D, a checker C and a recovery circuit R that is triggered when the checker detects an error. Even before looking into the semantics of the above construct, just the fact that the original design, the checker and the recovery circuits have been separated out is of value. Clearly this separation is not required to specify these circuits, any hardware description language today can be used for that purpose. However, the language construct provide a clear separation of these and thus an ease in design management at the least. Further, given the clear separation, the compiler can handle the components differently if need be.

Beyond the minimal benefit of the separation of the checking and recovery circuits from the main design, the above construct can also provide us with the following basic hardware implementation template through its semantics. The checking circuit C operates in parallel with the design D. It is allowed to examine the state of D (possibly limited). On detecting an error, (some or all) computation in D is stalled, i.e. (some or all) state updates in D are disabled. Also, recovery R is triggered. R is permitted to change (possibly limited) state in D. At the end of recovery, state updates in D are enabled once again. Fig. 4 shows the flow of information and the relationship between these components.

The above semantics provide for the design D operating with no performance overhead in the absence of any failures detected by C. Also, the recovery R is silent in the absence of failures detected by C. Further, it provides for automated handling of stalls in case of an error. The stalling mechanism of disabling state updates needs to

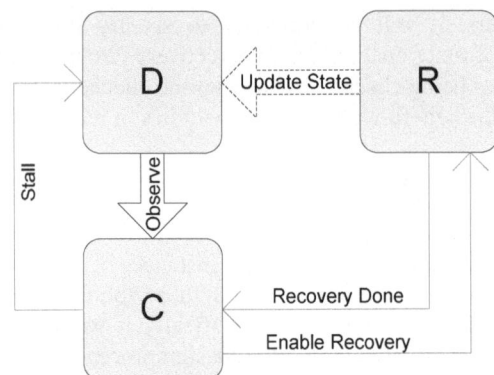

Fig. 4. General Design Template

be carried through with design composition and hierarchy. If a module interacts with another module and it stalls, other modules may need to stall to since they may be producing inputs for this module or may be consuming the outputs from this module. A static analysis of the design can determine these relationships. The overall language semantics are for these stalls to be enforced as based on the compositional relationships. Note that this is an important aspect of the overall design that the designer has been freed from. Traditionally determining the transitive effect of stalls and implementing them correctly is non-trivial for most designs. Providing for it in the semantics of the language and off-loading the responsibility of this to the synthesis process makes this aspect completely transparent to the designer.

There are other features that may be of interest in such a language. For example, for backward recovery through rollback, check-pointing of design state may be required. This could be specified through a check-pointed register data type which is used to indicate through parameters the frequency and depth of check-pointing of a given register.

Based on the language features and semantics, the synthesis process handles all the hardware implementation tasks. These will include: the analysis and implementation of the stalling mechanisms in case of an error, the implementation of checkpointed registers, using different constraints for the design, checking and recovery circuits (for example, the recovery need not be performance constrained, and needs to be electrically more robust). Thus, the burden on the designer is limited to the intellectual aspects of specifying the design, checking and recovery circuits. All aspects of their implementation are managed by the synthesis process.

Such a design methodology could easily be built by augmenting existing design languages such as Verilog, VHDL or System-C. For example, an initial implementation of this can be done by extending Verilog, and then preprocessing this extended specification to a Verilog description with appropriate design constraints. This can then be used in a standard Verilog flow.

5 Summary

This paper makes a case for runtime validation of hardware for functional errors. It argues that this is likely to occur as a confluence of two trends – the lack of scalability of current verification techniques and consequently a larger number of bugs escaping the verification process, as well as introduction of runtime validation for operational failures. It demonstrates how functional failures dominate both operational failures and manufacturing defects, and thus their solutions can provide for a unified attack for all failure modes. Some recent specific instances of runtime validation are also discussed. These include specific micro-architectural solutions for uniprocessors as well as SMT processors, as well as validation for temporal properties using hardware checkers possibly combined with offline model checking. Finally, it discusses a RT level design methodology with language and synthesis support for runtime validation.

Acknowledgements

This research is supported by a grant from the Microelectronics Advanced Research Consortium (MARCO) through the Gigascale Systems Research Center (GSRC). It includes inputs from Kaiyu Chen, Ali Bayazit and Yogesh Mahajan.

References

[1] Brian Bailey, "A New Vision for Scalable Verification," EE Times, March 18, 2004
[2] Tom Schubert, "High-Level Formal Verification of Next Generation of Microprocessors," in, *DAC' 03: Proceedings of the 40th ACM/IEEE Design Automation Conference,* 2003.
[3] Gregory S. Spirakis, "Opportunities and Challenges in Building Silicon Products in 65nm and Beyond," in, *DATE'04: Proceedings of the Design, Automation and Test in Europe Conference and Exposition,* 2004.
[4] Todd Austin, Valeria Bertacco, David Blaauw and Trevor Mudge, "Opportunities and Challenges for Better than Worst-Case Design," in, *ASPDAC'05: Proceedings of the Asia-Pacific Design Automation Conference,* 2005
[5] T. M. Austin, "DIVA: A reliable substrate for deep-submicron microarchitecture design," in *MICRO'99: 32nd Annual International Symposium on Microarchitecture,* Nov. 1999.
[6] Maher Mneimneh, Fadi Aloul, Chris Weaver, Saugata Chatterjee, Karem Sakallah, Todd Austin, "Scalable hybrid verification of complex microprocessors," in, *DAC'01: Proceedings of the 38th IEEE/ACM conference on Design Automation,* 2001.
[7] Kaiyu Chen and Sharad Malik, "Runtime Validation of Multithreaded Processors," Technical Report, Dept. of Electrical Engineering, Princeton University, May 2005. Available by email from kchen@princeton.edu.
[8] H. Foster, A. Krolnik, and D. Lacey, *Assertion Based Design.* Kluwer Academic Publishers, 2003.
[9] Y. Abarbanel, I. Beer, L. Glushovsky, S. Keidar, and Y. Wolfsthal, "Focs: Automatic generation of simulation checkers from formal specifications," in *CAV'00: Proceedings of the 12th International Conference on Computer Aided Verification.* London, UK: Springer-Verlag, 2000, pp. 538–542.

[10] S. Berezin, S. V. A. Campos, and E. M. Clarke, "Compositional reasoning in model checking," in *COMPOS'97:Revised Lectures from the International Symposium on Compositionality: The Significant Difference*. London, UK: Springer-Verlag, 1998, pp. 81–102.

[11] Ali Bayazit and Sharad Malik, "Complementary Use of Runtime Validation and Model Checking," in *ICCAD'05: Proceedings of the IEEE/ACM International Conference on Computer-Aided Design*, 2005.

Assertion-Based Verification for the SpaceCAKE Multiprocessor – A Case Study

Milind Kulkarni and Benita Bommi J.

Philips Research India,
#1, Murphy Road, Ulsoor, Bangalore, India
Milind.Kulkarni@philips.com
Benitabommi@rediffmail.com

Abstract. This paper presents a case study of the application of assertion-based verification to a multi-million-gate design of the SpaceCAKE architecture with shared L2 cache. SpaceCAKE L2 cache is highly configurable and implements Distributed Shared Memory (DSM) architecture. This paper discusses the issues faced during the functional verification of this architecture. A number of techniques are employed to verify the design. The paper serves as a case study for verification of such a complex architecture. A description of the different techniques that were used to verify this architecture and an assessment of using a comprehensive coverage-driven verification plan that exploits the benefits of the traditional simulation techniques through the use of assertions is presented. We have found that the tools, currently provided by the market, for assertion-based static verification approach need more maturity. A 50% reduction in debug time has been achieved through the use of assertions.

1 Introduction

Verification of SoCs through self-checking testbenches, directed tests, regression tests and simulation-based verification are the order of the day. The effort required to implement the above mentioned techniques of verification for a multi-million gate is monumental. In spite of the huge time spent on verifying the design using these techniques, the ability of the techniques to uncover the source of error is limited. The ability to present sources of errors is very necessary while verifying massive designs, as it is practically impossible to locate the source of error in such designs with reasonable effort. This necessity paved the way for implementing assertion-based verification for this design.

The paper discusses the areas where assertions can be applied without increasing the simulation time overhead to a prohibitive level while still facilitating verification of the complete functionality of the design. It also discusses the use of assertions in acquiring coverage information that identify functionalities that have not been verified.

The organization of the paper is as follows. Section 2 is a description of the architecture verified. Section 3 discusses the need of assertion-based verification for multi-million gate designs. Section 4 is an account of the process of identification and implementation of the assertions.

S. Ur, E. Bin, and Y. Wolfsthal (Eds.): Haifa Verification Conf. 2005, LNCS 3875, pp. 43 – 55, 2006.
© Springer-Verlag Berlin Heidelberg 2006

Section 5 is a note on the implemented static verification of assertions. Section 6 is a description of the test set-up used to dynamically verify this architecture. Section 7 is a discussion of the results obtained through the use of this methodology followed by the conclusion in section 8.

2 The SpaceCAKE Architecture

SpaceCAKE is a homogenous, tile-based, DSM architecture [1][2]. The tile is a multiprocessor system on a chip consisting of a number of programmable processor cores with an integrated L1 caches with Harvard architecture (i.e., separated instruction and data caches), a snooping bus-based interconnection network (ICN) which connects the cores, L2 cache, and the memory management unit interface to the off-chip DDR memory as shown in Fig.1. The tiles communicate via high-speed routers. Cache coherence within the tile and across tiles is implemented so that synchronization is taken out of software programmers botheration.

The features of this architecture are as follows:

Distributed Shared Memory architecture.
Level-1 and Level-2 Cache coherency.
Shared resource reservation framework.
Software controlled cache operations and debug functionality.
Performance measurement based run-time adaptations.
Multiple outstanding requests with optional out of order processing.

The interconnection network and the processor cores support hardware cache coherence, which greatly simplifies the programming model. There are multiple memory banks to increase the concurrency and improve throughput. Since the snooping interconnection network supports concurrent transactions, multiple CPUs

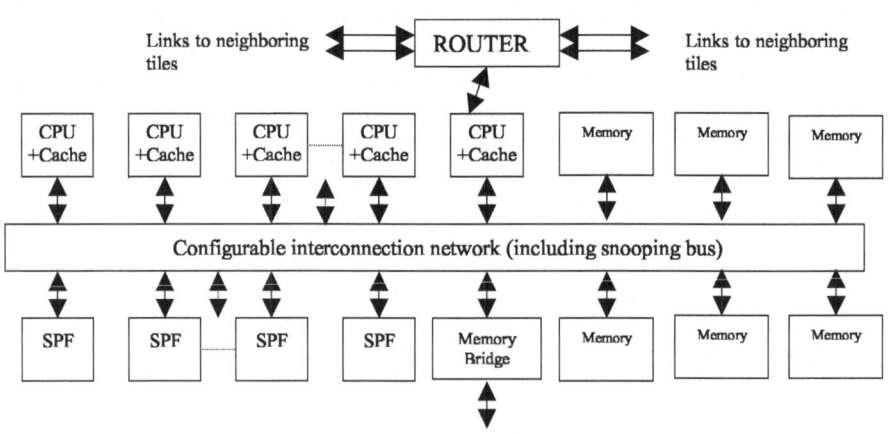

Fig. 1. SpaceCAKE Tile

can be busy with memory transactions at the same time, as long as they are accessing different memory banks. The blocks labeled SPF represent the special purpose hardware functions that are key to the computational efficiency of SpaceCAKE chip. All SPFs execute under control of a task running on any of the processors. Inter-tile communication takes place through the routers. The interconnect network consists of a set of multi-purpose registers which can be software configured to implement Distributed Shared Memory (DSM) functionality. Due to the multiple configurations possible, high concurrency and advanced features with respect to cache coherency and snooping protocols the verification effort required is considerably high. Although, the focus of verification effort was restricted to the verification of interconnect network, as all other design entities are off-the-shelf pre-verified IPs.

3 The Verification Impasse

Verifying the functionality of the design is a key requirement. An effort to independently verify each individual module in the design using traditional simulation based verification will require Herculean effort and time. However, creating a top-level testbench for the verification of the architecture will lead to loss in the design visibility and hence the prolonged debug time. This impasse is resolved by the use of an assertion-based technique for the verification of the SpaceCAKE architecture. The assertion-based technique is implemented by embedding assertions into the design.

4 Implementation of Assertions

The different verification hot spots in the design were identified, and assertions were written to verify the hot spots. Verification hot spots denote design structures or features of the design that are difficult to verify. Critical modules of the design that are highly depended upon by other modules are considered as hot spots since the total verification of the module is crucial for the operation of other modules. Simulation-based verification methodologies do not adequately verify hot spots because a typical verification hot spot processes too many combinations of events to be simulated exhaustively. A systematic verification methodology that employs exhaustive assertion-based dynamic and static tools is used to achieve verification signoff for the design.

4.1 Assertion Specification

Assertions expressed in Property Specification Language (PSL) [3] and Open Verification Library (OVL) [4] were embedded in the design. Though OVL provides ease of use because it comes as a predefined library, adapting the OVL to suit the needs of the design is complex. PSL provides the flexibility to express the functionality of the design explicitly and hence was used to a greater extent. PSL assertions are implemented in an external file, which is then explicitly bound to the modules of the design. This facilitates non-obtrusive use of the assertion-based technique with any verification set-up that is used.

4.2 Identifying Hot Spots

The hot spots identified can be classified into four categories.

1. Hot spots in complex design structures.
2. Hot spots in reusable protocol interface.
3. Hot spots in interface of different modules.
4. Hot spots in protocols and policies specific to the design.

4.3 Hot Spots in Complex Design Structures

SpaceCAKE consists of a number of design structures such as FIFOs, buffers, memories, arbiters, FSMs and pipelines. A few properties verified and the coverage information obtained in the above design structures are as follows.

FIFO
A queue is a FIFO structure. The FIFO structure has many corner cases that needs to be verified and is hence considered as a hot spot.
   The properties verified for a FIFO include:

1. The FIFO does not accept data once full and does not refuse to accept data if not full.
2. Normal FIFO operation.
3. Data overwriting does not occur on FIFO overflow.

   The coverage information obtained for a FIFO include:

1. Is FIFO size over-designed?
2. Is FIFO size under-designed?

BUFFER
The buffer design construct is mainly used for storing data temporarily before it is passed on to the next stage. Due to the high frequency of use of this construct, it is considered a hot spot.
   The properties verified for a buffer include:

1. Simultaneous read and write to the same location of buffer is not performed.
2. Buffer indexing for reads and writes are consistent.
3. Width of data written into buffer is consistent with the width of the buffer.
4. Overwriting of data at a location, before a read on the location, is prohibited.

   The coverage information obtained for a buffer include:

1. Are all the locations of the buffer read from and written into?
2. How many times is the buffer filled/emptied?
3. The numbers of read and write transactions.

MEMORY
Since the SpaceCAKE architecture is configurable, there are many memories that initialise the device operating conditions on boot and reflect the operation of the

device over the course of time. Also, memories that facilitate table lookup operations are present.

These facts make memories an important hot spot to verify.

The properties verified for memories include:

1. The value of the memory on reset is as per specification.
2. Illegal/unknown values are not stored in memory.
3. Appropriate accessing of memory for read and write requests.
4. Configurable memories are initialized before certain time duration.

The coverage information obtained for memories include:

1. Reads and writes to all possible locations of the memory.
2. Number of read and write transactions.

ARBITER

An arbiter is an element that allows the bus to be shared by multiple devices. It is commonly used in all the SoC designs and provides many corner cases that should be verified.

The properties verified for an arbiter include:

1. The different arbitration schemes that can be implemented by the arbiter.
2. The request that has the highest priority is granted.
3. The grant to a request is within a certain number of cycles to the arrival of the request.

The coverage information obtained for an arbiter include:

1. Is every arbitration channel exercised?
2. How many simultaneous requests are made to the arbiter?

FSM

FSMs are design structures consisting of many legal paths. Each of these paths has to be verified to give the confidence of complete verification. This makes FSMs a hot spot for verification.

The properties verified for FSMs include:

1. Are the state transitions legal?
2. Has the FSM deadlocked or live-locked in any state?

The coverage information obtained for FSMs include:

1. Whether each state in the FSM has been exercised?
2. Whether every specified arc of the FSM is exercised?

PIPELINE

Pipelines form a crucial part of the present day designs as they help to increase throughput. Pipelines serve several concurrent actions that may be interdependent. This might lead to stalls in the pipeline. Therefore, the six-stage pipeline that forms a part of the SpaceCAKE design is a very crucial construct that has to be verified thoroughly.

The properties verified for pipelines include:

1. Every request is processed to completion.
2. Data corruption does not occur in the pipeline.
3. No live-locks/deadlocks in the pipeline.
4. Handshaking of signals between each stage.
5. Handling of pipeline stalls.

The coverage information obtained include:

1. Number and type of requests passed through the pipeline.
2. Number of pipeline stalls.

4.4 Hot Spots in Protocol Interface

The data transfer protocol interfaces, used in the SpaceCAKE architecture, are the MTL [5], DTL [6], and AXI [7] protocols. Inclusion of assertions that completely defines the protocol at the interface ensures that any block that connects to the interface will be automatically checked for protocol compliance. Assertions to check the read/write operation of the protocol, address/data relationships, cycle latency and event scheduling time in the protocol (i.e. the time within which the interface should respond to a request) are used to define the MTL, DTL and AXI interfaces of the design. e Verification Components [8] that come along with these protocol checks were also used to verify these protocols.

4.5 Hot Spots in Inter-module Interfaces

Interdependent modules are used in the design. Therefore, the verification of each module operating as a stand-alone unit is not sufficient. Use of interconnected modules can lead to deadlocks, if the modules are highly interdependent. Assertions that check the transmit/receive of valid data from each module, handshake behaviour modelling, the behaviour of communication between the two modules, and timeout conditions (indicate that a given module has not responded to a request within a particular duration) are implemented at the inter-module interfaces.

4.6 Hot Spots in Protocols and Policies Specific to the Design

The DSM policy of the SpaceCAKE architecture is implemented as a lookup table that is programmed with the desired cache mode. This table lookup action is part of a pipeline and is therefore verified thoroughly. The two important design specific protocols are the cache coherency protocol and the snooping protocol. The cache coherency protocol implemented in the SpaceCAKE design is the MESI protocol [9]. The different cache modes possible are un-cached, bypass on miss, fetch on miss, and allocate on write. The snooping protocol allows four modes of read operations namely shared read, exclusive read, bus upgrade, bus invalid and two modes of write operation namely normal write and coherence write. Snooping ensures that all caches see and react to all bus events. Assertions to check whether each of these modes of operation is exercised by the design and whether the expected output is provided by the design once these different modes are activated are implemented.

The properties verified include:

1. Reads/writes in different cache modes.
2. Appropriate tag updates.
3. Cached transactions with snooping.

The coverage information obtained include:

1. The number of cache hits and misses.
2. Cache modes/snooping protocol coverage
3. The number of requests pertaining to the cache mode exercised.

5 Verifying Assertions Using Static Techniques

The static checking tool [12] was used to further verify the assertions specified. The assertions that passed the static check are completely verified for all possible input sequences. The staticverilog tool is used to initially compile the design and then statically check the design using design-partitioning techniques. The main limiting factor of this style of verification, for such a large complex design, is the time factor. Time consumed to verify the assertions statically is more than ten times the time taken to dynamically verify the assertion.

6 Dynamic Verification Using Assertions

Dynamic verification involves generation of test stimulus to the design model. Generating test stimulus for the many individual modules of the SpaceCAKE design is a time consuming activity. Therefore, a top-level testbench is created. Though the utilization of a top-level testbench leads to loss of design visibility, the presence of embedded assertions alleviates this problem. The details of the SpaceCAKE test suite set up are given below.

6.1 Test Suite Set-Up

The SpaceCAKE architecture is highly parameterised and configurable. Due to this fact the testbench set up for the design should be highly reusable. The top-level test environment set up for the architecture is as shown in Fig.2. Testbenches in C language are created to generate addresses and data corresponding to local and remote tiles. These C testbenches create Verilog tests that are applied to bus functional models of the processors. These are used to transfer the data to the SpaceCAKE interconnect network.

The tests issue read and write requests to the memory through a Bus Functional Model (BFM) of the processors. A BFM is a model that emulates the processor bus cycles. It does not execute the processor instruction set nor does it maintain the internal cache. The addresses generated by the tests are constrained to either fall in the local memory or in the remote memory that is part of the distributed shared memory space. The data generated by the tests is written at a particular address of the

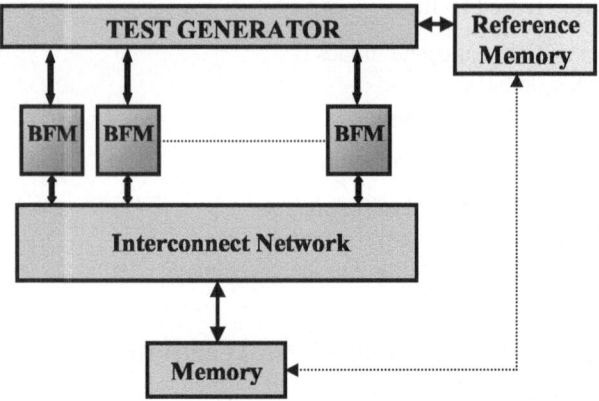

Fig. 2. Test set-up

chip memory through the interconnect network and simultaneously into the reference memory. When a read request is generated to an address at which data was previously written into, the read data obtained by the read transaction is compared with the data stored in the reference memory. If the data matches then the write/read transactions have been performed correctly by the interconnect network. These tests are self-checking and trigger an error if any discrepancy is found in the data transfer transactions. This self-checking mechanism helps verify design at a global level; perfectly complementing the local assertions to locate problems.

The test set up was run on the design before and after the addition of assertions in the design. Tests that passed the design simulation run earlier, triggered assertions after the inclusion of assertions. This was attributed to factors such as presence of speculative read operations (i.e. read operations for which data correctness checks are not done), programmer-transparent address translations for L2 cache etc.. This can potentially trigger the exposure of more bugs in the design. Also, the effort required to locate the bug once a test failed had considerably improved due to the high visibility provide by the assertions. The above set-up will check for acceptable write into memory and read from the memory operations for a single configuration of the interconnect network.

6.2 Activating Different Configurations

Typically the interconnect network can be configured in different ways as follows.

1. With remote or/and local memory access.
2. With and without debug functionality.
3. With/without different cache modes - fetch on miss, allocate on write, bypass on miss.
4. With/without snooping - exclusive read, bus upgrade, shared read, bus invalid.
5. With and without victimization/refill.
6. With DTL, MTL or AXI traffic.
7. With and without cache reservations.

Considering the high configurability of the interconnect network, a method of running the test suite using every configuration is needed to ensure that the interconnect functions correctly in all the different possible configurations. The manual selection of these configurations is time consuming and random selection may not yield meaningful values and sufficient confidence. Therefore, the different possible combinations were obtained by use of the Orthogonal Array Technique (OAT) tool [10][11]. The possible parameters that can be configured along with the possible values that can be acquired by the parameter were provided as input to the tool. The tool generates the different possible configurations using the OAT method. The OAT technique was used so that the configurations obtained are well spread out and covers the state space uniformly without being redundant.

6.3 Regression Environment

A regression environment was set up to run the test suite for different configurations. The environment also stores the waveforms, assertion firing reports, code and functional coverage reports pertaining to each test. Code coverage reports include information about the block, expression, and path coverage; while functional coverage indicates the coverage of essential properties of design as defined by the architect. This provides an automated way of running the different test suites. The code and functional coverage information reports obtained are merged to provide collective coverage information.

6.4 Coverage Closure

Coverage closure is the process of finding areas of the HDL code not exercised by a set of tests so as to create additional tests to increase coverage by targeting holes, and determining a quantitative measure of coverage. The methodology depicted in Fig. 3 targeted coverage closures for the SpaceCAKE design.

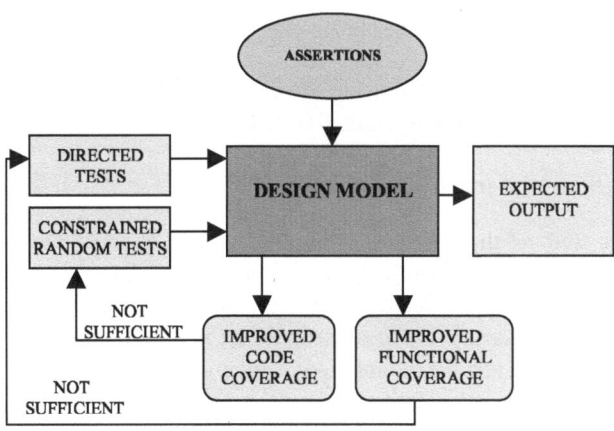

Fig. 3. Coverage closure

Once a collective code coverage report is available the areas of code that are not covered are manually checked for:

1. Non-accessibility of code with all configurations.
2. Presence of dead code.
3. Presence of bugs.

If the code coverage reports are satisfactory then the functional coverage reports are checked to ensure 100% functional coverage. Constrained random testing is used to increase the code coverage and directed tests are used to cover the caveats reported by the functional coverage tool. Constrained random tests are good to improve code coverage but it is often not enough to achieve 100% code coverage. Directed tests to cover the most important function coverage goals were needed.

7 Results

The top-level test suite in Fig. 2 was used to simulate the design before and after the addition of assertions. The bug detection rate increased since the same tests that had passed the design simulation run earlier triggered assertions, after the inclusion of assertions. Also, the task of locating the bug for the failed test considerably improved due to the high visibility provided by the assertions. The following results were obtained after the test set-up was run with the embedded assertions.

Bugs uncovered in the critical module of the design
    Unassigned outputs.
    Non-accessible code.
    Illegal output values.
    Errors in the lookup table.
50% reduction in debugging time.
10% of assertions completely verified using static checking. Static checking is unable to support part indexed select operator construct from Verilog2001 which is widely used in the design implementation. Also, restrict statements are not supported which leads to illegal input values being checked for. Due to the above limitations of the tool static checking could not be used extensively to verify the design.
62% of assertions were verified using dynamic verification.
Code coverage of 95% for DSM functionality.

7.1 Observations: Assertion Based Technique

A comparative study of the different techniques adopted to verify the design, based on the results obtained, is given in this section. The use of assertion-based verification improved the productivity of the verification effort since it reduced the debug time by 50%. Increasing the density of the assertions in the design code will further reduce the debug time, however the simulation time overhead increases. This dilemma requires that the density of assertions be kept in check. The proficient placing of the assertions uniformly throughout the design implementation with increased concentration at the hot spots facilitated the resolving of the dilemma. The overhead while using assertions is depicted in Fig. 4.

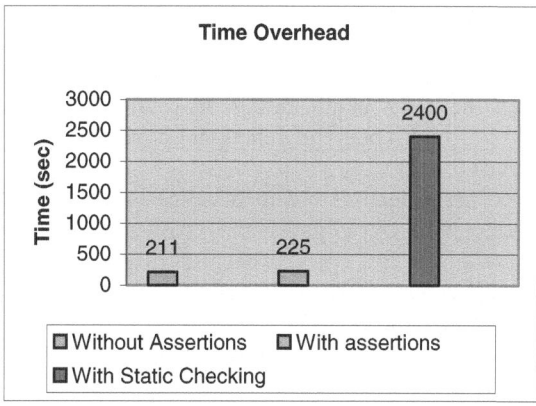

Fig. 4. Time Overhead

The addition of assertions in the design has caused only a 6% increase in the simulation time while using dynamic verification. This overhead can be overlooked considering the immense visibility provided to the design by the addition of assertions. The time overhead for static checking of assertions was more than ten times the time taken to verify the assertions using simulation-based techniques. The advantage of using static checking though is to ensure the fact that the functionality has been verified completely.

7.2   Observations: Coverage Metrics

Two different coverage metrics namely functional coverage and code coverage were obtained. The two coverage metrics complement each other. The graph in Fig. 5 depicts the code and functional coverage obtained. The effort used to obtain functional coverage, is comparatively less while using assertions than using

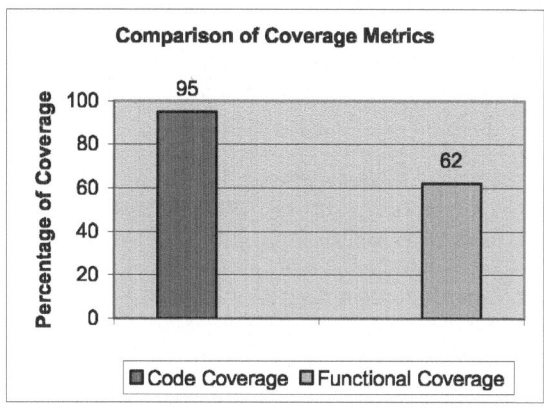

Fig. 5. Comparison of coverage metrics

post-processing tests that perform sequence checking [13] etc. The graph emphasizes the fact that though 100% code coverage can be obtained for modules, the functional coverage reports give a true picture of the functionalities covered by the test suite. This is because all the paths in the design may not be covered with 100% code coverage. Only with a high code and functional coverage report can the design be released for tape-out.

Analyzing the coverage metrics obtained indicates that for the design, presently, certain corner cases, multi-cycle paths and cross-correlated paths are yet to be verified.

Therefore, the verification effort needs to be directed towards verifying the corner cases and different paths of the design.

8 Conclusions

The key contribution provided by the use of the assertion-based technique, for verifying the SpaceCAKE architecture, is the remarkable increase in design visibility and therefore, a marked reduction in debug time. The assertions implemented can be used in concurrence with other verification techniques, like directed and pseudo random tests, to continuously run a background check and find unexpected bugs. Also, the use of the assertion-based functional coverage metric complemented with the code coverage metric provided a deeper insight into the areas of the design that needed further verification effort. This facilitated the verification effort to be directed towards verifying the key functionalities of the design.

To summarize, moving verification to the functional level, by focusing on protocol monitors, verification hot spots and critical coverage points provided a substantial return of investment. Assertion-based verification is an important methodology for the future generation complex chips. It is important for the EDA companies to make stronger tools around this methodology so that the verification effort is kept under control especially for large complex designs that are here to stay. As a methodology, it is a comprehensive start-to-finish approach, which not only integrates static, dynamic and coverage tools; it executes them in perfect coordination to accomplish a well-specified common goal.

References

1. Stravers. P, Hoogerbrugge. J, *Homogeneous Multiprocessing and the Future of Silicon Design Paradigms,* Proceedings of the IEEE International Symposium on VLSI technology, Systems and Applications, April 2001, pp. 184-187.
2. Stravers. P, *Homogeneous Multiprocessing for the Masses,* IEEE Workshop on Embedded Systems for Real-Time Multimedia, September 2004, pp. 3.
3. *Property Specification Language Version 1.1, Reference Manual,* Accellera, June 2004.
4. *Open Verification Library, Assertion Monitor Reference Manual,* Accellera, June 2003.
5. Philips Semiconductors, *CoReUse 3.2.1 Memory Transaction Level (MTL) Protocol Specifications.* Sept 2002.
6. Philips Semiconductors, *CoReUse 3.1.5 Device Transaction Level (DTL) Protocol Specifications.* Dec 2001.

7. *AMBA Advanced eXtensible Protocol v1.0 Specification.*
8. Cadence Specman Elite v4.3.4.
9. John Hennessy, David Patterson, *Computer Architecture and Organization,* Morgan Kaufmann Publishers, 3rd edition, 2003.
10. Sloane, Neil J. A. *A Library of Orthogonal Arrays.* Information Sciences Research Center, AT&T Shannon Labs. 9 Aug. 2001
11. *Jenny Tool.* http://burtleburtle.net/bob/math/jenny.html
12. *Incisive Static Assertion Checking Guide,* Product Version 5.1, October 2003.
13. M. Kantrowitz, Lisa M. Noack, I'm Done Simulating: Now What? Verification Coverage Analysis and Correctness Checking of the DECchip 21164 Alpha Microprocessor, Proceedings of the 33rd Design Automation Conference, June 1996, pp. 325-330.

Simultaneous SAT-Based Model Checking
of Safety Properties

Zurab Khasidashvili[1], Alexander Nadel[1,2], Amit Palti[1], and Ziyad Hanna[1]

[1] Design Technology Solutions, INTEL Corporation
{zurab.khasidashvili, alexander.nadel, amit.palti,
ziyad.hanna}@intel.com
[2] Department of Computer Science,
Tel Aviv University, Ramat Aviv, Israel

Abstract. We present several algorithms for simultaneous SAT (propositional satisfiability) based model checking of safety properties. More precisely, we focus on Bounded Model Checking and Temporal Induction methods for simultaneously verifying multiple safety properties on the same model. The most efficient among our proposed algorithms for model checking are based on a simultaneous propositional satisfiability procedure (SSAT for short), which we design for solving related propositional objectives simultaneously, by sharing the learned clauses and the search. The SSAT algorithm is fully incremental in the sense that all clauses learned while solving one objective can be reused for the remaining objectives. Furthermore, our SSAT algorithm ensures that the SSAT solver will never re-visit the same sub-space during the search, even if there are several satisfiability objectives, hence one traversal of the search space is enough. Finally, in SSAT all SAT objectives are watched simultaneously, thus we can solve several other SAT objectives when the search is oriented to solve a particular SAT objective first. Experimental results on Intel designs demonstrate that our new algorithms can be orders of magnitude faster than the previously known techniques in this domain.

1 Introduction

Bounded Model Checking (BMC) [BCCZ99, BCC+03] is a SAT (or satisfiability) [DLL62] based verification technique, well suited for finding counter-examples to a given safety property P, in a transition system. It arose as a complementary approach to BDD-based [Bry86] *Symbolic Model Checking* technique [McM93], and is increasingly adopted by the industry [PBG05]. The idea of BMC is to unroll the transition system to k time steps, and search using a SAT solver for a state transition path of length less or equal to k, starting with an initial state and ending in a state violating the property. We recall that a SAT solver searches for a satisfying assignment to a Boolean formula written in CNF form; such a formula is represented as a set of clauses, a clause being a disjunction of literals, where a literal is a Boolean variable or its negation.

S. Ur, E. Bin, and Y. Wolfsthal (Eds.): Haifa Verification Conf. 2005, pp. 56–75, 2006.

We restrict ourselves to safety properties written as *AGp* in Computation Tree Logic (CTL, [CGP99]). Such properties are often called *invariants*. Proving a property P using BMC technique means showing that there is no counter-example to P of the length less or equal to the diameter of the system, i.e., the maximum length of a shortest (thus loop-free) path between any two states. From practical point of view, BMC is an incomplete technique in that it can rarely prove a property arising from an industrial application of software or hardware verification, since the diameter for such systems is too large to handle by current SAT solvers. A practical, complete SAT-based model checking method was proposed by Sheeran.et. al. [SSS00] as a (temporal) induction method, allowing proving safety properties by means of unrolling to much lower depths than the diameter. Roughly, BMC in this method corresponds to the base of temporal induction, and the induction step, at depth m, attempts to prove that there is no state transition path $s_0,...,s_m, s_{m+1}$ such that P holds in all but the last state (here s_0 needs not be an initial state). Once such an m is found, and it has been shown in the base of induction that there is no counter-example to the property of length m or less, the property P is proven valid at all states reachable from the initial state.

Several usability enhancements have been proposed in the literature to the above methods, boosting the capacity to handle larger systems and complex properties, and faster. These enhancements are very important for successful application of the methods in practice. Here we review briefly two enhancements that are most relevant to our work. For more information, we refer the reader to a recent survey of SAT-based model checking [PBG05].

In a BMC run, to avoid unnecessary unrolling of the transition relation, one starts with low bounds k, and if no counter example is found for the property of length smaller or equal to k, the bound k is increased, repeatedly, till it reaches the diameter or a maximal user given value for the bound. Therefore a BMC run involves a number of calls to the SAT solver. Similarly, proving the induction step in temporal induction method needs several calls to the SAT solver, with increasing bounds m. These SAT instances are closely related, and the idea of *incremental SAT solving* in BMC and induction (as well as in other SAT applications), proposed independently by Strichman [Str04] and Whittemore et al [WKS01], is in re-using *pervasive* learned conflict clauses across consecutive calls to the SAT solver. Here pervasive learned clauses are logical consequences of all involved SAT instances, thus adding them to the clause set permanently is safe. Eén and Sörensson [ES03] extended this approach to temporal induction, and proposed a simple interface to a SAT solver enabling incremental BMC and induction schemes where all conflict clauses are pervasive and can be re-used.

In typical industrial model checking applications, one needs to prove a number of properties on the same model. Since several properties may share the "cone of influence" in the model, (dis)proving several properties in one model-checking session may yield a significant speedup. To the best of our knowledge, Fraer et al. [FIK+02] were the first to propose an extension to the classic BMC and the induction method allowing to *simultaneously* check a number of safety properties $P_1,...,P_n$ on the same model.

Here we propose a number of new algorithms for simultaneous SAT-based model checking of multiple safety properties, which strengthen the method of [FIK+02].

Incrementality through verification depths is one source of incremental SAT-based model checking [Str04, WKS01, ES03]. Our algorithms are *double-incremental*, meaning that learned clauses of the SAT solver can be reused across depths, as well as across the properties at every depth. The most efficient among our algorithms use a *simultaneous SAT solver* (SSAT), which is able to resolve several objectives related to the same instance *in one traversal* of the search space. In SSAT, besides a selected, *currently watched* objective, one actually watches all unresolved objectives as well, and can falsify or prove them valid during the search oriented to solve the currently watched objective. Because of these "one traversal" and "all watched" principles, our algorithm is more efficient than previous approaches to fully incremental SAT solving which, like SSAT, allowed reusing *all* learned conflict clauses [GN01, ES03]. We will discuss these approaches in detail in a related work section and will provide experimental results to demonstrate the superiority of the SSAT approach.

The paper is organized as follows: in the next section, we will give a short introduction into modern DPLL-based algorithms. In Section 3, we describe SSAT algorithm and its implementation on top of a DPLL-based propositional SAT solver. In Section 4, we compare the SSAT algorithm with previous approaches to incremental solving of related satisfiability objectives. As one can see, sections 2-4 are dedicated to propositional satisfiability checking. In Section 5, we propose several new, double-incremental methods for simultaneous model checking of safety properties based on SAT algorithms described in sections 2-4. In Section 6, we present experimental results demonstrating the usefulness of the SSAT approach on series of benchmarks originating from formal property verification and formal equivalence verification of Intel designs. Conclusions appear in Section 7.

2   The Basic DPLL Algorithm in Modern SAT Solvers

The DPLL algorithm [DP60, DLL62] is the basic backtrack search algorithm for SAT. We briefly describe the functionality of modern DPLL-based SAT solvers, referring the interested reader to [LM02] or [Nad02] for a more detailed description.

Most of the modern SAT solvers enhance the DPLL algorithm by the so-called Boolean constraint propagation (BCP) [ZM88], conflict driven learning [MS99], [ZMM+01] and search restarts [GSK98]. The SAT solver receives as input a formula in Conjunctive Normal Form (CNF), represented as a set of clauses, each clause being a disjunction of literals, where a literal is a Boolean variable or its negation. The solver builds a binary search tree until it either finds an assignment satisfying all the clauses—a *model*, in which case the formula is *satisfiable*; or it explores the whole search space and finds no model, in which case the formula is *unsatisfiable*. Note that some of the variables in a model may be *don't cares*, meaning that any assignment to these variables still yields a model of the CNF formula.

At each node of the search tree the solver performs one of the following steps:

1. It chooses and assigns the next decision literal and propagates its value using BCP. A unit clause is a clause having all but one literal assigned *false*, while the remaining literal l is unassigned. Observe that l must be assigned *true* in order to satisfy the formula; this operation is often referred to as the *unit clause rule* and

$l = true$ is referred to as an *implied* assignment. BCP identifies unit clauses and repeatedly applies the unit clause rule until either:

- No more unit clauses exist. In this case, the solver checks whether all the clauses are satisfied. If they are, we have found a model and the formula is satisfiable, otherwise the solver is looking for the next decision literal;
- A variable exists that must be assigned both *false* and *true* in two different unit clauses, in which case we say that a *conflict* is discovered.

2. If a conflict is discovered, the solver adds one or more *conflict clauses* to the formula. A conflict clause is a new clause that prevents the set of assignments that lead to the conflict from reappearing again during the subsequent search [MS99]. Then, if a literal y exists such that it is sufficient to flip its value in order to resolve the conflict, the solver backtracks and flips the value of y; otherwise the formula is unsatisfiable. The former case is referred to as a *local conflict* and the latter case is referred to as a *global conflict*. For our purposes, it is also important to mention that during conflict analysis one or more literals may be discovered to be *globally true*, that is, they must be assigned *true* independently of other variable values. This happens every time when a new conflict clause containing exactly one literal l is learned. The literal l as well as all the literals assigned as a result of BCP following the assignment $l = true$ are globally true.

3. Once in a while the solver restarts the search, keeping all or some of the learned conflict clauses [GSK98].

We demonstrate the above concepts on a simple example taken from [Str04].

Example 1: Consider the following set of clauses $\{c_1, c_2, c_3, c_4\}$, where

$$c_1 = \neg x_1 \vee x_2$$
$$c_2 = \neg x_1 \vee x_3 \vee x_5$$
$$c_3 = \neg x_2 \vee x_4$$
$$c_4 = \neg x_3 \vee \neg x_4$$

Assume the current assignment is $x_5 = false$ (*i.e.*, x_5 is assigned *false*), and a new decision assignment is $x_1 = true$. The resulted implication graph is shown in Figure 1: applying BCP leads to conflicting assignments $x_4 = true$ and $x_4 = false$. The clauses $\neg x_1 \vee \neg x_3$; $\neg x_1 \vee x_5$ are examples of conflict clauses, and a subset of conflict clauses is kept as *learned* conflict clauses.

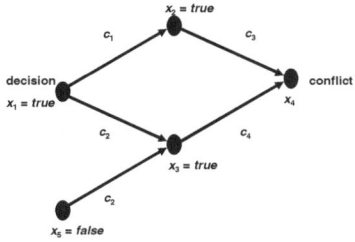

Fig. 1. An implication graph

3   SSAT Implementation Within a DPLL-Based SAT Solver

Now we describe the simultaneous propositional satisfiability algorithm, denoted as *SSAT*. In addition to the input formula (the SAT instance, or CNF instance), SSAT receives as a parameter a list of proof objective literals (PO literals, or POs for short). The POs must be proven *falsifiable* or *valid*. We require the variables of the PO literals to occur (positively or negatively, or both) in the CNF instance.[1]

Example 1 (continued): Assume our SAT instance consists of the same four clauses $\{c_1, c_2, c_3, c_4\}$, and assume our PO literals (or simply POs) are $PO_1 = \neg x_1$, $PO_2 = x_5$, and $PO_3 = x_2$. One can verify that all of the POs are falsifiable – there is a model for the SAT instance (i.e., an assignment satisfying the instance) where $\neg x_1$ is assigned *false* (thus x_1 is assigned *true*), there is a model where x_5 is assigned *false*, and a model where x_2 is assigned *false*. For example, the (partial) assignment $x_1 = false$, $x_3 = false$, $x_4 = true$, $x_5 = false$ is a model for $\{c_1, c_2, c_3, c_4\}$ in which PO_2 is falsified. Note that x_2 is unassigned in the model – x_2 is a *don't care variable* since assigning any of the truth values to x_2 yields a model for $\{c_1, c_2, c_3, c_4\}$. Assigning $x_2 = false$ yields a model in which PO_3 is falsified. The partial assignment $x_1 = false$, $x_3 = false$, $x_4 = true$ is another model (with more don't cares) for $\{c_1, c_2, c_3, c_4\}$.

A straightforward way to implement SSAT is as follows: (1) proceed with a regular DPLL-based search; (2) when a model is discovered, mark all the POs that are assigned *false* in the model as *falsifiable*; (3) as soon as all the search space is explored, mark all the unmarked POs *valid* and exit. However, there is a major problem with this solution: the number of models might be very large and therefore it is extremely inefficient to visit each of them during the search. Moreover, if a SAT solver uses search restarts (as do most of the state-of-the-art solvers), the algorithm might never finish, since the same models could be rediscovered after each restart. One solution could be adding clauses preventing the rediscovery of each model, but this might lead to memory explosion. We propose the following solution to this problem.

We always maintain a PO literal that we are trying to falsify, called the *currently watched PO (CWPO)*. At the beginning of the search we set *CWPO* to be any PO literal. At any stage of the search, prior to invoking a generic decision heuristic, we assign *CWPO* the value *false*, if not already assigned. The *CWPO* ceases to be the currently watched PO under two circumstances: (1) When a model containing *CWPO* = *false* is discovered, in which case we mark as *falsifiable* the *CWPO* as well as all the POs that are assigned *false* (or are don't care literals) in the model; (2) When *CWPO* is discovered to be globally *true*, in which case we mark the *CWPO* as well all other globally valid POs (if any) as *valid*. On both occasions, we check whether there exists a PO l that has not been discovered *valid* or *falsifiable*, in which case we set *CWPO* to l, otherwise the algorithm halts. This simple adjustment ensures that: (a) the number of discovered models is at most the number of POs; (b) a model is never rediscovered even if search restarts are used. Indeed, after encountering a model we

---

[1] We expect that POs are related to the instance; the definition of the POs in terms of variables in the instance can be included as a part of the CNF instance; thus the above requirement is not a restriction from application point of view.

always choose a *CWPO* that has not been *false* under any model and assign it the value *false*. This guarantees that any model will be different from all the previously discovered models. In addition, since the number of *CWPO*s is at most equal to the number of POs, the number of discovered models is at most equal to the number of POs. Also, our algorithm ensures that every PO is visited during new *CWPO* selection and thus every PO is marked *valid* or *falsifiable* after SSAT terminates.

The SSAT algorithm is presented in Figure 2. First, SSAT chooses a *CWPO*. Then, it enters a loop that terminates only when all the POs are proven to be either *falsifiable* or *valid*. Within the loop, SSAT first checks whether the current *CWPO* has already received a value. If it has, then a new *CWPO* is selected and assigned *false*. If all POs are resolved, the algorithm terminates. If the current *CWPO* has not been resolved yet, a new decision literal is picked using a generic decision heuristic. At the next stage, a conflict analysis loop is entered. After BCP, SSAT marks any PO that was found to be globally *true* as *valid*. Then, *SSAT* checks what the status of the formula is after BCP. If a global conflict has been discovered, that is, all the assignment space has been explored, the algorithm marks all the unmarked POs as

```
SSAT ([PO₁,..,POₙ], cnf_instance) {
   Literal CWPO = any PO literal;
   while (1) {
      if (CWPO is valid or falsifiable) {
         if (all the POs are valid or falsifiable) Return;
         CWPO = any PO literal that is neither valid nor falsifiable;
         Assign CWPO = false;
      } else {
         Assign choose_decision_literal();
      }
      do {
         status = BCP();
         Mark any PO literal that is discovered to be globally true as valid;
         if (status == global_conflict) {
            Mark all unmarked PO literals valid; Return;
         }
         if (status == model) {
            Mark any falsified and don't care PO literal falsifiable;
            Unassign all the literals that are not globally true;
         }
         if (status == local_conflict) {
            Add a conflict clause; Backtrack;
            Assign literal that must be flipped following conflict analysis;
         }
      } while (status == local_conflict);
   }
}
```

Fig. 2. SSAT pseudo-algorithm

valid (since after exploring the whole search tree, we discovered that no model falsifies them) and halts. If a model has been discovered, SSAT marks as *falsifiable* all the POs that are assigned *false* or are don't cares in the model and unassigns all the literals except ones that are globally *true*. Observe that in this case the algorithm exits the conflict analysis loop and picks the next *CWPO* during a new iteration of the global loop. Finally, if a local conflict has been encountered, SSAT backtracks and flips the value of a certain literal. Observe that in this case the algorithm goes on with the conflict analysis loop. Notice that it is safe to use restarts in the SSAT algorithm.

4 Comparing Simultaneous SAT Algorithm with Previous Work

The incremental satisfiability technique proposed in [MS97, WKS01, Str04] is based on identifying and reusing the *pervasive conflict clauses* encountered by the SAT solver during the search for a satisfying assignment to a given CNF formula. When one is trying to solve related SAT problems, the clauses occurring in the CNF formulas that are to be checked for satisfiability, which we will call the satisfiability objectives, can be divided into two classes: the clauses that are common to all satisfiability objectives will be called *pervasive clauses*, and the remaining clauses will be called *temporal clauses*. Then the conflict clauses that can be derived solely from the pervasive clauses are *pervasive conflict clauses*, and can be used for resolving each satisfiability objective. Experimental results in [WKS01, Str04] amply demonstrate that pervasive conflict clauses can significantly accelerate solving families of related SAT objectives. We will refer to this approach as PISAT approach.

To understand the differences between our SSAT approach and the PISAT approach, here we explain on an example the definition of pervasive conflict clauses and their usage in incremental SAT solving, as proposed in [Str04].

Example 1 (continued): Suppose again we have the same SAT formula consisting of clauses $\{c_1, c_2, c_3, c_4\}$. Further, define clauses $c_5 = x_1$, $c_6 = \neg x_5$, and $c_7 = \neg x_2$, and assume we are interested in solving the following three SAT instances:

(1) $\{c_1, c_2, c_3, c_4, c_5\}$
(2) $\{c_1, c_2, c_3, c_4, c_6\}$
(3) $\{c_1, c_2, c_3, c_4, c_7\}$

The incremental SAT solving approach proposed in [Str04] is as follows: One observes that clauses $\{c_1, c_2, c_3, c_4\}$ are common to all three SAT problems, and clauses c_5, c_6 and c_7 are unique to particular SAT instances (1), (2) and (3), respectively. When solving the instance (1), one marks clauses c_1, c_2, c_3, and c_4, and for every conflict clause c encountered during the SAT search, if all clauses leading to the conflict are already marked, then one marks c as well. Note that all pervasive conflict clauses are logical consequences of $\{c_1, c_2, c_3, c_4\}$, thus the satisfiability of (2) and (3) will remain unaffected if the pervasive conflict clauses are added to instances (2) and (3).

Suppose when solving instance (1), the SAT solver chooses first the assignment $x_5 = false$.[2] From this assignment, using clause c_5, BCP will force implied assignment $x_1 = true$, and further iterations of unit clause rule in BCP will lead to the discovery of

---

[2] Most of the modern SAT solvers would start with BCP, and BCP in our example would find a model. We have chosen to start with assignment $x5 = false$ for demonstration purposes, and this allows us to reuse example from [Str04] (and to keep the presentation simple).

a conflict clause $\neg x_1 \lor x_5$ (as shown before, see Figure 1). Since clause c_5 is responsible for that conflict, the conflict clause will not be marked as pervasive, and its usage is not allowed during SAT search for instance (2) (and instance (3)). When trying to resolve instance (2), the SAT solver may again choose to assign $x_5 = false$ and then $x_1 = true$, and discover the same conflict clause again – a duplication of work, which is desired to avoid.

The reader may have noticed that solving SAT problems (1), (2) and (3) corresponds to solving the validity of POs $PO_1 = \neg x_1$, $PO_2 = x_5$, and $PO_3 = x_2$, respectively, from our running Example 1. In the *SSAT* algorithm, there is no need to distinguish between pervasive and other conflict clauses – all conflict clauses are re-usable till the end of the *SSAT* search. Thus any conflict clause can be added to the original clause set without affecting any of the POs' status, and no such conflict clause will be encountered twice in a *SSAT* search. The following is a possible scenario of a *SSAT* run on our running example: Suppose *SSAT* selected PO_2 as the first currently watched PO. Then x_5 is assigned *false*. BCP yields no implied assignments, and *SSAT* may chose $x_1 = true$ as the next decision. BCP will then discover the conflict clause $\neg x_1 \lor x_5$. A clever decision here is to flip the conflicting assignment of x_1, and assign $x_1 = false$. This assignment satisfies clauses c_1 and c_2. *SSAT* may then choose $x_4 = true$ as the next decision. This assignment will satisfy the clause c_3, and BCP will force implied assignment $x_3 = false$ to satisfy c_4 as well. Thus we got a model $x_5 = false$, $x_1 = false$, $x_3 = false$, $x_4 = true$. Variable x_2 is a don't care variable for the discovered model, thus PO_3 can also be declared *falsifiable*. Thus SSAT is left with PO_1; it chooses PO_1 as *CWPO* and assigns it *false* – thus $x_1 = true$. BCP will use the previously discovered conflict clause $\neg x_1 \lor x_5$ to force assignment $x_5 = true$ (here we have used a conflict clause that is not pervasive in the sense of [Str04]); BCP will also imply assignments $x_2 = true$, $x_4 = true$ and $x_3 = false$. SSAT has thus discovered a counter model to validity of $PO_1 - x_1 = true$, $x_5 = true$, $x_2 = true$, $x_4 = true$, $x_3 = false$. *SSAT* will report PO_1 as *falsifiable* and exit.

We have mentioned that SSAT can declare a PO *valid* during the search when it discovers that the PO is *globally true*. This happens when a conflict clause is encountered in which the PO is the unique literal. This can also happen during BCP. In the experimental results section we give data on the valid POs proved in such situations – *all* POs proved valid in SSAT are such POs. Note that in the PISAT approach, a PO can be proved valid if the instance where the PO is assigned *false* is *unsatisfiable* – thus it is necessary to cover the entire search space, while in *SSAT* the POs can be proved valid after a partial traversal of the search space.

It is worth reiterating that in SSAT it is possible to falsify several POs based on the same model. We have seen above a toy example where three POs are falsified based on two models. In the next section we will give experimental data on this as well. This simultaneous falsification feature significantly accelerates the SSAT algorithm. When one works with POs, a similar feature can also be implemented for the PISAT approach based on pervasive conflict clauses. Such a simultaneous falsification feature was indeed activated in the benchmark runs reported in the next section.

Re-usage of certain conflict clauses is the essence of the incremental approach of [MS97, WKS01, Str04]. However, PISAT allows one to reuse only pervasive conflict clauses; hence any conflict whose associated conflict clause is temporal may reappear while solving the next instance. In another context, Goldberg and Novikov [GN01] proposed a method, which we refer to as GN, allowing one to reuse all conflict clauses when solving multiple POs for a given single propositional instance – thus tracking pervasive clauses becomes redundant. Similarly to our approach, GN maintains a currently watched PO (*CWPO*). It assigns *CWPO* the value *false* prior to assigning values to other variables. From then on, GN treats *CWPO* as a normal decision variable and proceeds with a DPLL-style, backtrack search. If a model is found while exploring the subspace, where *CWPO* is assigned *false*, *CWPO* is *falsifiable*, otherwise it is *valid*. After GN completes checking a certain *CWPO*, it augments the initial formula with all or some of the recorded conflict and uses the described above method to determine the status of the remaining POs.

In contrast with PISAT, GN allows one to reuse every conflict clause recorded while checking a certain PO. Indeed, CWPOs are treated as internal assumptions and every recorded conflict clause is guaranteed to be independent of internal assumptions. This feature is common with our SSAT algorithm; still there are important differences between the GN and SSAT algorithms. Most importantly, SSAT is oriented towards simultaneous solving of the POs – it watches all POs and tries to decide other POs while checking the CWPO, whereas GN treats one PO – the CWPO, at a time. To begin with, suppose that a model, falsifying all the POs, is discovered. In this case, SSAT declares all the POs falsifiable and exits, whereas GN falsifies only the *CWPO* and continues to work to falsify other POs. Generally, each time a model falsifying more than one PO is discovered, GN falsifies only the *CWPO*. Another advantage of SSAT is that it can never rediscover the same model, because it always chooses as a *CWPO* only POs that have not been previously falsified by any model. This allows SSAT to prune the search space in a much more efficient manner. In contrast, GN can reach the same models again and again while checking different POs. We will demonstrate empirically that SSAT is more efficient than GN. It is worth mentioning that [GN01] was not written in the context of BMC. Experimental data section of [GN01] contains benchmarks having only a few hundred variables and clauses. Modern BMC benchmarks are larger by 3-5 orders of magnitude, and it is interesting to see how GN performs compared with PISAT and SSAT on such instances.

Conflict clause re-usage was also proposed by Eén and Sörensson [ES03]. The basic idea is the same as in [GN01]. The POs are considered to be internal assumptions, thus every conflict clause is guaranteed to be globally correct. However, [ES03] treats only the case where one should prove a single PO. The conflict clauses are passed between formulas corresponding to different base and step depths. Roughly, the enhanced API proposed by [ES03] corresponds to our SSAT API. However, [ES03] were concerned with solving one objective at a time and their approach lacks the "one traversal" and "all watched" principles of SSAT. We will refer to the fully incremental approaches of [GN01] and [ES03] as FISAT – indeed, their aim (as stated in the respective papers) is to achieve a maximal re-usage of conflicts, rather that *simultaneous solving* of related objectives.

5 Methods for Simultaneous Bounded Model Checking and Induction

Previous sections were concerned with simultaneous solving of propositional objectives. In this section we propose several new methods for simultaneous model checking of multiple safety properties, using the proposed propositional algorithms. Since BMC corresponds to the base of temporal induction, we will mainly discuss induction algorithms. Let us first briefly recall the induction method of [SSS00]. Let $path(s_0,\ldots,s_k)$, $base(P,k)$, $step(P,k)$, and $loopFree(k)$ denote the following formulas:

$$path(s_0,\ldots,s_k) = Tr(s_0,s_1) \wedge Tr(s_1,s_2) \wedge \ldots \wedge Tr(s_{k-1},s_k) \tag{1}$$

$$base(P,k) = I(s_0) \wedge path(s_0,\ldots,s_k) \wedge P(s_0) \wedge \ldots \wedge P(s_{k-1}) \wedge \neg P(s_k) \tag{2}$$

$$step(P,k) = path(s_0,\ldots,s_{k+1}) \wedge P(s_0) \wedge \ldots \wedge P(s_k) \wedge \neg P(s_{k+1}) \tag{3}$$

$$loopFree(k) = path(s_0,\ldots,s_k) \wedge (\bigwedge_{0 \leq i < j \leq k} (s_i \neq s_j)) \tag{4}$$

where Tr is a transition relation between states s_0, s_1, ... of a Finite State Machine M, and $I(s_0)$ denotes that s_0 is an initial state of M. Then a pseudo-code for the basic induction algorithm for an invariant property P looks as follows:

```
BASIC-TEMPORAL-INDUCTION (P, max_depth) {
 k = 0;
   while ( k ≤ max_depth ) {
       If ( satisfiable (base(P,k) ) {
        Return "P is falsifiable (counter-example length is k)";
       }
       If  ( unsatisfiable (step(P,k) && loopFree(k) ) {
         Return "P is valid";
       }
       k++;
   }
 Return "P has no counter-example of length max_depth or less";
}
```

Fig. 3. Basic temporal induction scheme

Checking P in the BMC style consists of finding a k such that $base(P,k)$ is *satisfiable*. We then can generate a counter-example (CE) of length k, which is an *error trace* for P. The above induction scheme for verifying P consists of finding a k such that either $base(P,k)$ is *satisfiable* (and a CE will be generated) or $base(P,i)$ and $step(P,k)$ are *unsatisfiable* for $0 \leq i \leq k$, in which case P is *valid* at all reachable states. The *loopFree* condition is needed for the completeness of the algorithm, but the proofs obtained without this condition remain sound. In the algorithms below, for the simplicity of presentation we omit this condition.

Work [ES03] discusses several variations of this basic induction algorithm. There are several ways to combine base checks with step checks, for example. Further, in the basic algorithm the depth k is incremented by 1, while larger increments are possible by slight modification of the base and step formulas. In the next section where we describe versions of temporal induction algorithms for simultaneous verification of safety properties, for the simplicity of presentation we will only consider a combination of base and step parts in the style of the basic induction algorithm above, and we will only consider increment 1 in base and step depths. Variations similar to those discussed in [ES03] are possible also for this basic simultaneous induction algorithm.

5.1   The Previous Work on Simultaneous Induction

Fraer et al. [FIK+02] proposed a method for *simultaneously* checking a number of safety properties $P_1,...,P_n$ on the same model. Their idea is to form a *conjunction P* from the properties P_i. If P is *false* at depth 0, a CE to P is a CE for a number of properties P_i. These P_i are reported *falsifiable* (at depth 0), and remaining properties will form a conjunction P'. The same process will be applied to P', repeatedly, till the maximal subset of properties whose conjunction is not *falsifiable* at depth 0 is found. To perform BMC, such properties must be checked for depth 1, and so on. The BMC check terminates when all properties are falsified or the depth limit is reached. For the step, the idea is similar: The aim is to find a maximal subset (which actually is *the* maximal subset) V of yet unresolved properties whose conjunction P^* can be proven at current depth k (that is, the corresponding step formula $step(P^*,k)$ must be *unsatisfiable*). The subset V is found after several iterations of SAT-checking of conjunctions of unresolved properties and eliminating properties that cannot be proven at depth k, by inspecting the models returned by the SAT solver.

The next figure describes a basic induction algorithm for multiple safety properties; here and in the remaining algorithms below, U will denote the list of safety properties to be resolved. Furthermore, in these algorithms we normally use callbacks to report the status of the properties (the callbacks are activated during the run, or after the algorithm terminates). The callbacks may or may not be mentioned explicitly. All algorithms return the list of unresolved properties (remaining from the input list).

```
SIMULTANEOUS-INDUCTION(U, max_depth) {
  k = 0;
    while ( k ≤ max_depth && U != ∅) {
        U = simultaneous_base(U,k);
        If ( U != ∅)
          U = simultaneous_step(U,k);
        k++;
    }
  Return U;
}
```

Fig. 4. A basic simultaneous induction scheme for multiple safety properties

In the induction scheme above, for a depth k, the algorithm *simultaneous_ base(U,k)* checks which of the properties in U are *falsifiable* in the instance unrolled to depth k. This can clearly be done in different ways. The *BASE_CONJUNCTION* algorithm below corresponds to the method in [FIK+02] for performing simultaneous base on properties in U:

```
BASE_CONJUNCTION(U,k) {
P = ∧U; // the conjunction of all formulas in U
  While ( U != ∅) {
    if( satisfiable( base(P,k) ) ) {
      U = base_conj_callback( M );
      P = ∧U; }
    else {
      break; }
  }
  Return U;
}
```

Fig. 5. The conjunction method for simultaneous base at depth k

Here M is the model returned by the SAT solver, and *base_conj_callback* checks M: all properties P_j in U whose representative variables at depth k are false in M are reported to the user as *falsifiable* at depth k; the list of remaining P_j is saved as U.

Similarly, the *STEP_CONJUNCTION* algorithm below corresponds to the way *simultaneous_step* procedure is performed in [FIK+02]:

```
STEP_CONJUNCTION(U,k) {
  V = U; // properties we may still prove valid at depth k
  U = ∅; // properties we already know cannot be proven at depth k
    While ( V != ∅ ) {
    P = ∧V;
    if ( satisfiable( step(P,k) ) {
      (V,U) = step_model_callback(M); }
    else {
      Break;   }
    }
    valid_callback(V);
    Return U;
}
```

Fig. 6. The conjunction method for simultaneous step at depth k

Here M is the model returned by the SAT solver, and *step_model_callback* checks which of the variables representing properties in V at depth $k + 1$ are false in M; such properties are moved from V to U, as we know they cannot be proven at depth k; *valid_callback* will report all properties in V valid to the user (the list V may be empty after *STEP_CONJUNCTION* terminates).

5.2 SSAT-Based Induction

In this subsection we propose several new methods for simultaneous temporal induction for multiple safety properties.

The following *BASE_SSAT(U,k)* procedure is a way to perform the *simultaneous_base(U,k)* procedure in the *SIMULTANEOUS-INDUCTION* scheme of Figure 4; it uses the SSAT algorithm:

```
BASE_SSAT(U, k) {
   U = SSAT⁺( U, base_ssat_callback );
   Return U;
}
```

Fig. 7. The SSAT method for simultaneous base at depth k

Here $SSAT^+$ starts by running SSAT; the callback *base_ssat_callback* updates the user every time a property P_j from U gets falsified; finally, the list of remaining properties (properties, proved *valid* by SSAT) will be assigned to U.

To describe simultaneous step algorithms for multiple safety properties, let us define:

$$\underline{step\_ssat}(U,k) = [\neg step(P_1,k),..., \neg step(P_n,k)] \tag{5}$$

$$\underline{step\_hybrid}(U,k) = [step_2(U,k,1),...,step_2(U,k,n)] \tag{6}$$

where both $\underline{step\_hybrid}(U,k)$ and $\underline{step\_ssat}(U,k)$ are lists of formulas; $step_2(U,k,l) = path(s_0,...,s_{k+1}) \wedge P(s_0) \wedge ...\wedge P(s_k) \rightarrow P_l(s_{k+1})$, P_l is a property in $U= \{P_1,...,P_n\}$, and $P =\bigwedge U$. Then two methods of performing simultaneous step are described by *STEP_SSAT* and *STEP_HYBRID* algorithms below:

```
STEP_SSAT(U, k) {
   (U, V) = SSAT*(step_ssat(U, k) )
   valid_callback(V);
   Return U;
}
```

Fig. 8. SSAT method for simultaneous induction step at depth k

Here we assume that $SSAT^*$ runs $SSAT$ and returns a pair of lists, where the first list contains all properties P_j from U whose corresponding formulas $\neg step(P_j,k)$ get falsified by SSAT solver, and the second list consists of the remaining properties P_j from U; *valid_callback* reports all formulas P_j in V valid to the user. Indeed, for all such P_j from V, the corresponding step formula $step(P_j,k)$ is *unsatisfiable*, and since P_j was not falsified till depth k, it is *valid* according to the temporal induction scheme in [SSS00].

```
STEP_HYBRID (U, k) {
  V = U; // properties we may still prove valid at depth k
  U = Ø; // properties we already know cannot be proven at depth k
  fixpoint_reached = false;
  While ( ! fixpoint_reached ) {
        U_old = U;
        (U, V) = SSAT**( step_hybrid(U, k) )
        if ( U == U_old )
           fixpoint_reached = true;
  }
  valid_callback(V);
  Return U;
}
```

Fig. 9. Hybrid method for simultaneous induction step at depth k

Here $SSAT^{**}$ procedure runs $SSAT$ and updates U and V as follows: it moves from V to U all formulas P_j whose corresponding step formulas $step_2(U,k,j) = path(s_0,...,s_{k+1}) \wedge P(s_0) \wedge ... \wedge P(s_k) \rightarrow P_j(s_{k+1})$ from the list step_hybrid(U, k) get falsified in SSAT. When there are no such formulas, the while loop stops – *fixpoint_reached* is assigned *true*. Note that in such a case $path(s_0,...,s_{k+1}) \wedge P(s_0) \wedge ... \wedge P(s_k) \wedge \neg P(s_{k+1})$ is *unsatisfiable*, and it is the step formula for the conjunct P, thus P (and all its conjuncts) are valid according to the temporal induction scheme in [SSS00]. The callback *valid_callback* reports all properties in V valid to the user.

Notice the differences between *STEP_SSAT* and *STEP_HYBRID* algorithms. *STEP_SSAT* needs one call to SSAT, thus in general is faster than *STEP_HYBRID* which may require more calls to SSAT. However, *STEP_HYBRID* works more like *STEP_CONJUNCTION* in that it can find the maximal subset of U whose conjunction can be proved at a given depth k. On the other hand, in *STEP_SSAT*, each unresolved property P_j is proved "without help of other properties", meaning that P_j at depth $k +$ 1 is attempted to prove based on assuming P_j valid at depths 0 to k, while *STEP_HYBRID* and *STEP_CONJUNCTION* use stronger assumptions that the conjunction of all unresolved properties in U is valid at depths 0 to k. Thus *STEP_HYBRID* and *STEP_CONJUNCTION* may in general prove more properties at a given depth than *STEP_SSAT*. The difference between *STEP_HYBRID* and *STEP_CONJUNCTION* is that the latter uses a SAT solver rather than SSAT; therefore the number of calls to the SAT solver depends on the returned models, and since in these models normally not all the properties falsifiable under the current step assumption come up *false*; in general *STEP_CONJUNCTION* needs much more iterations than *STEP_HYBRID*. This is one of the main advantages of *STEP_HYBRID* over *STEP_CONJUNCTION*.

Several variations of *simulatenous_base* and *simultaneous_step* are also possible. For example, one may choose to use the GN algorithm instead of SSAT in simultaneous base and step schemes proposed in this section. We have already mentioned that variations are possible in combining the base and step parts of induction, and increments to the depth other than 1 can easily be allowed by slight modification of the base and step formulas.

All base and step procedures described in this section can be made incremental in various ways. The conjunction method (and implementation) proposed in [FIK+02] is non-incremental, but it can easily be made incremental in the PISAT style of [WKS01, Str04] which needs tracking of pervasive learned clauses, or in the FISAT style of [ES03] where all learned clauses are pervasive. In the former approach, variables used to define the involved base and step formulas can be soundly removed from the instances, while for the latter option one needs to keep them. Since in SSAT all learned clauses are pervasive, it is safe to use them across all calls to SSAT solver at the same or different depths, as long as no variables and clauses used to define the involved base and step formulas are removed; indeed, since no temporal assumptions are made before SSAT calls, all learned clauses are logical consequences of the unrolled instances. From our experience, the extra defining clauses and variables of the base and step formulas are not a significant overhead to the SSAT solver.

6   Experimental Results

In this section, we report experimental results on some important applications in formal hardware verification domain where simultaneous and incremental SAT solving is very beneficial. In particular, we will compare the performance of the basic simultaneous induction algorithm when different simultaneous base and step procedures and different incremental schemes are used. All benchmarks originate from Intel designs. The performance results were generated on a 3.2 GHz machines with 4GB memory.

Since there are many variations of simultaneous temporal induction, there is a choice to be made here. As a base line, we choose the non-incremental conjunction method of [FIK+02]. We will refer to it as **conj**. We will consider a double-incremental version of it, in the PISAT style – we will refer to it as **dincr_conj**; this method is double incremental as we transfer the learned clauses from iteration to iteration at the same depth, as well as from depth to depth. It will allow us to measure the effect of pervasive learning on simultaneous induction. To measure the effect of fully incremental approach FISAT, we will consider SSAT-based schemes where the simultaneous falsification feature (which requires watching of all objectives) is disabled; we will consider two options: **incr_gn**, where learned clauses are not transferred from low to higher depths; and **dincr_gn**, where learned clauses are transferred to higher depths (note that the [GN01] approach corresponds to **incr_gn** rather than **dincr_gn**, since they did not consider incremental learning for related instances – rather, they considered the same instance for multiple objectives). We then consider the SSAT-based induction schemes – *BASE_SSAT* as *simultaneous_base* algorithm, and both *STEP_SSAT* and *STEP_HYBRID* as *simultaneous_step*. We will consider the double-incremental versions for both schemes, and refer to them as **dincr_ssat** and **dincr_hybr**, respectively. Furthermore, to measure the effect of double-incremental learning in SSAT-based induction as well, we will disable transferring the learned clauses from low to higher depths in **dincr_ssat**; we call the resulting scheme **incr_ssat**. We will not consider the scheme precisely corresponding to [ES03], as in the majority of our benchmarks we have tens or hundreds of properties in the same session, and even non-incremental simultaneous induction schemes are superior to repeated application of incremental induction for single safety properties, when solving multiple properties.

The benchmark Tables 1-2 below originate from simultaneous SAT-based model checking of 543 invariant properties. The pruned model (that includes only the "cone of influence" of the properties) contains 2723 state elements (latches), 37159 logic gates and 3767 inputs. The first table gives data of a BMC run using both SSAT and PISAT algorithms at depths 6-15 (the lower depths took less than a second each to complete; the depth count in our algorithm starts from 0). And the second table gives similar data for the step part of the induction algorithm. The combinational instances at each depth are represented as $\wedge$ / $\neg$ graphs (and-inverter graphs, or AIGs, allow for a compact representation of Boolean formulas, see e.g. [KGP01]) and then are translated to the CNF representation to run PISAT or SSAT algorithm. The PISAT algorithm requires multiple calls to the DPLL algorithm, to resolve each (non-trivial) PO. For each PO, the corresponding cone of influence is built, and then translated to CNF representation. Usually, lots of optimizations are performed when translating an AIG representation into a CNF representation, allowing one to reduce significantly the amount of variables in the CNF instance. The downside is that performing such optimizations for each PO separately may be quite an overhead in some cases, especially when the "cones of influences" of the POs have a significant overlap. Usually, vectors of invariant properties are formed so that the properties share large chunks of common logic (otherwise there would be no reason for performing simultaneous model checking). This is the case in the benchmarks reported below.

Table 1. BMC at depths 6-15

BMC depth	# of POs	# of gates	# of inputs	# of literals	# of clauses	PISAT (sec)	GN (sec)	SSAT (sec)
6	543	98288	25383	93784	254145	174.36	9.28	2.42
7	494	113938	28885	108488	295157	245.51	5.41	4.68
8	473	132372	33352	125745	342993	210.47	8.14	3.96
9	450	150565	37454	142720	390432	316.93	2.61	2.61
10	450	170016	42072	160938	440968	305.79	11.79	6
11	435	189670	46529	179233	492157	508.97	14.61	11.6
12	418	209885	51380	198212	544750	364.76	7.68	6.69
13	417	229883	55880	216769	596763	576.3	5.72	5.71
14	417	250285	60745	235896	649809	424.04	11.14	11.38
15	415	270393	65243	254571	702148	686.75	7.96	8.05
Total						3813.88	84.34	63.1

Tables 1-2 show the size of the entire instance both in its AIG representation as well as in CNF representation. In these runs, around 40% of total runtime was spent on the pruning (i.e., relevant cone formation and CNF re-generation) part of the PISAT algorithm (the reported PISAT run times include the pruning times). And the PISAT algorithm spends more time in DPLL search than the SSAT algorithm even if the pervasive learned conflicts are re-used in the PISAT algorithm (we know however that SSAT allows sharing of more conflict clauses). Overall, these tables demonstrate that the SSAT algorithm can perform orders of magnitude faster than the PISAT algorithm, at least on many practical designs. Furthermore, even if the time-consuming pruning is not performed in the GN algorithm, the benchmarks show that overall SSAT is significantly faster than GN.

Table 2. Induction step at depths 1-10

Step depth	# of POs	# of gates	# of inputs	# of literals	# of clauses	PISAT (sec)	GN (sec)	SSAT (sec)
1	543	52786	8026	45811	132490	87.42	30.3	1.99
2	433	73274	9024	62397	184397	117.71	44.6	2.2
3	433	94927	10065	79932	239277	170.86	62.13	3.26
4	433	117160	11128	97964	295687	230.31	78.97	3.92
5	433	140251	12196	117095	355143	291.87	97.73	5.03
6	384	161723	13135	134894	410561	323.46	109.5	15.93
7	260	182905	14182	152888	465867	273.91	117.62	10.36
8	236	204098	15210	170834	521100	292.09	117.87	8.56
9	236	225355	16241	188819	576470	321.43	131.28	11.03
10	221	246438	17261	206653	631386	340.79	139.34	9.24
Total						2449.85	929.34	71.52

Data in Tables 3-4 originate from compositional formal sequential equivalence verification runs [KSKH04]. The specification and implementation circuits are divided into corresponding sub-circuits and verified separately. Equivalence of the circuits can be derived from the equivalence of the corresponding sub-circuits. The properties that are checked state that the corresponding pairs of sub-circuit outputs in the specification and implementation models have same values for any input vector sequence of the respective sub-circuits.

We report experimental data on 6 vectors of safety properties. In order to show the overall impact on the run-time of the end tool (in this case, a SAT-based model checker), in this and other tables below we give total runtimes (BMC or Induction till a certain bound), and do not report a breakdown of times spent at each depth. Tables 3 and 4 both originate from the same model-checking runs (simultaneous induction).

In Table 3 we report the data showing the impact of several advanced features of the SSAT algorithm that do not exist in PISAT and FISAT. The POs in SSAT represent the base and step formulas at different depths. We report the amount of models discovered by SSAT and the number of POs falsified using these models – on average, each model was sufficient to falsify 8-9 POs. We also report the amount of POs proved valid based on global assignments before the end of search (i.e., based on a partial search) – all other POs were trivial to resolve, and they were not passed to the SAT solvers (their validity was discovered during translating the AIG representation to CNF). More interestingly, we give also data on the ratio of pervasive conflicts in the PISAT algorithm, and the amount of conflicts in the SSAT algorithm.

Table 3. Full proof with induction: conflicts, globally true, and simultaneous faslification data

# properties	#PISAT pervasive conflicts	#PISAT conflicts	#SSAT conflicts	#SSAT globally true	#SSAT models	#SSAT false	Bound reached
9	6170	35225	13615	126	21	189	22
9	8549	40680	15709	135	21	189	22
9	8258	37814	14206	135	21	189	22
9	8488	39945	14276	135	21	189	22
9	7056	35370	14251	135	21	189	22
9	6968	13257	13257	135	21	189	22

Table 4. Full proof induction

#properties	Conj	dincr-conj	dincr-gn	incr-ssat	dincr-ssat	dincr-hybr
9	189.57	72.67	66.65	35.05	29.84	29.32
9	200.13	73.23	61.72	40.46	27.76	29.09
9	222.29	66.11	67.17	35.84	26.06	27.83
9	246.51	67.85	62.33	37.22	28.24	29.5
9	253	68.14	59.55	39.04	28.66	30.13
9	215.09	70.25	60.5	35.52	26.7	27.86
Total (sec)	**1326.59**	**418.25**	**377.92**	**223.13**	**167.26**	**173.73**

Table 5. BMC run times (sec) till different depth

#properties	BMC depth	conj	dincr-conj	incr-gn	dincr-gn	incr-ssat	dincr-ssat
32	50	32.95	5.4	536.3	8.31	533.87	8.14
32	50	32.85	5.46	543.15	8.42	534.85	8.15
3	50	318.87	108.54	3041.72	46.17	3064.57	46.05
3	50	360.67	464.32	3760.78	210.15	3747.52	210.45
3	50	310.64	367.93	3653.52	50.23	3612.3	50.06
3	50	242.4	231.59	3337.25	199.96	3330.65	199.46
8	50	78.8	30.69	681.68	27.46	685.82	27.18
8	50	78.14	30.77	680.2	26.97	682.88	26.91
8	50	18.53	3.28	162.84	9.98	157.66	10.03
8	50	18.46	3.35	157.25	10.23	157.69	10.18
543	15	139.82	41.18	51.26	32.78	36.08	18.91
172	30	145.75	52.13	76.83	40.5	63.59	28.06
1035	3	2478.61	406.2	1743.56	1618.58	386.93	229.28
Total run times:		**4256.49**	**1750.84**	**18426.34**	**2289.74**	**16994.41**	**872.86**

There is a clear correlation between these conflict counts and the runtimes of the **dincr_conj** and **dincr_hybr** algorithms reported in Table 4. In Table 4, one can also see the advantage of the double-incremental verification. And furthermore, the SSAT-based algorithms **dincr-ssat** and **dincr-hybr** are the fastest among all simultaneous induction algorithms discussed in previous sections.

In Table 5 we report the model checking runtimes for various methods of simultaneous BMC on test cases originated from formal property verification as well as formal equivalence verification of Intel designs. Again, the SSAT-based algorithm **dincr_ssat** is clearly superior.

In Table 6, we compare the main double-incremental schemes, and show the speedup of **dincr_ssat** and **dincr_hybr** compared with **dincr_conj** (columns speedup 1 and speedup 2 respectively). The superiority of the SSAT based schemes is evident. We also present data on the number of properties, and number of properties proved valid or falsifiable. In the corresponding column, for the proved properties we give two figures: properties proved in **dincr_hybr** and properties proved in **dincr_ssat**; we already know that the former scheme may prove more properties than the latter.

Table 6. Full induction till bounds 10-30

#PO/#false/#valid	depth	conj	dincr_conj	dincr_ssat	dincr_hybr	speedup 1	speedup 2
172/106/65	30	2777.54	1718.51	200.69	237.66	8.56	7.23
543/128/(275:214)	30	3978.52	3287.89	312.08	233.97	10.54	14.05
100/0/(73:41)	10	18040.21	6390.95	830.61	1384.08	7.69	4.62
249/32/(70:64)	14	31338.58	7730.62	3710.35	2148.58	2.08	3.60

7 Conclusions

We presented an incremental propositional satisfiability technique allowing one to solve simultaneously and efficiently multiple satisfiability problems for related formulas. Insignificant modification to a (regular) DPLL-based SAT solver is sufficient to implement our Simultaneous SAT algorithm. Further, we presented several novel techniques for simultaneous SAT-based model checking of multiple safety properties. We provided experimental results demonstrating that the SSAT algorithm may be orders of magnitude faster in solving related SAT problems than the previous incremental SAT solving approaches that require multiple calls to a SAT solver, and that double-incremental simultaneous model checking of related safety properties employing the SSAT algorithm can accelerate verification significantly.

References

[BCCZ99] Biere A., A. Cimatti, E. Clarke, Y. Zhu, Symbolic model checking without BDDs, Tools and Algorithms for the Construction and Analysis of Systems, TACAS 1999.

[BCC+03] Biere, A., A. Cimatti, and E. Clarke, O. Strichman, Y. Zhu, Bounded Model Checking, Chapter in Advances in Computers, vol. 58, 2003.

[Bry86]   Bryant R.E., Graph-based algorithms for Boolean function manipulation, IEEE
             Trans. Computers, C-35(8), 1986.
[CGP99]   Clarke E.M., O. Grumberg, D.A. Peled, *Model Checking*, MIT Press, 1999.
[DLL62]   Davis M., G. Logemann, D. Loveland, A machine program for theorem proving. In
             Communications of the ACM, (5):394-397, 1962.
[DP60]   Davis M., H. Putnam, A computing procedure for quantification theory, J. ACM,
             vol 7, 1960.
[ES03]   Eén N, N. Sörensson, Temporal induction by incremental SAT solving,
             International Workshop on Bounded Model Checking, BMC 2003.
[FIK+02]   Fraer, R., S. Ikram, G. Kamhi, T. Leonard, A. Mokkedem, Accelerated verification
             of RTL assertions based on satisfiability solvers, HLDVT, 2002.
[GN01]   Goldberg E., Y. Novikov, An efficient learning procedure for multiple implication
             check. In Design, Automation, and Test in Europe (DATE '01), 2001.
[GSK98]   Gomes C.*P.*, B. Selman, H. Kautz, Boosting combinatorial search through
             randomization, National Conference on Artificial Intelligence, 1998.
[KGP01]   Kuehlmann A., M.K. Ganai, *V.* Paruthi, Circuit-based Boolean reasoning, DAC
             2001.
[KSKH04]   Khasidashvili, Z., M. Skaba, D. Kaiss, Z. Hanna, Theoretical framework for
             compositional sequential hardware equivalence verification in presence of design
             constraints, ICCAD'04, 2004.
[LM02]   Lynce I., J. Marques-Silva, Building state-of-the-art SAT solvers, European
             Conference on Artificial Intelligence (ECAI), 2002.
[MS97]   Marques-Silva J.*P.*, K.A. Sakallah, Robust search algorithm for test pattern
             generation, IEEE Fault-Tolerant Computing Symposium, 1997.
[MS99]   Marques-Silva J.*P.*, K.A. Sakallah, GRASP: A search algorithm for propositional
             satisfiability, IEEE Transactions on Computers, vol. 48, 1999.
[McM93]   McMillan, K.L., *Symbolic Model Checking*, Kluwer, 1993.
[Nad02]   Nadel A. Backtrack search algorithms for propositional satisfiability: Review and
             Innovations, Master Thesis, the Hebrew University of Jerusalem, 2002.
[PBG05]   Prasad M., A. Biere, A. Gupta, A survey of recent advances in SAT-based formal
             verification, Int. Journal on Software Tools for Technology Transfer (STTT), vol.
             7, number 2, 2005.
[SSS00]   Sheeran, M., S. Singh, G. Stalmarck, Checking safety properties using induction
             and a SAT-solver, FMCAD, 2000.
[Str04]   Strichman, O., Accelerating bounded model checking of safety properties, Formal
             Methods in System Design, vol, 24, 2004.
[ZM88]   Zabih R., D.A. McAllester, A rearrangement search strategy for determining
             propositional satisfiability, National Conference on Artificial Intelligence, 1988.
[ZMM+01]   Zhang, L., C.F. Madigan, M.H. Moskewicz, S. Malik, Efficient conflict driven
             learning in a boolean satisfiability solver. International Conference on Computer-
             Aided Design (ICCAD'01), 2001.
[WKS01]   Whittemore, J., K. Kim, K. Sakallah, SATIRE: A new incremental satisfiability
             engine, DAC, 2001.

HaifaSat: A New Robust SAT Solver

Roman Gershman[1] and Ofer Strichman[2]

[1] Computer Science, Technion, Haifa, Israel
[2] Information Systems Engineering, Technion, Haifa, Israel
gershman@cs.technion.ac.il, ofers@ie.technion.ac.il

Abstract. The popular abstraction/refinement model frequently used in verification, can also explain the success of a SAT decision heuristic like Berkmin. According to this model, conflict clauses are abstractions of the clauses from which they were derived. We suggest a clause-based decision heuristic called Clause-Move-To-Front (CMTF), which attempts to follow an abstraction/refinement strategy (based on the resolve-graph) rather than satisfying the clauses in the chronological order in which they were created, as done in Berkmin. We also show a resolution-based score function for choosing the variable from the selected clause and a similar function for choosing the sign. We implemented the suggested heuristics in our SAT solver HaifaSat. Experiments on hundreds of industrial benchmarks demonstrate the superiority of this method comparing to the Berkmin heuristic. There is still room for research on how to explore better the resolve-graph information, based on the abstraction/refinement model that we propose.

1 Introduction

A SAT solver can be thought of as a search engine based on *enumeration* of solutions, but also as a *proof engine* based on inference through the resolution rule. Traditionally the first view was dominant, hence the emphasis in designing SAT solvers and explaining their success was on pruning search spaces. Decision heuristics and learning schemes can all be interpreted as aiming at this goal. Yet the harder and larger the CNF instances are, pruning alone cannot account for the success of modern SAT solvers. It is their ability as proof engines that makes them succeed. This distinction has practical implications, too. For example, for many years decision heuristics gave higher priority to variables in shorter clauses, and to learning shorter conflict clauses. The reasoning was that such clauses can potentially prune larger search-spaces. Although this claim is true, all modern decision heuristics (VSIDS [4], VMTF [3], Berkmin [2]) ignore the length of the clauses, after reaching empirically the conclusion that there are more important considerations. Ryan experimented in his thesis [3] with first-UIP and all-UIP learning schemes, and although the latter generate on average shorter clauses, the former is empirically better. He hypothesized that the learning scheme should be geared towards resolution rather than for pruning. In this article we extend this approach by looking on clause-learning and the decision

S. Ur, E. Bin, and Y. Wolfsthal (Eds.): Haifa Verification Conf. 2005, LNCS 3875, pp. 76–89, 2006.

heuristic as one complete mechanism and refer to a SAT solver as a prover rather than as a search engine. It turns out, empirically, that when conflict clauses are effective, which is the case in all real-world instances, this is the right way to go.

Conflict clauses are derived through a process of resolution (see, for example, [6] and [1] for a more formal treatment of this subject). If a clause c is derived by resolution from a set of clauses $c_1 \ldots c_n$ then

$$c_1 \wedge \cdots \wedge c_n \rightarrow c$$

while the other direction does not hold. This means that we can see c as an over-approximating abstraction of the resolving clauses $c_1 \ldots c_n$. Attempting to satisfy c first, therefore, can be seen as an attempt to satisfy the abstract model first. Like any abstraction/refinement technique, a successful assignment to c is one that satisfies the concrete model (the $c_1 \ldots c_n$ clauses) as well. And an unsuccessful assignment leads to a refinement step, or, in our case, to derivation of new conflict clauses which further constrain the abstract model. According to this model, Berkmin is only one of many possible strategies to refine the abstract model. In Section 3 we suggest one such alternative clause-based decision heuristic called Clause-Move-To-Front (CMTF), which attempts to follow the order of the clauses in the *resolve-graph* [7] rather than their chronological order in which they were created. In Section 4 we also show a resolution-based score function for choosing the variable from the selected clause and a similar function for choosing the sign. In Section 5 we report experimental results on hundreds of industrial benchmarks that prove the advantage of our approach.

2 Background

The explanation of our methods and the analysis of various heuristics later on will require some basic definitions.

Abstraction and refinement of formulas. While the typical use of the terms *abstraction* and *refinement* refer to models and programs, here we define them in the context of formulas.

We say that a formula $\hat{f}$ is a conservative abstraction (over-approximation) of another formula f if

$$f \rightarrow \hat{f}$$

A refinement process of $\hat{f}$ with respect to f finds an intermediate formula $\hat{f}_1$ such that

$$f \rightarrow \hat{f}_1 \quad \text{and} \quad \hat{f}_1 \rightarrow \hat{f}$$

Abstraction-Refinement is an iterative process in which one begins with some abstract formula $\hat{f}$ of a concrete formula f and gradually refines it through a series of formulas $\hat{f}_1, \ldots, \hat{f}_n$ until proving or disproving the desired property of f (in the worst case $\hat{f}_n = f$).

Restricting our attention to CNF formulas, suppose we have two sets of clauses, S and $\hat{S}$, s.t. $\hat{S} \subseteq S$. It is straight forward that $S \rightarrow \hat{S}$ and, therefore, $\hat{S}$ is an abstraction of S.

2.1   Conflict Clauses and Resolution

The well-known binary resolution rule is:

$$\frac{a_1 \vee \ldots \vee a_n \vee \beta \qquad b_1 \vee \ldots \vee b_m \vee (\neg\beta)}{a_1 \vee \ldots \vee a_n \vee b_1 \vee \ldots \vee b_m}$$

where $a_1, \ldots a_n, b_1, \ldots b_m, \beta$ are literals. β is known as the *resolution variable* (also known as the *pivot variable*) of this derivation. Clauses $(a_1, \ldots, a_n, \beta)$ and $(b_1, \ldots, b_n, \bar{\beta})$ are called *resolving clauses* and clause $(a_1, \ldots, a_n, b_1, \ldots b_n)$ is a *resolvent*. It follows by the soundness of the rule, that the resolvent is always implied by its resolving clauses and can therefore be thought of as an abstraction of the clauses that participated in the derivation.

Algorithm 1. The First-UIP resolution algorithm

---

    procedure ANALYZECONFLICT(Clause: conflict)
2:     $currentClause \leftarrow conflict$;
      $ResolveNum \leftarrow 0$;
4:     $NewClause \leftarrow \emptyset$;
      repeat
6:        **for** each literal $lit \in currentClause$ **do**
           $v \leftarrow var(lit)$;
8:        **if** v is not marked **then**
           Mark v;
10:         **if** $dlevel(v) = CurrentLevel$ **then**
             $++ ResolveNum$;
12:         **else**
             $NewClause \leftarrow NewClause \cup \{lit\}$;
14:         **end if**
        end if
16:        **end for**
        $u \leftarrow$ last marked literal on the assignment stack;
18:        Unmark $var(u)$;
        $-- ResolveNum$;
20:        $ResolveCl \leftarrow Antecedent(u)$;
        $currentClause \leftarrow ResolveCl \setminus \{u\}$;
22:     **until** $ResolveNum = 0$;
      Unmark literals in $NewClause$;
24:     $NewClause \leftarrow NewClause \cup \{\bar{u}\}$;
      Add $NewClause$ to the clause database;
26: **end procedure**

We now show why the process of generating conflict clauses indeed can be seen as a sequence of resolution steps. Algorithm 1 shows a simple and efficient implementation of the First-UIP resolution scheme, which is implemented in most competitive SAT solvers, including our solver HAIFASAT. We will refer to this algorithm simply as the *resolution* algorithm. First, a conflicting clause is set to be the current resolved clause. The main loop processes literals in the current clause. All literals from the previous decision levels are gathered into *NewClause* at line 13 and marked. Literals from the current level are marked in order to resolve on them (i.e., use them as resolution variables) further. In every iteration a new marked (yet unprocessed) literal u is chosen in line 17. This literal must be from the current decision level. The algorithm resolves on u by setting *currentClause* to be the antecedent clause without u.

ResolveNum counts the number of the marked literals from the current decision level that still have to be processed. When $ResolveNum = 0$ at line 22, then u is the *FirstUIP* or the *asserted* literal. The negation of this literal is added to the *NewClause* causing u's value to be flipped after backtracking. For more details on the resolution algorithm see [4,3].

We will use the following definition in order to denote the initial state of *NewClause*:

Definition 1 (Asserting clause). *Suppose a new conflict clause C was created in Alg. 1 with asserted literal u. Suppose also that the solver backtracks after the conflict to level dl. Then C becomes an* asserting *clause when it implies $\overline{u}$ for the first time at level dl, and stops being asserting when the solver backtracks from dl.*

It follows from definition that every conflict clause becomes asserting exactly once.

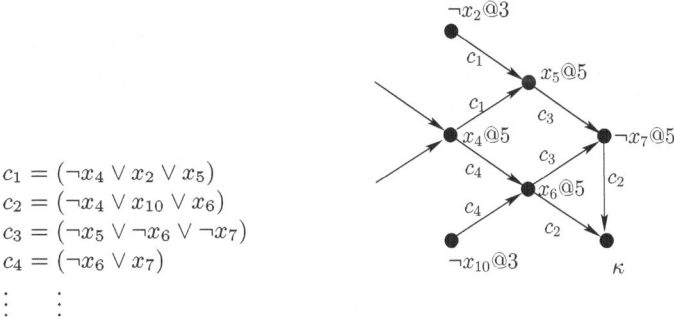

$$c_1 = (\neg x_4 \lor x_2 \lor x_5)$$
$$c_2 = (\neg x_4 \lor x_{10} \lor x_6)$$
$$c_3 = (\neg x_5 \lor \neg x_6 \lor \neg x_7)$$
$$c_4 = (\neg x_6 \lor x_7)$$

Fig. 1. A partial implication graph and set of clauses demonstrate ANALYZECONFLICT. x_4 is the *FirstUIP*, and $compl x_4$ is the asserted literal.

Example 1. Consider the following partial implication graph [5] and set of clauses Denote by $Resolve(s, t, x)$ the binary resolution of clauses s and t with the resolution variable x. Then the conflict clause $c_5 : (x_{10}, x_2, \neg x_4)$ is computed through a series of binary resolutions, starting from the conflicting clause c_4, and going backwards on the implication graph until all literals in the conflict clause are either from previous decision levels or the *firstUIP*.

$$Resolve(Resolve(Resolve(c_4, c_3, x_7)), c_2, x_6), c_1, x_5) = (x_{10}, x_2, \neg x_4)$$

Algorithm 1 implicitly performs these resolution steps while computing the conflict clause c_5.   □

NewClause is derived through a series of binary resolutions that can be seen as a tree: every time the solver reaches line 21, an intermediate clause (consisting of all marked literals) is resolved with the antecedent clause of the chosen resolution variable. We can treat this process as one atomic action of *Hyper-resolution* (resolution between more than two clauses). Since each conflict clause is derived from a set of other clauses, we can keep track of this process with a *Resolve-Graph* [7]. Here we define a variation of the well-known resolve-graph that distinguished between two types of resolutions:

Definition 2 (Colored Resolve Graph). *A* Resolve Graph *is a directed acyclic graph where each node corresponds to a clause, and there is an edge* (u, v) *if and only if v participated in the Hyper-resolution of u as a CurrentClause at line 6 of Alg 1.*
   The color *of the edge* (u, v) *is defined to be blue if v was an asserting (conflict) clause during the resolution and red otherwise.*

In this graph, edges come from the resolvent to its resolving clauses. The leafs of the graph correspond to the original clauses in the formula. Notice that since a conflict at level dl necessarily implies that the solver backtracks from dl and unassigns all the variables that were resolved on, any asserting clause which participated in the resolution will stop being asserting. Therefore for any conflict clause there can be at most one incoming blue edge. The original clauses do not have outgoing edges, and only red incoming edges.

Example 2. Consider once again the implication graph in Figure 1. Since $c_1 \ldots c_4$ participate in the resolution of c_5, the corresponding resolve-graph is as appears in Figure 2. Assuming that $c_1 \ldots c_4$ are original clauses, then all the edges in this graph are red, because original clauses cannot be asserting.

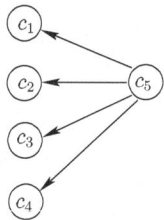

Fig. 2. A resolve-graph corresponding to the implication graph in Figure 1

Now consider a similar case in which c_2 is not an original clause, and at the time when $x_4@5$ is assigned it does not yet exist (the notation $l@i$, adopted from [5], means that literal l is satisfied at decision level i). The implication graph at this stage appears in Figure 3. Now assume that due to further decisions and implications in deeper decision levels a conflict is encountered, the solver creates

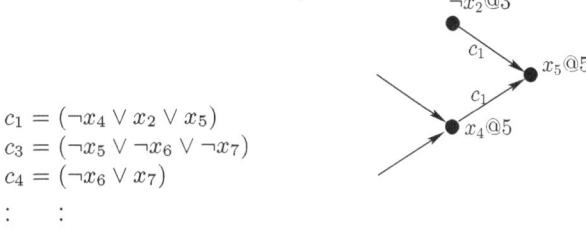

$c_1 = (\neg x_4 \lor x_2 \lor x_5)$
$c_3 = (\neg x_5 \lor \neg x_6 \lor \neg x_7)$
$c_4 = (\neg x_6 \lor x_7)$

Fig. 3. A partial implication graph corresponding to c_1, c_3, c_4 and the decision $x_4@5$

the new conflict clause c_2, backtracks to decision level 5 and asserts $x_6@5$. This, in turn, completes the implication graph to the way it looks in Figure 1. But now, since c_2 asserts x_6, we consider its edge on the resolve-graph from c_5 as blue.   □

The distinction between the two type of edges is important because a blue edge (u, v) indicates that the solver had to create u in order to later create v^1.

2.2   The Berkmin Decision Heuristic

We describe shortly the Berkmin Decision heuristic, since not only that it is considered to be one of the best known, but also because it is a clause-based decision heuristic, like HAIFASAT's heuristic, and therefore convenient for comparison.

Berkmin [2] pushes every new conflict clause to a stack, and makes a decision by choosing an unassigned variable from the last unsatisfied conflict clause in this stack (if there is more than one such variable, it uses the VSIDS score system[4]). If all the conflict clauses are satisfied, it continues with a different heuristic.

In Fig. 4(a) we show a sketch of the progress of Berkmin, which is helpful in understanding why this process can be seen as abstraction-refinement. Clauses $c_1, \ldots, c_{100}$ are conflict clauses ordered by their creation time (c_1 is first). Berkmin tries to satisfy these clauses from last to first, i.e. from right to left. Suppose that all clauses $c_{51} \ldots c_{100}$ are already satisfied, and now Berkmin focuses on c_{50}. We refer to $S = \{c_{51}, \ldots, c_{100}\}$ as our current abstract formula of the original formula φ (it is abstract because each of the clauses in S is derived by a resolution chain from the clauses of φ). Clauses in S must be satisfied, since the decision heuristic reached c_{50}. Therefore, S is an abstract formula of φ. Berkmin now makes a decision on a variable from c_{50} which leads to a conflict and learning of a new clause. The decision heuristic backtracks to the clauses on the end of the list, until finally, through possibly additional iterations of conflicts and added clauses, it reaches c_{50} again while all the clauses to its right are satisfied. Denote by S' the clauses to the right of c_{50} at this point, e.g. $S' = \{c_{51} \ldots, c_{110}\}$. Clearly $S \subseteq S'$ and S' is an abstraction of φ. We can therefore say that S' is a refinement of S with respect to φ.

---

[1] By this we do not mean that this is the only way to create v.

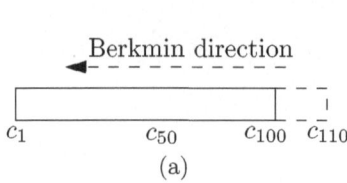
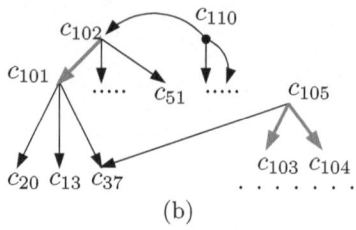

(a)   (b)

Fig. 4. Berkmin's decision heuristic can be thought of as an abstraction-refinement, where a range of the conflict clauses from the right end until c_i represents an abstract model of the clauses on the left of c_i. (a) Berkmin clauses stack: after encountering a conflict, the new resolved clauses are added on the right end. By the time the solver returns to c_{50}, it will have a partial assignment that satisfies a refined model, i.e. the clauses $c_{51} \ldots c_{110}$ (b) The resolve sub-graph of some newly created clauses. Grey thick edges denote the blue edges in the graph.

This view of the process possibly explains why a strategy of giving absolute priority to variables in a specific clause is empirically better than previous approaches like VSIDS that used only a score function.

Fig. 4(a) shows a 'linear' view of the conflict clauses in the order that they are added, which is also the order in which they are considered by Berkmin. The Berkmin heuristic never tries to satisfy a clause before satisfying its resolvents and thus mimics a gradual process of refinement.

A different view of conflict clauses considers their partial order in the Resolve Graph. Fig.4(b) presents a possible Resolve sub-Graph corresponding to the same set of clauses. After the conflicts, Berkmin starts from satisfying c_{110}. c_{102} is a resolving clause that can potentially refine the initial model, however Berkmin first passes through $c_{105}, c_{104}, c_{103}$ to which c_{110} is not connected at all. Therefore Berkmin is dispersed trying to refine several abstractions. Such unfocused behavior can lead to longer proofs. This problem is exactly what our decision heuristic CMTF attempts to solve, as we soon show.

Our SAT solver HAIFASAT makes a decision in three steps: it chooses an unsatisfied clause according to the CMTF heuristic, it then chooses an unassigned variables from this clause, and finally gives it a value. The next sections describe in detail these decision steps.

3   The Clause-Move-To-Front (CMTF) Decision Heuristic

The description above of Berkmin's decision heuristic, and the alternative view of the conflict clauses as being part of a resolve-graph, hints towards the process which is described in Figure 5. In the first line $roots(ResolveGraph)$ refer to resolvent clauses that did not resolve other clauses. Note that in this general scheme a clause is processed only if at least one of its abstractions (its resolvent clauses) has already been processed. It is easy to see that Berkmin is an

1: $S = roots(ResolveGraph)$;
2: Choose an unsatisfied clause (vertex) $v \in S$; If there is no such clause, exit;
3: Process v;   ▷ Processing a clause, among other things, satisfies it.
4: $S = S \cup children(v)$;
5: Goto 2

Fig. 5. A Resolve-Graph Based decision heuristic

instantiation of the scheme. In fact, Berkmin is more strict and processes a clause only if *all* its abstractions are satisfied.

CMTF is a method that instantiates this scheme in a different way. It causes the decision heuristic to be more focused on the current refinement path, i.e. to satisfy children of the currently satisfied clause s. It works as follows:

- All the conflict clauses are stored in a list.
- During the resolution in Alg 1, a bounded number of resolving conflict clauses which are processed at line 6 are moved to the front (front corresponds to the right end of Fig 4(a)). The newly created clause *NewClause* is also added to the list (can be done at line 25).
- Clauses are processed from right to left in the list, while ignoring satisfied clauses. If all the conflict clauses are satisfied then the original VMTF strategy (from Siege [3]) is applied.

The idea of this strategy is to keep clauses that participate in resolution adjacent to their resolvents (at least until the next time they participate in a resolution, a case in which they can be moved to a new location).

CMTF shows a big improvement on many industrial problems comparing to the Berkmin heuristic. Both are specific instantiation of the scheme showed above. The advantages of CMTF is its simplicity and the fact that the explicit storage of the resolve-graph is not required. However, it seems that there is still room for future research on how to use the general scheme. For example, classic AI search methods like best-first-search can be used to decide on the exploration order of nodes in S at line 2. It may happen that partial or full storage of the resolve-graph will improve the performance.

4   Resolution-Based-Scoring

In the previous section we showed how HAIFASAT decides which clause to satisfy first. Given a clause c there can still be several ways to satisfy it. HAIFASAT computes dynamically an *activity score* for each variable and then chooses the variable with the maximal score. Then another *sign score* is used to determine its Boolean value.

We define a scoring heuristic based solely on the resolution algorithm (Algorithm 1). The idea, intuitively, is to give higher weights to variables that were frequently resolved on recently, while distinguishing between resolutions that were necessary for the progress of the solver, and those that were made due to the imperfection of the decision heuristic. We will need several definitions and lemmas to explain this heuristic more precisely.

Suppose that every time the solver makes a decision or processes a conflict it writes into a log the event $a_i = (dl, e)$ where dl is the decision level where the event occurred and $e \in Conflicts \cup Decisions$ is either a conflict event or a decision event. The global index i is incremented every time the event happens. We call the sequence $\{a_i\}_1^N$ the *flat log* of the solver's run. We will denote by $DL(a_i)$ the decision level of the event. We consider only the case in which $dl > 0$. All conflict events other, potentially, than the last one in an unsatisfiable instance are included by this definition[2]. It must hold that for any conflict c there exists a decision d at the same level as c. In such a case, we say that d is refuted by c. More formally:

Definition 3 (Refuted decision by a conflict). *Let $a_j = (dl, c)$ be a conflict event. Let $a_k = (dl, d)$, $k < j$, be the last decision event with decision level dl preceding a_j (note that for $i \in [k+1, j-1] : DL(a_i) > dl$). We say that d is the refuted decision of the conflict c, and write $\mathbf{D(a_j)} = a_k$.*

Note that because of non-chronological backtracking the opposite direction does not hold: there are decisions that do not have conflicts on their levels that refute them.

For any conflict event a_j, the range $(D(a_j), a_j)$ defines a set of events that happened after $D(a_j)$ and led to the conflicts that were resolved into the conflict a_j which, in turn, refuted $D(a_j)$. These events necessarily occurred on levels deeper than $DL(D(a_j))$.

Definition 4 (Refutation Sequence and sub-tree events). *Let a_j be a conflict event with $a_i = D(a_j)$. Then the (possibly empty) sequence of events $a_{i+1}, \ldots, a_{j-1}$ is called the* Refutation Sequence *of a_j and denoted by $\mathbf{RS}(a_j)$. Any event $a_k \in RS(a_j)$ is called a* sub-tree event *of both a_j and a_i.*

Example 3. Consider the conflict event $a_j := (27, c_{110})$ in Fig. 6. For every event a_i that follows decision $D(a_j) = (27, d_{202})$ until (but not including) the conflict c_{110} it holds that $a_i \in RS(a_j)$. Note that the solver can backtrack from deeper levels to level 27 as a result of conflict events. However no event between a_i and a_j occurred on levels smaller or equal to 27.   □

The number of resolutions for each variable is bounded from above by the number of sub-tree conflicts that were resolved into the current conflict. However, not all sub-tree conflict clauses resolve into the current refuting conflict. Some of them could be caused by the imperfection of the decision heuristic and are therefore not used at this point of the search. Our goal is to build a scoring system that is based solely on those conflicts that contribute to the resolution of the current conflict clause. In other words, we compute for each variable an *activity score* which reflects the *number of times it was resolved-on in the process of generating the relevant portion of the refutation sequences of recent conflicts*. We *hypothesize that this criterion for activity leads to faster solution times*.

---

[2] A conflict that occurs at level 0 proves that the instance is unsatisfiable.

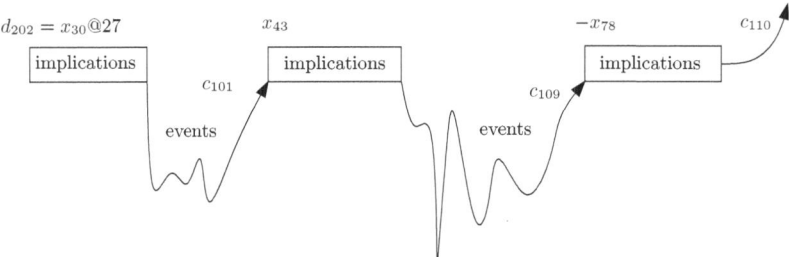

Fig. 6. A possible scenario for the flow of the solver's run. After deciding x_{30} at decision level 27 the solver iteratively goes down to deeper decision levels and returns twice to level 27 with new asserted literals x_{43} and $\overline{x}_{78}$. The latter causes a conflict at level 27 and the solver backtracks to a higher decision level. Implications in the boxes denote assignments that are done during BCP after implying decision or asserting literal.

The information in the colored resolve-graph can enable us to compute such a score.

Definition 5 (Asserting set). *Let $G = (V, E)$ be a colored resolve-graph, and let $v \in V$ be a conflict clause. The Asserting set $B(v) \subset V$ of v is the subset of (conflict) clauses that v has a blue path to them in G.*

The following theorem relates between a resolve-graph and sub-tree conflicts.

Theorem 1. *Let e_v be the conflict event that created the conflict clause v. Then the asserting set of v is contained in the refutation sequence of e_v, i.e. $B(v) \subseteq RS(e_v)$. In particular, since conflict events in $B(v)$ participate in the resolution of v by definition, they necessarily correspond to those sub-tree conflicts of e_v that participate in the resolution of v.*

Note that $B(v)$ does not necessarily include *all* the sub-tree conflicts that resolve into v, since the theorem guarantees containment in only one direction. Nevertheless, our heuristic is based on this theorem: it computes the size of the asserting set for each conflict.

In order to prove this theorem we will use the following lemmas.

Lemma 1. *Denote by $stack(a_j)$ the stack of implied literals at the decision level $DL(a_j)$, where a_j is a decision event. Suppose that a literal t is asserted and entered into $stack(a_j)$, where a_j is a decision event. Further, suppose that t is asserted by the conflict clause cl (cl is thus asserting at this point) which was created at event a_i. Then it holds that $j < i$, i.e cl was created after the decision event a_j occurred.*

Proof. Right after the creation of cl, the DPLL algorithm backtracks to some level dl' with a decision event $a_k = (dl', d)$ and implies its asserted literal. It holds that $k < i$, because the solver backtracks to a decision level which already exists when cl is created. By the definition of an asserting clause, cl can be asserting exactly once, and since cl is asserting on level dl', it will never be asserting after the DPLL algorithm will backtrack from dl'. Therefore it must hold that $a_k = a_j$ $(k = j)$ and $dl' = dl$. ☐

Lemma 2 (Transitivity of RS). *Suppose that a_i, a_j are conflict events s.t. $a_i \in RS(a_j)$. Then, for any event $a_k \in RS(a_i)$ it follows that $a_k \in RS(a_j)$.*

Proof. First, we will prove that $D(a_i) \in RS(a_j)$, or, in other words, that $D(a_i)$ occurred between $D(a_j)$ and a_j. Clearly, $D(a_i)$ occurred before a_i and, therefore, before a_j. Now, falsely assume that $D(a_i)$ occurred before $D(a_j)$. Then the order of events is $D(a_i), D(a_j), a_i$. However, this can not happen since $D(a_j)$ occurred on shallower (smaller) level than a_i and this contradicts the fact that all events between $D(a_i)$ and a_i occur on the deeper levels. Therefore, both $D(a_i)$ and a_i occurred between $D(a_j)$ and a_j. Now, since a_k happened between $D(a_i)$ and a_i it also happened between $D(a_j)$ and a_j and from this it holds that $a_k \in RS(a_j)$. ☐

Using this lemma we can now prove Theorem 1.

Proof. We need to show that any blue descendant of v is in $RS(e_v)$. By Lemma 2 it is enough to show it for the immediate blue descendants, since by transitivity of RS it then follows that for any blue descendant. Now, suppose that there exists a blue edge (v, u) in the resolve-graph. By the definition a blue edge, clause u was asserting during the resolution of v. On the one hand, u was resolved during the creation of v and, therefore, was created before v. On the other hand, by Lemma 1 it was created after $D(e_v)$. Therefore, $e_u \in RS(e_v)$. ☐

Definition 6 (Sub-tree weight of the conflict). *Given a resolve-graph $G(V, E)$ we define for each clause v a state variable $W(v)$:*

$$W(v) = \begin{cases} \sum_{(v,u) \in E} W(u) + 1 & v \text{ is asserting} \\ 0 & \text{otherwise} \end{cases}$$

The function $W(v)$ is well-defined, since the resolve-graph is acyclic. Moreover, since the blue sub-graph rooted at v forms a tree (remember that any node has at most one incoming blue edge), $W(v)$ equals to $|B(v)| + 1$. Our recursive definition of $W(v)$ gives us a simple and convenient way to compute it as part of the resolution algorithm. Algorithm 2 is the same as Algorithm 1, with the addition of several lines: in line 5 we add $W \leftarrow 1$, at line 24 we add $W += W(ResolveCl)$ and, finally, we set $W(NewClause) \leftarrow W$ at line 29. We need to guarantee that $W(C)$ is non-zero only when C is an asserting clause. Therefore, for any antecedent clause C, when its implied variable is unassigned we set $W(C) \leftarrow 0$.

Computing the scores of a variable. Given the earlier definitions, it is now left to show how activity score and sign score are actually computed, given that we do not have the resolve-graph in memory. For each variable v we keep two fields: $activity(v)$ and $sign\_score(v)$. At the beginning of the run $activity$ is initialized to $\max\{lit\_num(v), lit\_num(\overline{v})\}$ and $sign\_score$ to $lit\_num(v) - lit\_num(\overline{v})$. Alg. 2 shows the extended version of the resolution algorithm which computes the weights of the clauses and updates the scores. Recall that any clause weight is reset to zero when its implied variable is unassigned, so that any clause weight is contributed at most once. In order to give a priority to recent resolutions we occasionally divide both activities and sign scores by 2.

Our decision heuristic chooses a variable from the given clause with a biggest activity and then chooses its value according to the sign score: TRUE for the positive values and FALSE for the negative values of the sign score.

Algorithm 2. First-UIP Learning Scheme, including scoring

---

    procedure ANALYZECONFLICT(Clause: conflict)
2:    $currentClause \leftarrow conflict$;
        $ResolveNum \leftarrow 0$;
4:    $NewClause \leftarrow \emptyset$;
        wght $\leftarrow$ **1**;
6:    **repeat**
        for each literal $lit \in currentClause$ **do**
8:            $v \leftarrow var(lit)$;
            if v is not marked **then**
10:               Mark v;
               if $dlevel(v) = CurrentLevel$ **then**
12:                   $++ ResolveNum$;
               else
14:                   $NewClause \leftarrow NewClause \cup \{lit\}$;
               end if
16:            **end if**
        end for
18:        $u \leftarrow$ last marked literal on the assignment stack;
        Unmark $var(u)$;
20:        **activity(var(u)) += wght**;
        sign_score(var(u)) −= wght · sign(u);
22:        $-- ResolveNum$;
        $ResolveCl \leftarrow Antecedent(u)$;
24:        **wght += W(ResolveCl)**;
        $currentClause \leftarrow ResolveCl \setminus \{u\}$;
26:    **until** $ResolveNum = 0$;
    Unmark literals in $NewClause$;
28:    $NewClause \leftarrow NewClause \cup \{\overline{u}\}$;
    W(NewClause) $\leftarrow$ **wght** ;
30:    Add $NewClause$ to the clause database;
    end procedure

---

5 Experiments

Figure 7 shows experiments on an Intel 2.5Ghz computer with 1GB memory running Linux, sorted according to the winning strategy, which is CMTF combined with the RBS scoring technique. The benchmark set is comprised of 165 industrial instances used in various SAT competitions. In particular, *fifo8*, *bmc2*, *CheckerInterchange*, *comb*, *f2clk*, *ip*, *fvp2*, *IBM02* and *w08* are hard industrial benchmarks from SAT02; *hanoi* and *hanoi03* participated in SAT02 and SAT03; *pipe03* is from SAT03 and *01_rule*, *11_rule_2*, *22_rule*, *pipe-sat-1-1*, *sat02*, *vis-bmc*, *vliw_unsat_2.0* are from SAT04 each instance was set to 3000 seconds. If an instance could not be solved in this time limit, 3000 sec. were added as its solving time. All configurations are implemented on top of HaifaSat, which guarantees that the figures faithfully represent the quality of the various heuristics, as far as these benchmarks are representative. The results show that using CMTF instead of Berkmin's heuristic for choosing a clause leads to an average reduction of 10% in run time and 12-25% in the number of fails (depending on the score heuristic). It also shows a 23% reduction in run time when using RBS rather than VSIDS as a score system, and a corresponding 20-30% reduction in the number of fails. The differences in run times between HAIFASAT running the berkmin heuristic and Berkmin561 are small: the latter solves these instances in 210793 sec. and 53 timeouts. We also ran zChaff2004.5.13 on these formulas: it solves them in 210395 sec, and has 53 timeouts.

Benchmark	instances	BERKMIN+RBS		BERKMIN+VSIDS		CMTF+RBS		CMTF+VSIDS	
		time	fails	time	fails	time	fails	time	fails
hanoi	5	389.18	0	530.62	0	130.72	0	74.55	0
ip	4	191.02	0	395.52	0	203.24	0	324.27	0
hanoi03	4	1548.25	0	1342.1	0	426.87	0	386.28	0
CheckerI-C	4	1368.25	0	3323.16	0	681.56	0	3457.78	0
bmc2	6	1731.96	0	1030.9	0	1261.97	0	1006.94	0
pipe03	3	845.97	0	6459.62	2	1339.29	0	6160.12	1
fifo8	4	1877.57	0	3944.31	0	1832.65	0	3382.61	0
fvp2	22	1385.64	0	8638.63	1	1995.17	0	11233.7	3
w08	3	2548.62	0	5347.62	1	2680.96	0	4453.28	0
pipe-sat-1-1	10	1743.23	0	3881.49	0	3310.41	0	6053.84	0
IBM02	9	7083.55	1	9710.52	1	3875.64	0	7163.95	0
f2clk	3	4389.04	1	5135.25	1	4058.62	1	4538.15	1
comb	3	3915.15	1	3681.45	1	4131.05	1	4034.53	1
vis-bmc	8	15284.45	3	7905.9	1	13767.52	3	10119.34	2
sat02	9	17518.09	4	22785.77	5	17329.64	4	21262.25	4
01_rule	20	22742.11	4	33642.33	9	19171.5	2	23689.37	5
vliw_unsat_2.0	8	16600.67	4	24003.62	8	19425.41	5	22756.03	7
11_rule_2	20	31699.69	8	34006.97	10	22974.7	6	28358.05	6
22_rule	20	28844.07	8	33201.87	10	27596.78	8	30669.91	8
Total:	165	161706.5	34	208967.7	50	146193.7	30	189125	38

Fig. 7. A comparison of various configurations, showing separately the advantage of CMTF, the heuristic for choosing the next clause from which the decided variables will be chosen, and RBS, the heuristic for choosing the variable from this clause and its sign

6   Summary

We presented an abstraction/refinement model for analyzing and developing SAT decision heuristics. Satisfying a conflict clause before satisfying the clauses from which it was resolved, can be seen according to our model as satisfying an abstract model before satisfying a more concrete version of it. Our Clause-Move-To-Front decision heuristic, according to this model, attempts to satisfy clauses in an order associated with the resolve-graph. CMTF does not require to maintain the resolve-graph in memory, however: it only exploits the connection between each conflict clause and its immediate neighbors on this graph. Perhaps future heuristics based on this graph will find a way to improve the balance between the memory consumption imposed by saving this graph and the quality of the decision order. We also presented a heuristic for choosing the next variable and sign from the clause chosen by CMTF. Our Resolution-Based-Scoring heuristic scores variables according to their involvement ('activity') in refuting recent decisions. Our experiments show that CMTF and RBS either separately or combined are better than Berkmin and the VSIDS decision heuristics.

Acknowledgments

We thank Maya Koifman for helpful comments on an earlier version of this paper.

References

1. P. Beame., H. Kautz, and A. Sabharwal. Towards understanding and harnessing the potential of clause learning. *Journal of Artificial Intelligence Research*, 22:319–351, 2004.
2. E. Goldberg and Y. Novikov. Berkmin: A fast and robust sat-solver. In *Design, Automation and Test in Europe Conference and Exhibition (DATE'02)*, page 142, Paris, 2002.
3. L.Ryan. Efficient algorithms for clause-learning SAT solvers. Master's thesis, Simon Fraser University, 2004.
4. M. Moskewicz, C. Madigan, Y. Zhao, L. Zhang, and S. Malik. Chaff: Engineering an efficient SAT solver. In *Proc. Design Automation Conference (DAC'01)*, 2001.
5. J.P.M. Silva and K.A. Sakallah. GRASP - a new search algorithm for satisfiability. Technical Report TR-CSE-292996, Univerisity of Michigen, 1996.
6. L. Zhang, C. Madigan, M. Moskewicz, and S. Malik. Efficient conflict driven learning in a boolean satisfiability solver. In *ICCAD*, 2001.
7. L. Zhang and S. Malik. Extracting small unsatisfiable cores from unsatisfiable boolean formula. In *Theory and Applications of Satisfiability Testing*, 2003.

Production-Testing of Embedded Systems with Aspects

Jani Pesonen[1], Mika Katara[2], and Tommi Mikkonen[2]

[1] Nokia Corporation, Technology Platforms,
P.O.Box 88, FI-33721 Tampere, Finland
jani.p.pesonen@nokia.com
[2] Tampere University of Technology, Institute of Software Systems,
P.O.Box 553, FI-33101 Tampere, Finland
Tel: +358 3 3115 {5512, 5511}; Fax: +358 3 3115 2913
mika.katara@tut.fi, tommi.mikkonen@tut.fi

Abstract. A test harness plays an important role in the development of any embedded system. Although the harness can be excluded from final products, its architecture should support maintenance and reuse, especially in the context of testing product families. Aspect-orientation is a new technique for software architecture that should enable scattered and tangled code to be addressed in a modular fashion, thus facilitating maintenance and reuse. However, the design of interworking between object-oriented baseline architecture and aspects attached on top of it is an issue, which has not been solved conclusively. For industrial-scale use, guidelines on what to implement with objects and what with aspects should be derived. In this paper, we introduce a way to reflect the use of aspect-orientation to production testing software of embedded systems. Such piece of a test harness is used to smoke test the proper functionality of a manufactured device. The selection of suitable implementation technique is based on variance of devices to be tested, with aspects used as means for increased flexibility. Towards the end of the paper, we also present the results of our experiments in the Symbian OS context that show some obstacles in the current tool support that should be addressed before further case studies can be conducted.

1   Introduction

A reoccurring problem in embedded systems development is the maintenance and reuse of the test harness, i.e., software that is used to test the system. Although code implementing the tests is not included in the final products, there is a great need to reuse and maintain it especially in the context where the system-under-test is a part of an expanding product family. However, as the primary effort goes into the actual product development, quality issues concerning the test harness are often overlooked. This easily results in scattered and tangled code that gets more and more complex each time the harness is adapted to test a new member of the product family.

S. Ur, E. Bin, and Y. Wolfsthal (Eds.): Haifa Verification Conf. 2005, LNCS 3875, pp. 90–102, 2006.

Aspect-oriented approaches provide facilities for sophisticated dealing with tangled and crosscutting issues in programs [1]. Considering the software architecture, aspects should help in modular treatment of such code, thus facilitating maintenance and reuse, among other things. With aspects, it is possible to weave new operations into already existing systems, thus creating new behaviors. Moreover, it is possible to override methods, thus manipulating the behaviors that already existed.

With great power comes great responsibility, however. The use of aspect-oriented features should be carefully designed to fit the overall system, and ad-hoc manipulation of behaviors should be avoided especially in industrial-scale systems. This calls for an option to foresee functionalities that will benefit the most from aspect-oriented techniques, and focus the use of aspects to those areas. Unfortunately, case studies on the identification of properties that potentially result in tangled or scattered code, especially in testing domain, have not been widely available. However, understanding the mapping between the problem domain and its solution domain, which includes both conventional objects as well as aspects, forms a key challenge for industrial-scale use.

In this paper, based on our previous work [2], we analyze the use of aspect-oriented methodology in production testing of a family of embedded systems. In this domain, common and device-specific features correspond to different categories of requirements that can be used as the basis for partitioning between object-oriented and aspect-oriented implementations. The way we approach the problem is that the common parts are included in the object-oriented base implementation, and the more device-specific ones are then woven into that implementation as aspects. As an example platform, we use Symbian OS [3,4] based mobile phones.

The rest of this paper is structured as follows. Section 2 gives an overview of production testing and introduces a production-testing framework for the Symbian platform. Section 3 discusses how we apply aspect-orientation to the framework. Section 4 provides an evaluation of our solution including the results of our practical experiments. Section 5 concludes the paper with some final remarks.

2   Production Testing

Production testing is a verification process utilized in the product assembly to measure production line correctness and efficiency. The purpose is to evaluate the correctness of devices' assembly by gathering information on the sources of faults and statistics on how many errors are generated with certain volumes. In other words, production testing is the process of validating that a piece of manufactured hardware functions correctly. It is not intended to be a test for the full functionality of the device or product line, but a test for correct composition of device's components. With high volumes, the production testing involves test harness software that must be increasingly sophisticated, versatile, cost-effective, and adapt to great variety of different products. In software, the most successful way of managing such variance is to use product families [5,6].

2.1   Overview of Symbian Production Testing Framework

Individual design of all software for a variety of embedded system configurations results in an overkill for software development. Therefore, product families have been established to ease their development. In such families, implementations are derived by reusing already implemented components, and only product-specific variance is handled with product-specific additions or modifications. In this paper, we will be using a running example where a product family based on Symbian OS [3,4] is used as the common implementation framework. Symbian OS is an operating system for mobile devices, which includes context-switching kernel, servers that manage devices' resources, and rich middleware for developing applications on top of the operating system. According to the Symbian home page, there are almost 40 million installations of the operating system and over 50 different phone models running the system currently on the market from seven licensees worldwide [4].

The structure of a production-testing framework in Symbian environment follows the lines of Figure 1 and consists of three subsystems: user interface, server and test procedures. Test procedure components (Test 1, Test 2, etc.) implement the actual test functionalities and together form the test procedures subsystem. These components form the system's core assets by producing functionality for basic test cases. Furthermore, adding specializations to these components produces different product variants hence dedicating them for a certain specific hardware, functionality, or system needs. In other words, the lowest level of abstraction created for production test harness is composed of test procedure components that only depend on the operating system and underlying hardware. As a convenience mechanism for executing the test cases, a testing server has been implemented, which is responsible for invoking and managing the tests. This server subsystem implements the request-handling core and generic parts of the test cases, which are abstract test procedure manifestations, as test proxies.

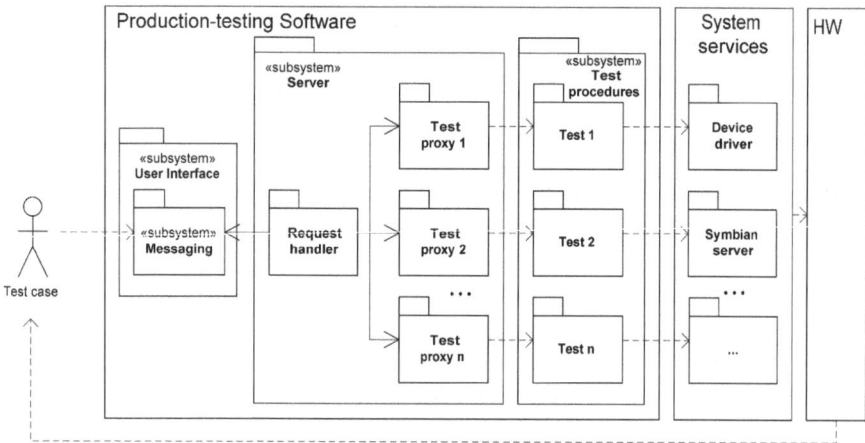

Fig. 1. Production-testing framework

Finally, a user interface is provided that can be used for executing the test cases. The user can be a human user or a robot that is able to recognize patterns on the graphical user interface, for instance. The user interface subsystem implements the communication between the user and the production-testing system.

2.2 Variability Management

From the viewpoint of production testing, in the case of Symbian, the most important pieces of hardware are Camera, Bluetooth, Display and Keyboard. In addition, also other, less conventional pieces of equipment in mobile phones can be considered, like WLAN for instance. The test harness on the target is then composed of components for testing the different hardware and device driver versions, which are specific to the actual hardware. When composing the components, one must ensure that concerns related to a certain piece of hardware are taken into account in relevant software components as well. For instance, more advanced versions of the camera hardware and the associated driver allow higher resolution than the basic ones, which needs to be taken into consideration while testing the particular configuration. Since the different versions can provide different functional and non-functional properties, the harness must be adapted to the different configurations. For example, the execution of a test case can involve components for testing display version 1.2, Bluetooth version 2.1 and keyboard version 5.5. Moreover, possible manufacturer-specific issues need to be taken into consideration. The particular display version may suggest using a higher resolution pictures as test data than previous versions, for instance. To complicate matters further, the composition of hardware is not fixed. All the hardware configurations consist of a keyboard and a color display. However, some configurations also include a camera or Bluetooth, or both. Then, when testing a Symbian OS device with a camera but without Bluetooth, for instance, Bluetooth test procedure components should be left out from the harness.

To manage the variability inherent in the product line, the test harness is assembled from components pertaining to different layers as follows. Ideally, the basic functionality associated with testing is implemented in the general test procedure components that only depend on the operating system or certain simple, common test functionality of generic hardware. However, to test the compatibility of different hardware variants, more specialized test procedure components must be used. Moreover, to cover the particular hardware and driver versions, suitable components must be selected for accessing their special features. Thus, the harness is assembled from components, some of which provide more general and others more specific functionality and data for executing the tests.

3 Applying Aspect-Oriented Techniques to Production Testing

In traditional object-oriented designs, there are usually some *concerns*, i.e. conceptual matters of interest that cut across the class structure. The implementation of such concerns is *scattered* in many classes and *tangled* with other parts inside the classes [7]. Scattering and tangling may suggest emergence of problems concerning traceability, comprehensibility, maintainability, low reuse, high impacts of changes and reduced concurrency in development [8].

Aspect-oriented programming provides means for modularizing crosscutting concerns. The code that would normally be scattered and tangled is written into its own modules, usually called aspects. There are basically two kinds of aspect-oriented languages, asymmetric and symmetric. The former ones assume an object-oriented base implementation to which the aspects are woven either statically or at run time. The latter treat the whole system as a composition of aspects. In the following, we assess the possibilities of applying aspect-oriented techniques to production testing by identifying the most important advantages of the technique in this problem domain.

3.1 Identifying Scattering and Tangling

Strive for high adaptability and support for greater variability implies implementations that are more complex and a large number of different product configurations. Attempts to group such varying issues and their implementations into optimized components or objects using conventional techniques make the code hard to understand and to maintain. This leads to heavily loaded configuration and large amounts of redundant or extra code, and complicates the build system. Thus, time and effort are lost in performing re-engineering tasks required to solve emerging problems. Hence, for industrial-scale systems, such as production-testing software, this kind of scattered and tangled code should be avoided in order to keep the implementation cost-effective, easily adaptable, maintainable, scalable, and traceable.

Code scattering and tangling is evident in test features with long historical background and several occasional specializations to support. The need for maintaining backwards compatibility causes the implementation to be unable to get rid of old features, whereas the system cannot be fully optimized for future needs due to the lack of foresight. After few generations, the test procedure support has cluttered and complicated the original simple implementation with new sub-procedures and specializations. As an example, consider testing a simple low-resolution camera with a relatively small photo size versus a megapixel camera with an accessory flashlight. In this case, the first generation of production-testing software had simple testing tasks to perform, perhaps nothing else but a simple interface self-test. However, when the camera is changed the whole testing functionality is extended, not only the interface to the camera. In addition to new requirements regarding the testing functionality, also some tracing or other extra tasks may have been added. This kind of expansion of hardware support is in general fully supported by the product-line, but several one-shot products rapidly overload the product-line with unnecessary functionalities. While the test cases remain the same, the test procedures subsystem becomes heavily scattered and tangled piece of code.

Another typical source of scattered and tangled code is any additional code that implements features not directly related to testing but still required for all or almost all common or specialized implementations. These are features such as debugging, monitoring or other statistical instrumentation, and specialized initializations. Although the original test procedure did not require any of these features, apart from specialized products, and certainly should be excluded in software in use in mass production, they provide useful tools for software and hardware development, research, and manufacturing. Hence, they are typically instrumented into code using precompiler macros, templates, or other relatively primitive techniques.

In object-oriented variation techniques, such as inheritance and aggregation, the amount of required extra code for proper adaptability could be large. Although small inheritance trees and simple features require only a small amount of additional code, the amount expands rapidly when introducing test features targeted for not only one target but for a wide variety of different, specialized hardware variants. Redundant code required for maintaining such inheritance trees and objects is exhaustive after few generations and hardware variants. Hence, the conserved derived code segments should provide actual additional value to the implementation instead of gratuitous repetition. Furthermore, these overloaded implementations easily degrade performance. Hence, the variation mechanism should also promote light-weighted implementations, which require as little as possible extra instrumentation.

Intuitively, weaving the aspects into code only after preprocessing, or precompiling, does not add complexity to the original implementation. However, assigning the variation task to aspects does only move the problem into another place. While the inheritance trees are traceable, the aspects and their relationships, such as interdependencies, require special tools for this. Hence, the amount of variation implemented with certain aspects and grouping the implementations into manageable segments forms the key asset in avoiding scattering and tangling with at least tolerable performance lost.

3.2 Partitioning to Conventional and Aspect-Oriented Implementation

Symbian OS provides more abstract interfaces on upper and more specialized on lower layers. Hence, Symbian OS components and application layers provide generic services while the hardware dependent implementations focus on the variation and specializations. In order to manage this layered structure in implementation a distinction between conventional and aspect-oriented implementation is required. However, separating features and deciding which to implement as aspects and which using conventional techniques is a difficult task. On the one hand, the amount of required extra implementation should be minimized. On the other hand, the benefits from introducing aspects to the system should be carefully analyzed while there are no guidelines or history data to support the decisions.

We propose a solution where aspects instrument product level specializations into the common assets. Ideally, aspects should be bound into the system during linking. The common product-specific, as well as architecture and system level, test functionalities comply with conventional object-oriented component hierarchy. However, certain commonalities, such as tracing and debugging support, should be instrumented as common core aspects, and hence be optional for all implementations. Thus, we identify two groups of aspects: test specialization aspects and general-purpose core aspects. The specialization aspects embody product-level functionalities, and are instrumented into the lowest, hardware related abstraction level. These aspects provide a managed, cost-effective and controlled tool for adding support for any extraordinary product setup. Secondly, the common general-purpose aspects provide product-level instrumentation of optional system level features.

In this solution, we divided the implementation based on generality; special features were to be implemented using aspect-oriented techniques. These are all strictly product-specific features for different hardware variants clearly adding dedicated implementations relevant to only certain targets and products. On the contrary, however, the more common the feature is to all products, it does not really matter whether it is implemented as part of the conventional implementation or as a common core aspect. The latter case could benefit from smaller implementation effort but suffer from lack of maintainability, as it would not be part of the product-line. Hence, common core aspects are proposed to include only auxiliary concerns and dismiss changes to core implementation structures and test procedures.

Our solution thus restricts the use of aspects to implementations that are either left out from final products or dedicated to only one specific product. In other words, conventional techniques are utilized to extend the product-line assets, while the aspect-orientation is used in tasks and implementations supporting product peculiarities as well as research and development tasks. Hence, the product-line would remain in proper shape despite the support for aggressive and fancy products with tight development schedules.

3.3   Camera Example

We demonstrate the applicability of aspect-orientation in production-testing domain with a simple example of an imaginary camera specialization. In this example, an extraordinary and imaginary advanced mega-pixel camera with accessory flashlight replaces a basic VGA-resolution camera in a certain product of a product family. Since this unique hardware setup is certainly a one-time solution, it is not appropriate to extend the framework of the product family. Evidently, changes in the camera hardware directly affect the camera device driver and in addition to that, require changes to the test cases. New test procedure is needed for accessory flashlight and camera features and the old camera tests should be varied to take into account the increased resolution capabilities. Hence, enhanced camera hardware has an indirect effect on the production-testing software, which has to support these new test cases and algorithms by providing required testing procedures. Hence, camera related specialization concerns affect four different software components, which are all located on different levels of abstraction: the user interface, request handler, related test procedure component, and the device driver. Components requiring changes compared to the initial system illustrated in Figure 1 are illustrated in Figure 2 as grey stars.

From the figure, it is apparent that the required specialization cuts across the whole infrastructure and it will be difficult to comply with the extraordinary setup without producing excessive scattered or tangled code inside the product-line assets. In other words, maintaining the generality of the system core assets using exclusively conventional techniques is likely to be difficult when the system faces such requirements. In practice, the new set-up could involve new initialization values, adaptation to new driver interface, and, for example, introduce new algorithms. With conventional techniques, such as object-orientation, this would entail inherited specialization class with certain functionalities enhanced, removed or added. Furthermore, a system level parameter for variation must have been created in order to cause related changes also in the server and the user interface level, which is likely to bind the implementation of each abstraction level together. Hence, a dependency is created not only between the hardware variants but also between the subsystem variants on each abstraction level. These modifications would be tolerable and manageable if parameterization is commonly used to select between variants. However, since this enhancement is certainly unique in nature, a conventional approach would stress the system adaptability in an unnecessary heavy manner.

The crosscutting nature of this specialization concern makes it an attractive choice for aspects that group the required implementation into a nice little feature to be included only in the specialized products. These aspects, which are illustrated in Figure 2 as a bold black outline, would then implement required changes to user interface, request handler, test procedure component, and device driver without intruding the implementation of the original system. Hence, the actual impact of the special hardware is negligible to the framework and the example thus demonstrates aspect-orientation as a sophisticated approach of incorporating excessive temporary fixes.

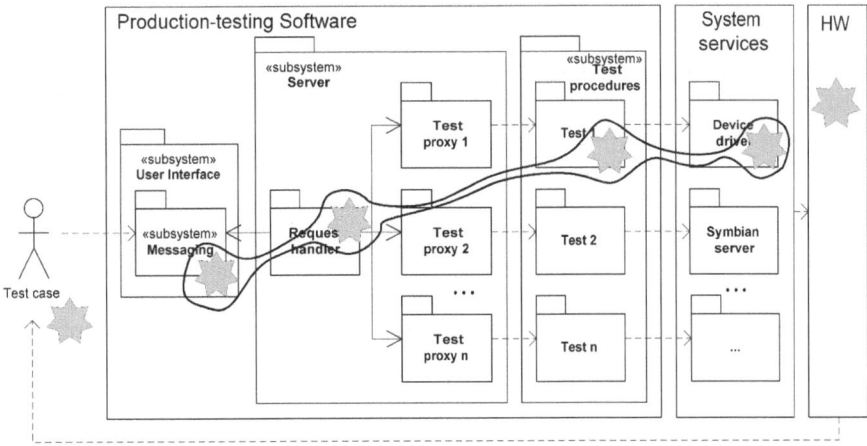

Fig. 2. An aspect capturing specialization concern in production-testing framework

4   Evaluation

In order to gather insight into the applicability of aspects to production-testing system, we assessed the technique against the most important qualities for such system. These include traceability and maintainability of the implementation as well as performance. In the following, we describe the experiments we carried out and provide an evaluation.

4.1   Practical Experiments

To investigate further the possibilities of aspect-oriented techniques in this context, we executed a set of practical experiments on an industrial production-testing system. Our first experiment was to infiltrate simple tracing support into an existing testing component and then, according to the results, actually provide an aspect-oriented solution to an example case described in Section 3.3. In the specialization experiment, a one-time system variant was branched out from the product-line by introducing camera related specialization aspects. These aspects introduced all the required modifications to the existing system in order to provide testing support in this special case. The benefit from the aspect approach was, as anticipated, the grouping of specialization concerns.

The prominent aspect implementation in C++ environment is AspectC++ [9,10] supporting the asymmetric view of aspect-orientation. The tool does a source-to-source compilation adding templates and namespaces, among other things, to the original C++ source. When using the AspectC++ compiler (version 0.9.3) we encountered some tool related issues that we consider somewhat inconvenient but solvable: Firstly, the compiler did not support Symbian OS peculiarities in full in its current form, and hence the tool required a lot of parameterization effort prior to being able to instrument code correctly. Another problem related to the AspectC++ compiler itself was its performance, since instrumenting even a small amount of code took a considerable amount of time on a regular PC workstation. This could have been acceptable, however, if the output would have been the final binary or even pre-compiled code.

The pre-processing of source code prior to normal C++ compilation added an extra step to the tool chain. This raised an issue on whether the aspects should be woven directly into the object files or the final binaries. Based on feedback and experiments, our conclusion is that industrial systems prefer cases where the source code is not available, but instead the aspects should be woven into already existing systems in binary format.

Other practical difficulties were related to the missing synergy between the used aspect compiler and C++ tool chain. Since the tool chain of the Symbian development is built around GCC [11] version 2.98, with some manufacturer-specific extensions needed in mobile setting [12], the AspectC++ compiler sometimes produced code that was not accepted by the GCC compiler. It was not evident whether this was caused by the GCC version or the nature of the Symbian style C++ code. Switching to other systems with different compilers not directly supported by AspectC++, for example the ARM RVCT (version 2.1) in

Symbian OS version 9.1 [13], revealed that this was a recurring problem. Furthermore, the reason why the instrumented code was unacceptable varied. These tool problems were finally solved by modifying the original source and correcting the resulting instrumented code by hand. This added another extra step that admittedly could be automated into the instrumentation process.

After the technical difficulties were solved as described above, our experiments with a simple tracing aspect revealed that the current AspectC++ implementation produced heavy overhead. A single tracing aspect woven into a dynamic link library (DLL) consisting of 22 functions with total size of 8kB doubled the size of the final DLL binary to 16kB. Unfortunately, this 100% increase in size is unacceptable in our case. Although weaving the tracing aspect into the whole component and each function was an overweighed operation, it revealed difficulties associated with the aspect implementation itself. This is a major drawback when considering testing of embedded systems, which are often delicate regarding memory footprint issues.

To summarize the above, a current feature of aspect-orientation in this context seems to be the tight relationship between the tools. This forces the organization to maintain a functional tool chain until a proven alternative is provided, which can become a burden for product families with a wide variety of different hardware platforms.

Moreover, the structure and architecture of the base system has a strong effect on how straightforward the aspect weaving is. Based on these experiments, we tend to believe that the highly adaptive framework actually enables such techniques as aspect-orientation to be adopted. However, this might not be the case with differently structured systems. The product-line architecture makes the adoption at least a bit more straightforward.

4.2   Evaluation Results

Since the production-testing system is highly target-oriented and should adapt easily to a wide variety of different hardware environments, the system's adaptability and variability are the most important qualities. We consider that by carefully selecting the assets to implement as aspects could extend the system's adaptability with still moderate effort. A convincing distinction between the utilization of this technique and conventional ones is somewhat dependent on the scope of covered concerns. Based on our experiments, while the technique is very attractive for low-level extensions, it seems to lack potential to provide foundation for multiform, large-scale implementations.

Including aspects in systems with lots of conventional implementations has drawbacks in maintenance and traceability. Designers can find it difficult to follow whether the implementation is in aspects or in the conventional code. As the objects and aspects have no clear common binding to higher-level operations, following the implementation and execution flow becomes more complex and difficult to manage without proper tool support. Aspects can be considered as a good solution when the instrumented aspect code is small in nature. In other words, aspects are used to produce only common functionalities, for example tracing, and do not affect the internal operation of the functions. That is, aspects do not disturb the conventional development. However, these deficiencies may be caused by the immaturity of the technique and hence reflect designers' resistance for changes. In addition, the lack of good understanding of the aspect-oriented technology and proper instrumentation and development tools tend to create skeptic atmosphere. Nevertheless, the noninvasive nature of aspects makes them a useful technique when incorporating one-time specializations.

Production-testing software should be as compact and effective as possible in order to guarantee highest possible production throughput. Thus, the performance of the system is a critical issue also when considering aspects. Although the conventional implementation can be very performance effective, the aspects could provide interesting means to ease the variation effort without major performance drawbacks. This, however, requires a careful definition and strictly limited expression of the woven aspect.

5   Conclusions

In this paper, we have described an approach for assembling production-testing software that is composed of components that provide test functionality and data at various levels of generality. To better support product-line architecture, we have described a solution based on aspects. The solution depends on the capability of aspects to weave in new operations into already existing components, possibly overriding previous ones. Thus, the solution provides functionality that is specialized for the testing of the particular hardware configuration.

One practical consideration, especially in the embedded systems setting, is the selection between static and dynamic weaving. While dynamic weaving adds flexibility, and would be in line with the solution of [14], static weaving has its advantages. The prime motivation for static weaving is memory footprint, which forms an issue in embedded systems. Therefore, available tool support [9,10] is technically adequate in this respect.

Unfortunately, we encountered some problems in practical experiments. Our first attempts indicate that using tools enabling aspects in the case of Symbian OS is not straightforward but requires more work. In principle we could circumvent the problem by using mobile Java and a more mature aspect language implementation such as AspectJ [15] to study the approach. However, hiding the complexities of the implementation environment would not be in accordance to the embedded systems development, where specialized hardware and tools are important elements. Furthermore, even a superficial analysis reveals that the problems related to size would be hardened, due to the size restrictions of MIDP Java [16].

Due to the reported problems, the results from our practical experiments in actual production testing are somewhat limited. It seems that the tool chain could be fixed to better support the target architecture, but more changes that are considerable would be needed to reduce the memory footprint. Defining the correct use cases and applications that could still benefit from aspect-orientation remains as future work. In addition, we would also like to investigate more on the possibilities aspects could have in the embedded systems domain. For instance, the compositionality of symmetrical aspects in the setting where platform-specific tools are needed is an open issue. Moreover, the effects of aspects to the configuration management need further research.

It should be possible to generalize our results further. Another experiment was actually inspired by the specialization experiment. Because of the noninvasive nature of aspects, the system and integration testing of the production test harness itself could benefit from aspects. However, this remains as a piece of future work.

References

1. Filman, R.E., Elrad, T., Clarke, S., Akşit, M.: Aspect-Oriented Software Development. Addison–Wesley (2004)
2. Pesonen, J., Katara, M., Mikkonen, T.: Evaluating an aspect-oriented approach for production-testing software. In: Proceedings of the Fourth AOSD Workshop on Aspects, Components, and Patterns for Infrastructure Software (ACP4IS 2005) in conjunction with AOSD'05,, Chicago, USA, College of Computer and Information Science, Northeastern University, Boston, Massachusetts, USA (2005)
3. Harrison, R.: Symbian OS C++ for Mobile Phones. John Wiley & Sons. (2003)
4. Symbian Ltd.: Symbian Operating System homepage. (At URL http://www.symbian.com/)
5. Bosch, J.: Design and Use of Software Architectures: Adopting and Evolving a Product-Line Approach. Addison–Wesley (2000)
6. Clements, P., Northrop, L.: Software Product Lines : Practices and Patterns. Addison–Wesley (2001)
7. Tarr, P., Ossher, H., Harrison, W., Sutton, Jr., S.M.: N degrees of separation: Multi-dimensional separation of concerns. In Garlan, D., ed.: Proceedings of the 21st International Conference on Software Engineering (ICSE'99), Los Angeles, CA, USA, ACM Press (1999) 107–119
8. Clarke, S., Harrison, W., Ossher, H., Tarr, P.: Subject-oriented design: towards improved alignment of requirements, design, and code. ACM SIGPLAN Notices **34** (1999) 325–339
9. Spinczyk, O., Gal, A., Schröder-Preikschat, W.: AspectC++: An aspect-oriented extension to C++. In: Proceedings of the 40th International Conference on Technology of Object-Oriented Languages and Systems (TOOLS Pacific 2002), Sydney, Australia (2002)
10. AspectC++: AspectC++ homepage. (At URL http://www.aspectc.org/)
11. Free Software Foundation: GNU Compiler Collection homepage. (At URL http://gcc.gnu.org/)
12. Thorpe, C.: Symbian OS version 8.1 Product description. (At URL http://www.symbian.com/)
13. Siezen, S.: Symbian OS version 9.1 Product description. (At URL http://www.symbian.com/)

14. Pesonen, J.: Assessing production testing software adaptability to a product-line. In: Proceedings of the 11th Nordic Workshop on programming and software development tools and techniques (NWPER'2004), Turku, Finland, Turku Centre for Computer Science (2004) 237–250
15. AspectJ: AspectJ WWW site. (At URL http://www.eclipse.org/aspectj/)
16. Riggs, R., Taivalsaari, A., VandenBrink, M.: Programming Wireless Devices with the Java 2 Platform, Micro Edition. Addison-Wesley (2001)

Assisting the Code Review Process Using Simple Pattern Recognition

Eitan Farchi[1] and Bradley R. Harrington[2]

[1] I.B.M. Research Laboratory, Haifa, Israel
[2] I.B.M. Systems and Technology Group, Austin, USA

Abstract. We present a portable bug pattern customization tool defined on top of the Perl regular expression support and describe its usage. The tool serves a different purpose than static analysis, as it is easily applicable and customizable to individual programming environments. The tool has a syntactic sugar meta-language that enables easy implementation of automatic detection of bug patterns. We describe how we used the tool to assist the code review process.

1 Introduction

Formal reviews in general and specifically formal code reviews are one of the most effective validation techniques currently known in the industry. Code reviews are known to find a defect every 0.2 hours and a major defect every 2 hours [3]. The down sides of formal reviews, especially formal code reviews, are that they are skill sensitive and often tedious.

Bug patterns and programming pitfalls are well known and effective programming tools that aid development, review, testing and debugging [5][1]. Bug patterns and pitfalls typically aid the review process through code review checklists or question catalogs (see appendix five in [4] for an example). As pointed out in [4], code review question catalogs must be kept small to be effective. As a result, a customization process is required - projects should create their own customized question catalog.

The developers of compilers and static analysis tools have attempted to address the skill sensitiveness, tediousness and need for customization of code reviews. Traditionally compilers could be directed to look for programming pitfalls. Notable examples of static analysis tools are lint [1], Beam[2], FindBugs[3] and the Rational code review feature. The user scenario in effect when applying these tools is to go over a list of warnings given by the tool and determine if a warning is actually a problem. The difficulty with this approach is false alarms, i.e., warnings which are not actually problems.

While all of the above tools provide customization options, these options are limited, as they only have a fixed list of pitfalls to choose from. Notably, Jtest[4]

---

[1] See also *http : //www.pdc.kth.se/training/Tutor/Basics/lint/indexframe.html.*

[2] See *http://www-cad.eecs.berkeley.edu/cad-seminar/spring02/abstract/krohm.htm*

[3] See *findbugs.sourceforge.net*

[4] See *www.parasoft.com*

S. Ur, E. Bin, and Y. Wolfsthal (Eds.): Haifa Verification Conf. 2005, LNCS 3875, pp. 103–115, 2006.

takes this a step further. Jtest for Java and for C has a GUI based language for specifying pitfalls.

Another difficulty with the static analysis approach is portability - static analysis tools require parsing of the programming language and additional information, such as symbolic tables, to perform their task. Such information is language specific. Today's organizations might work with a multitude of programming languages (we have seen instances of up to five different programming languages having different abstraction levels from assembly to C++ in the same organization). In addition, off the shelf tools will not always work as is, in specific environments and additional work will be required (up to 1 PY in one instance in our experience) to adapt an off the shelf tool that requires parsing of a language to a complex industrial development and testing environment.

When designing the language, we had several requirements in mind. It was important to use an easy to learn, easy to implement grammar, using off the shelf tools. Additionally, we felt that a declarative language will be better than an imperative language. The ready availability of a well known, portable, pattern recognition language for use as a building block was the most attractive option. The Perl language regular expression support perfectly fits this requirement. However, our example set, based on known review checklists [1] showed that the Perl regular expression support will not be sufficient to declaratively specify many bug patterns (see section 2.2 for some examples). Consequently, we identified that the following is required.

- Ability to find other patterns which follow or precede an identified pattern
- A hierarchical construct, enabling the developer to verify if a pattern does not match a line prior to checking the following line
- Identify repeated equal occurrences of a given pattern within an identified pattern

These requirements lead to the definition of a minimal syntactic sugar extension built on top of the Perl regular expression facility that enables the declarative definitions of patterns conforming to the above requirements.

The portable bug pattern customization tool serves a different purpose than static analysis, as it is easily applicable and customizable to individual programming environments. A given project has certain programming pitfalls that may not be applicable to other environments and may use a range of programming languages. Also, even though there is some overlap with static analysis, the tool catches different types of bugs by using specific project related information. Thus, the tool is meant to be used in concert with static analysis.

This work addresses the customization and portability requirements. We have prototyped a portable bug pattern customization tool to assist the Rephrase review process[5]. Bug patterns are defined using a simple pattern recognition language, building on top of Perl regular expressions. Thus, the solution is portable

---

[5] See *www2.umassd.edu/SWPI/NASA/figuide.html* for details on the Rephrase review technique.

and can be applied to any programming language or for that matter even to design documents. Once the tool is applied to a project, code lines are annotated with review questions. Next, during the review meeting, the annotated code is visible to all code reviewers, as the readers rephrase the code.

Whereas regular expression based tools have certain advantages, we recognize static analysis based tools (e.g., Jtest and Beam) may find problems regular expression based tools cannot. Specifically, static analysis tools operate on data flow (e.g., define-use graph) graphs and the control flow graph of the program. This type of information is not available when applying our regular expression based approach thus limiting its effectiveness. For example, data flow analysis enables aliasing analysis. As a result, two pointers that point to the same memory location can be identified and functions that have hazardous side effects can be identified. Another example is identifying variables that are initialized along some paths and are not initialized along some other paths in the program. This analysis requires both data flow and control flow information and is difficult to achieve using the regular expression approach.

The bug pattern customization tool is intended to be used as a supplement to existing source code review tools. The tool allows for the automatic detection of possible bug patterns during reviews.

This paper is organized as follows. First, we describe our simple pattern recognition review assisting tool. Next, typical bug pattern identification using the tool is discussed. We then describe how the tool is used to assist the review process. Finally, experience of applying the tool to real life software projects is presented.

2   A Simple Pattern Recognition Review Assisting Tool

In this section, an overview of our simple pattern recognition review tool is given. The user defines a set of patterns, (further explained in 2.1 and formally defined in A), applicable to the project at hand. The project code is then searched for any occurrence of the patterns defined by the user. When a pattern is found, a review question is inserted above the pattern instance. Thus, creating a new version of the code, annotated with review questions. The annotated code is then reviewed, instead of the original code.

2.1   Defining a Pattern

A bug pattern (see appendix A for a formal definition), or simply a pattern, is defined on top of the Perl regular expression facility [6]. Bug patterns are written by the project developers identifying possible problematic parts of the code under review. A pattern is implemented by using a set of Perl regular expressions. Using the syntactic sugar pattern language we have defined, it is feasible to check for a certain sequence of Perl regular expression matches and non-matches. We have found this syntactic sugar pattern language helpful in facilitating the definition of project specific bug patterns lists.

The bug pattern definition associates review questions with a bug pattern. Thus, a bug pattern definition includes the following elements.

- A definition of an interesting pattern to review
- A review question to be associated with each pattern instance found during the search (short question)
- A more detailed review question that is associated with each searched file for which at least one instance of the bug pattern was found

An example of a pattern definition follows. The pattern definition attempts to find a line that is not a trace line (a trace line logs information on the program to enable filed analysis), is probably a for loop line and has an assignment in its condition. Note that we are not too concerned if we will "find" instances that are not assignment in loop conditions as this will be easily ignored by the reviewers during the review process. Also note that in defining the pattern below we use project specific information - namely that traces are done using the *DEBUG_TRACE_PRINTF* macro. The example is annotated to make it self explanatory.

```
pattern
#The line is not a print statement
   nmatch
   DEBUG_TRACE_PRINTF
   next {
#The line contains an assignment within two semi column.
#It is probably something like for(i = 0;  i = 3; i++)
      match
      ;.*=.*;
      this {
#The part that was previously matched, e.g.,  ;i = 3; , is matched.
# .*\s+=\s+.* below means possibly an identifier (.*) followed by some
#spaces (\s+), an assignment (=), some additional spaces (\s+) and
#then another possible identifier (.*) which if matched looks like it is an
#assignment
         match
         .*\s+=\s+.*
      }
   }
short question: ASSIGNMENT IN CONDITION?
long question: Double check conditions to determine if = is used instead of ==
\pattern
```

We found writing bug patterns using this syntactic sugar pattern definition language easy. Within IBM it was quickly adopted by programmers from different organizations, with different skills and backgrounds and required a negligible learning curve. Three developers constructed a useful catalog of approximately 50 bug patterns within approximately 20 person hours.

2.2 Using the Tool to Search for Bug Patterns and Focus the Review

In preparation for the review meeting, the tool is applied to the project under review. Bug patterns can be grouped based on concerns. For a specific concern, (e.g., concurrency,) the tool identifies a subset of the project under review, (typically a set of files,) in which the occurrence of bug patterns related to the concern were found. The review can be further focused by briefing the reviewers to focus on the specific concern when reviewing the identified subset of the project.

2.3 The Annotated Code Used in the Review

After application of the tool, a new version of the code under review is produced with review questions to assist the review process. To illustrate, the pattern from the previous subsection will annotate the following code segment:

```
for(i = 2; i = j); i++)
for(i == 2; i = j); i++)
for(i = 2; i== j); i++)
for(i == 2; i== j); i++)
```

as follows:

```
/*REVIEW QUESTION(0) -  ASSIGNMENT IN CONDITION?  */
for(i = 2; i = j); i++)
/*REVIEW QUESTION(0) -  ASSIGNMENT IN CONDITION?  */
for(i == 2; i = j); i++)
for(i = 2; i== j); i++)
for(i == 2; i== j); i++)
```

Thus, warning of a possible assignment in a condition, in the proceeding code segment. Note that $for(i = 2; i == j); i++)$ and $for(i == 2; i == j); i++)$ are not flagged as they do not contain an assignment in the for condition.

3 Identification of Typical Bug Patterns

In this section we explain how to use the bug pattern customization tool to identify typical bug patterns. A detailed description of the bug pattern definition meta-language is formally defined in appendix A, briefly described in the previous section, and expanded upon here.

The meta-language consists of ten keywords built on top of Perl regular expressions. Perl regular expressions were chosen for their well known ability to allow for easy pattern matching, and grouping.

The meta-language is also source code independent, and could possibly be used to detect patterns in non-source code files such as design documentation. Using this language, we have been able to successfully detect bug patterns in the C, C++, Java and FORTH programming languages. A simple explanation of the ten keywords is as follows:

pattern, \pattern indicates the beginning and end of a bug pattern.

{} indicates a block of pattern detection code.

$dollarX where X is a number starting at 1, corresponds to the $<digits>Perl keywords.

this corresponds to the $MATCH and $& Perl keywords.

before corresponds to the $PREMATCH and $' Perl keywords.

after corresponds to the $POSTMATCH and $' Perl keywords.

match indicates the following line is a regular expression to be matched.

nmatch indicates the following line is a regular expression to be inversely (not) matched.

window(X) indicates the following pattern should check for a match on the next X source lines of code.

code indicates a free form block of Perl source code to be evaluated as a Boolean expression.

The following illustrates a few simple examples on how the language is used.

In a simple example, a project under review had a history of bugs in which the programmer unnecessarily uses floating point variables. The following pattern prints a simple statement above the variable declaration during the code review.

```
pattern
        match
        float
short question:Does the program unnecessarily use float or double?
long question:
\pattern
```

An original source code file being reviewed with this line:

```
float foobar;
```

will display this during the code review:

```
/* REVIEW QUESTION(0) Does the program unnecessarily use float or double ? */
float foobar;
```

In another example, a well known bug pattern occurs when resources are not released along error paths. General error path identification using static analysis is difficult, especially outside of specific programming constructs, such as the exception handling mechanism. In contrast, we can leverage a given project's conventional methods to easily identify error paths by identifying project specific constructs. On one program we worked with, the error paths occur when the return code from a function call stored in a variable named *rc* is not zero. Using this project specific information, team members where able to define the following bug pattern and use it effectively during review sessions.

```
pattern
        match
#Error path looks something like - if (rc != 0) {
        if\s*\(.*rc.*\)
short question: RESOURCE RELEASED ALONG ERROR PATH?
long question: Are obtained resources released along error paths?
\pattern
```

Another example, demonstrating the use of a sequence of matches and non-matches of Perl regular expressions, follows. During the code review session, the reviewers should ensure all dereferenced pointers are always verified to be non-null, at runtime. A review question associated with dereferencing pointers would remind reviewers of the risk of a null pointer exception. However, the tool should only identify sections of code, and not comments. This was important for the project under review, as there were many comments that included dereference statements. The bug pattern below ignores C++ style comment lines and then checks for a pointer dereference. Only then is the review question added to the code under review.

```
pattern
#check that this is not a comment line
        nmatch
        \/\/
        next {
#this is probably a dereference line
        match
        ->
    }
short question:  NULL DEREFERENCE?
long question: When dereferenced, can a pointer ever be null
\pattern
```

One obvious problem with this pattern is the possible presence of C-style comments. Due to the inherent limitations of regular expressions, it may be necessary for the project enforce a slightly more restrictive coding standard, or else risk false posititves. Conversely, the tool may be well suited for the enforcement of coding standards in the project itself. This is a topic for future exploration.

In a more complex example, perhaps the program has recently had a problem with C++ copy constructor methods.

```
pattern
        match
        (.*)::(.*)\(const (.*)\&
        code {
                dollar1 eq dollar2 && dollar1 eq dollar3
        }
short question:  COPY CONSTRUCTOR
long question:  Is a copy constructor method really required here?
\pattern
```

This pattern will match the following code:

```
Employee::Employee(const Employee& t)
```

Thus flagging the copy constructor method for special notice during the code review, giving the code the desired added attention.

In a more advanced example, perhaps the project team wanted to ensure all copy constructor methods are closely followed by corresponding destructor methods.

```
pattern
        match
        (.*)::(.*)\(const (.*)\&
        code {
                dollar1 eq dollar2 && dollar1 eq dollar3
        }
        window(20) {
                nmatch
                (.*)::~(.*)
                code {
                        dollar1 eq dollar2
                }
        }
short question: Is a matching destructor present?
long question:
\pattern
```

This pattern will alert that a matching destructor of the form:

```
Employee::~Employee()
```

might not be present.

It may also be useful to incorporate some or all of the contents of industry standard code inspection checklists into the tool. These are often very useful code review and programming guides, and are readily available. A bug pattern would be written to describe each item on the checklist, as applicable. One such checklist is Baldwin's Abbreviated C++ Code Inspection Checklist[2]. As part of our pilot, we have incorporated most of the patterns from Baldwin's list, to provide a baseline of bug patterns, which should be viable for most projects.

4   Using the Bug Pattern Customization Tool to Assist the Code Review Process

In preparation for adoption of the bug pattern customization tool by a project, a set of project specific bug patterns are created. As a starting point, general code review checklists, like the one provided in [4], are used. Project members use the meta-language to create a set of bug pattern definitions. The bug patterns are defined using knowledge of project specifics, generally previous problems, and are run against the project as sanity checks to see if the results make sense. Sometimes, at this stage, issues are revealed but this is not the objective of the stage, but a side benefit. Based on previous experience we estimate this work to be approximately one person week of work for a non-trivial middleware component (which could be amortized over a period of a few months).

Typically, code review checklists and pitfalls are categorized according to review concerns - for example, declarations, data items, initializations, macros, and synchronization constructs[4]. Code reviews should be performed with a mindset similar to testing, and should be driven by these concerns. When a certain concern drives a specific code review, the subset of bug patterns related to that concern are used, thus further focusing the review process. For example, if we are interested in problems related to the use of macros, then only the part of the project annotated with review questions related to macros is actually reviewed.

Next, as mentioned in previous sections, during the review meeting the annotated source code is reviewed instead of the original source code. When an issue is revealed during the review, the review team might write a bug pattern to search the code for additional occurrence of the same problem.

A concrete example follows. In a middleware project that extensively uses macros, it was found that a set of macros, $CA\_QUERY\_BIT\_SET$ being one of them (see below), should be used in such a way that the first parameter contains the word Flags and the second parameter contains the word FLAG. Once this is identified, a bug pattern can be written to check this for the entire code base.

```
if (!CA_QUERY_BIT_SET(pCsComTCB->BM.bmExtendedFlags,
                CS_BM_NVBM_EX_FLAG__CALLED_BY_DP)) {
```

5 Experience

In this section we report empirical instances of usage of the bug patterns customization tool. The initial results are encouraging and indicate that the tool is useful in assisting the code review process.

5.1 ConTest Code Reviews

ConTest[6] is a concurrent program testing tool, used extensively within IBM. As an intrusive test tool that modifies the object code of the program under test, it should meet high quality standards in order to be used safely and reliably. We found out that users have very little tolerance for false alarms that turn out to be bugs in ConTest itself.

As a result of the high quality requirements, the ConTest development team conducts regular code reviews. For the last two months, the code review sessions were conducted on code annotated by the bug pattern customization tool. The

---

[6] See *www.alphaworks.ibm.com/tech/contest* for details.

process of the code review included the owner going over the annotated source code projected on the screen and rephrasing the code.

Obtaining the annotated code turned out to be quick and did not require additional effort on the review team. In addition, the annotated review questions obtained, based on a customized set of bug patterns did not distract the reviewer. Finally, actual problems were identified with the help of the annotated review questions.

For example in one review instance it was decided to change the interface used in a function, $findTargetString()$, and pass it a pointer to one structure instead of passing it three separate pointers that are logically connected. The following is the annotated code segment with the review question that started the discussion and eventual code change.

```
/*REVIEW QUESTION(3) -  STORAGE INSTEAD? */
nTargetIndex = findTargetString(strNew.c_str(),&nIndexAfterTarget,
&callType);
```

5.2 Avoiding Field Escapes Through Guided Code Review

The following string buffer overflow bug, recently surfaced during an embedded software stack bringup. The offending code had been in place for several years, with no issues. The routine takes place at the hand-off point between two different embedded software components, where one component is about to terminate, and the other is about to begin. Thus, the component interaction is similar to an interprocess communication (IPC) scenario, as one component does not have much knowledge of the other's internal data structures.

The variable $glob\_fw\_vernum$ was defined as a 16 byte character array. The $fw\_version\_string$ variable points to a string of unknown length, because of the IPC-like interaction. The following is executed by the offending code:

```
strcpy(glob_fw_vernum,(char *)of_data_stackptr->fw_version_string);
```

The length of the string actually originated from the embedded software image, where it was originally generated by the build team. The build process changed because of new developments on the platform. Thus, the length of the string grew from 16 characters, to 40 characters. This increase in length created a buffer overflow situation, which was not detected until a long time after the offending code had been executed. Eventually, the embedded software stack crashed due to a data storage interrupt, where the platform was attempting to read from an address that no longer made sense, because it was overwritten by the new longer version string.

A simple bug pattern could have easily warned reviewers of this situation. Also, there is reason to believe the pattern would have been flagged, because of a long history of string problems within this project. The relevant bug pattern follows.

```
pattern
        match
        strcpy
short question: Is the buffer large enough?  Is overflow handled?
long question: Has enough space been allocated to hold the size of
the string to be copied, and if not are there an overflow precautions?
```

Had the bug pattern review tool been available during the development of this code, or during subsequent reviews of this portion of the firmware, there is reason to believe the bug would not have escaped the code review.

Another bug, this time a search routine error, was a unit test escape from an embedded software component. In this situation, a data structure search routine designed to be used, similar to the Standard C Library $fread()$ routine, is not properly utilized.

The search routine will only search the data structure beginning from the location of the most recent search. If the entire structure is to be searched the search routine must explicitly be instructed to do begin its search from the beginning. This is done with the $set\_start()$ routine. However, the bug would not be caught with a simple unit test, as the test would most likely succeed on at least the first pass, and would only fail after repeated runs of the test. The offending source code bug follows.

```
set_filter(&filter,(uint8 *) "PCI-DEV" , 1,
                S_Offset(PCI_DEV_NODE,uniqueId), drc_index);
node_ptr = find_node(&iter, &filter);
```

The repaired routine:

```
set_start(&iter, NULL);
set_filter(&filter,(uint8 *) "PCI-DEV" , 1,
                S_Offset(PCI_DEV_NODE,uniqueId), drc_index);
node_ptr = find_node(&iter, &filter);
```

A similarly simple bug pattern could have prevented this unit test escape, as this is a common pitfall in using this API. The relevant bug pattern is as follows.

```
pattern
        match
        set_filter
short question: Is set_start() called before set_filter()?
long question:
```

6   Conclusion

We have prototyped a portable bug pattern customization tool to assist the rephrase review process. Bug patterns are defined using a simple pattern recognition language, built on top of Perl regular expressions. Thus, the solution is portable and can be applied to any programming language or for that matter even to design documents. Once the tool is applied to a project, code lines are annotated with review questions. Next, during the review meeting, the annotated code is visible to all code reviewers, as the readers rephrase the code.

Empirical experience indicates that the bug pattern customization tool can be effectively used to assist the code review process.

Further research will focus on gaining a comprehensive experience in the usage to the tool and enhancing the pattern definition syntactic sugar language to meet different user's requirements.

References

1. Eric Allen. *Bug Patterns in Java*. Apress, 2002.
2. John T. Baldwin. An abbreviated c++ code inspection checklist, 1992.
3. Daniel Galin. *Software Quality Assurance*. Addison Wesley.
4. Brian Marick. *The Craft of Software testing*. Prentice Hall, 1995.
5. Scott Meyers. *Effective C++*. Addison-Wesley, 1997.
6. Randal L. Shwartz and Tom Phoenix. Learning perl. O'REILLY, 1993.

A   Pattern Language Semantics

Below is a description of the language used to define a bug pattern. For simplicity the semantics of the *window* and *code* blocks are not given. The description below assumes familiarity with Perl regular expressions[6]. Specifically, the meaning of $', $&, $', $1, $2, . . . is assumed and used.

A.1   Match Node (N)

A match node (N) is a node that contains a regular expression, N.R, to be matched,

    A N.$ ' (''before''), N.$& (''this'') and N.$' (''after''),

fields and a variable length list of optional fields N.$1, N.$2, . . . all pointing to either a match node or a don't match node.

The actual implementation of the language refers to

    N$' as before, N$& as this, N$' as after, N$1 as dollar1.

A.2   Don't Match Node (N)

A node containing a regular expression, N.R, that should not match and a pointer to either a match node or a don't match node N.next.

A.3   Pattern Definition

A pattern is a rooted tree of either match or don't match nodes. The Context of a Pattern P wrt a String S is next defined.

The context of each node in a pattern, P, is defined wrt to a string S. The context of the root tree of the pattern P is S. Given that the context of some match node, N, in the pattern P is defined to be the string S1 then the context of N.$`, N.$&, N$', N.$1, N.$2,.. are defined if N.R matches S1. They are defined exactly according to the semantic of Perl for these operators (i.e., N.$` is the part of S1 before the match, N.S& is the matched string, N$' is the part of S1 that is after the match, $1 is the string that matched the first sub regular expression that was enclosed in a parentheses in N.R, $2 matched the second one, etc). Given that the context of a don't match node N is S1 the context of N.next is defined only if S1 does not match N.R. In this case the context of N.next is S1.

The context of a pattern P wrt to S is the association of a context to each of its nodes as defined above.

A.4   A Match

A string S matches a pattern P if the association of each node with a context is a complete function to the set of strings (each node has a defined context).

An Extensible Open-Source Compiler Infrastructure for Testing

Dan Quinlan[1], Shmuel Ur[2], and Richard Vuduc[1]

[1] Lawrence Livermore National Laboratory, USA
{dquinlan, richie}@llnl.gov
[2] IBM Haifa
ur@il.ibm.com

Abstract. Testing forms a critical part of the development process for large-scale software, and there is growing need for automated tools that can read, represent, analyze, and transform the application's source code to help carry out testing tasks. However, the support required to compile applications written in common general purpose languages is generally inaccessible to the testing research community. In this paper, we report on an extensible, open-source compiler infrastructure called ROSE, which is currently in development at Lawrence Livermore National Laboratory. ROSE specifically targets developers who wish to build source-based tools that implement customized analyses and optimizations for large-scale C, C++, and Fortran90 scientific computing applications (on the order of a million lines of code or more). However, much of this infrastructure can also be used to address problems in testing, and ROSE is by design broadly accessible to those without a formal compiler background. This paper details the interactions between testing of applications and the ways in which compiler technology can aid in the understanding of those applications. We emphasize the particular aspects of ROSE, such as support for the general analysis of whole programs, that are particularly well-suited to the testing research community and the scale of the problems that community solves.

1 Introduction

Testing software involves a number of formal processes (*e.g.*, coverage analysis, model checking, bug pattern analysis, code reviews, deducing errors), which require accurate characterizations of the behavior of the program being tested. Deriving such characterizations for modern large-scale applications increasingly requires automated tools that can read, represent, analyze, and possibly transform the source code directly. These tools in turn often depend on a robust underlying compiler infrastructure. In this paper, we present the ROSE source-to-source compiler infrastructure for C, C++, and Fortran90 [1–3]. We believe that ROSE is well-suited to support testing tool development because it is easy to use, robust with respect to large and complex applications, and preserves the structure of the input source (including source file position and comment information), thereby enabling accurate program characterization.

S. Ur, E. Bin, and Y. Wolfsthal (Eds.): Haifa Verification Conf. 2005, LNCS 3875, pp. 116–133, 2006.
© Springer-Verlag Berlin Heidelberg 2006

There are three aspects of ROSE particularly relevant to developers of program testing and checking tools. First, the ROSE infrastructure can process complex, large-scale, production-quality scientific applications on the order of a million lines of code or more, which are being developed throughout the U.S. Department of Energy (DOE) laboratories. These applications use complex language and library features, and therefore possess all of the qualities of general large-scale industrial applications. Secondly, ROSE is designed to be accessible to developers who do not necessarily have a formal compiler background, enabling a broad community of tool builders to construct interesting tools quickly. Thirdly, ROSE is fully open-source, modularly designed, and extensible. This aspect is important because current and future static and dynamic analysis tools for testing have widely differing requirements of any underlying compiler infrastructure [4–6]. Collectively, these aspects make ROSE a suitable framework within which to build a variety of source-code analysis and transformation tools.

Automatic testing and defect (or bug) detection tools based on static analysis compose a large and growing body of recent work whose effectiveness demands accurate source code representation and analysis (see Section 5.2). For C++ in particular, accurately representing programs is particularly challenging due to complexity of both the language and commonly-used support libraries. The core ROSE intermediate representation (IR), used to form the abstract syntax tree (AST), is engineered to handle this complexity. Having been designed to support source-to-source translation, it preserves all the program information required to precisely reconstruct the original application in the same language (without language translation), and therefore represents the full type and structural representation of the original source, with only minor renormalization. As a result, ROSE has all of the information that may be required to support the kind of program analysis techniques needed, for instance, to keep a given tool's false positive reporting rates low.

The remainder of this paper reviews the interface between compilation and testing, with an emphasis on the ways in which compilers can support testing activities (Section 2). We discuss the ROSE infrastructure itself, highlighting the components of the infrastructure we believe will be particularly useful to testing tool builders (Section 3). We present a concrete example of an on-going collaboration between IBM and the ROSE project in the development of a testing tool for code coverage analysis, and discuss additional ideas for other testing tools that would require compiler support and be useful to the testing of large scale application (Section 4).

2   How Compilers Can Support Testing

Testing comprises many specific activities that can make use of compilation techniques. Although many useful tools can be built and used without using a full compiler infrastructure, compilers can provide stronger and deeper analysis to improve the quality of reports by, for instance, reducing the number of false positive reports in a defect detection tool. In this section, we review a number of important testing activities and show how a compiler infrastructure can assist.

2.1   Making Programs Fail

One important testing technique is to introduce instrumentation in a program in order to force more exhaustive coverage of control flow paths during execution. One example is ConTest, a tool developed at IBM, which introduces timing perturbation in multi-threaded applications at testing-time [7]. These perturbations force untested thread inter-leavings that might cause race conditions or deadlocks, for example. We are using ROSE to introduce this kind of instrumentation in C and C++ applications.

It may be possible to extend this idea further by building compiler-based tools to inject inputs automatically at various program input points, to force failures or perform security bug testing. Both the analysis to locate input points and the transformation to inject data at such points can best be supported using compiler-based tools that can generate correct code using an analysis of the context at an input point in the application.

2.2   Reproducing Bugs

Reproducing bugs generated from complex execution environments is the first step in fixing them, and knowing all the steps that lead to the bug is critical in understanding the context for the bug. Post-mortem analysis can be helpful, but either the support for this analysis is woven into the application by hand or automated tools are developed to support such transformations and enable the generation of a post-mortem trace. A compiler-based tool with full transformation capabilities should be relatively easy to build so that it can record limited tracing information and dump it to a file upon detection of internal faults. Such automated tools then work with any product application during development and provide useful information from failures in the field. This strategy accepts that bugs happen in production software and, where possible, provides a mechanism to detect the root cause or history of steps the led to uncovering the bug.

2.3   Extracting Simplifying Examples to Deduce Bugs

We can ease the process of deducing bugs by building an automated mechanism to remove irrelevant code within proximity of where an error occurs. Such a technique requires substantial compiler and program analysis support (dependence analysis) and is part of a traditional compiler-based program slicing analysis. Recent work within ROSE now supports reverse static slicing.

2.4   Bug Pattern Analysis

The goal in a bug pattern analysis is to find potential bugs by specifying a "pattern" (*e.g.*, a syntactic template) of code to be identified within the program source of interest, and then searching for instances of the pattern. To operate robustly against the numerous naming and language constructs, such matching should be done directly on the type-checked AST. Full compiler support, with no loss of source information, is required to perform this step robustly and with minimal false positives. Compiler-based tools can be built to accept bug pattern specifications (the development of which is a research issue) using grammar rules (a parser technology) and construct a pattern matching test on the AST of any input application. There are numerous fast search algorithms that could be employed to implement such tests on whole application projects via global analysis and with patterns of quite arbitrary complexity.

Hovemeyer and Pugh observe that bugs are often due simply to misuse of subtle or complex language and library features, and so they propose bug pattern analysis to identify these erroneous uses [5]. Although such features enable general purpose languages to support a broad audience of developers, they may require specialized knowledge to be used correctly. Indeed, individual development groups often develop explicit style guides to control the language feature and library use. Compiler-aided bug pattern analysis could be used to enforce such explicit guidelines, in addition to identifying actual incorrect usage.

2.5   Coverage Analysis

The main technique for demonstrating that testing has been thorough is called test coverage analysis [8]. Simply stated, the idea is to create, in some systematic fashion, a large and comprehensive list of tasks and check that each task is covered in the testing phase. Coverage can help in monitoring the quality of testing, assist in creating tests for areas that have not been tested before, and help with forming small yet comprehensive regression suites [9].

Instrumentation, in the source code or the object/byte code, is the foundation on which coverage tools are implemented. Coverage tools for different languages (*e.g.*, Java, C, C++) could have very similar front-ends to display the coverage results, but need different compiler-based tools for instrumentation. We are not aware of any easy-to-use coverage tool addressing the many different types of program coverage for large C++ applications, as are available for other languages such as Java; we are currently developing one based on ROSE.

2.6   Model Checking

Model checking is a family of techniques, based on systematic and exhaustive state-space exploration, for verifying program properties. Properties are typically expressed as invariants (predicates) or formulas in a temporal logic. Model checkers are traditionally used to verify models of software expressed in special modeling languages, which are simpler and higher-level than general-purpose programming languages. Manually producing models of software is labor-intensive and error-prone, so a significant amount of research is focused on abstraction techniques for automatically or semi-automatically producing such models. We review related work along these lines in Section 5.

To simplify the models, and fight the state space explosion, abstractions are used. The abstractions used today are performed after the model has been transformed into a net-list. New types of abstractions based on program transformation techniques have been suggested [10]. In this way, program transformation techniques can have a greater role in getting model checking to work with larger programs.

2.7 Code Reviews

Code reviews commonly use check lists to help reviewers express and address common problems.[1] For example, an item on the check list might be, "Should an array declaration containing a numerical value be replaced by a constant?" or perhaps, "If an if-statement contains an assignment, should the assignment actually be an equality comparison?" Though effective in practice, it can be very tedious to work with check lists and compare them against the source code to find where a given item is relevant. A better approach, suggested by Farchi [11], is to embed the review questions from the check lists in the code as comments. This requires two phases: one in which to check every item from the check list and see where it is reflected in the code (static analysis), and the second in which to annotate the code with the relevant comment (a simple code transformation). A compiler infrastructure is well-suited to both tasks.

2.8 Capture and Replay

Capture and replay tools enable test sessions to be recorded and then replayed. The test sessions might be edited, and then replayed repeatedly or with several test sessions running at the same time. Capture and reply is used in simulating load by helping to create many clones, in regression by automating re-execution of a test, and in debugging and testing of multi-threaded applications by helping to replay the scenario in which the bug appears. Most capture and replay tools enable the test sessions to be edited, parameterized, and generalized. Almost all of the tools have a facility to compare the expected results from a test run with those that actually occur.

One of the ways to implement capture and replay is by wrapping all the functions through which the applications communicate with the environment. Most commonly, this is done for GUI applications to record all the interactions of widgets. As another example, in scientific applications which use the Message Passing Interface (MPI) communication library, all the MPI calls could be to be recorded during the test execution; in the replay phase, the MPI calls would return the information collected during the testing session. It is expected that such techniques could greatly simplify isolating classes of bugs common to large parallel applications which occur only on tens of thousands of processors. Automatic wrapping of functions is another application for which a compiler framework is very suitable.

2.9 Identifying Performance Bugs

Of particular interest in high-performance scientific applications are *performance bugs*. These bugs do not represent computation errors, but rather result in unexpected performance degradation. Patterns of some performance bugs in parallel

---

[1] *E.g.*, see: http://ncmi.bcm.tmc.edu/homes/lpeng/psp/code/checklist.html

applications include synchronizing communication functions that can be identified from their placement along control paths with debugging code (often as redundant synchronization to simplify debugging) but without any use of their generated outputs. Control dependence analysis can be used to identify such bugs, but this analysis requires a full compiler support. Here the analysis of the AST must use the semantics of the communication library (*e.g.*, MPI, OpenMP) with the semantics of the language constructs to automatically deduce such bugs.

2.10 Aiding in Configuration Testing

One of the biggest problems in testing commercial code is *configuration testing*, *i.e.*, ensuring that the application will work not only on the test floor, but also at the customer site, which may have another combination of hardware and software. A configuration comprises many components, such as the type and number of processors, the type of communication hardware, the operating system and the compiler; testing all combinations is impractical.

Different compilers or even different versions of a compiler can vary in how they interact with a given application. For example, compilers often differ in what language features are supported, or how a particular feature is implemented when a language standard permits any freedom in the implementation. In addition, the same compiler can generate very different code depending on what compiler flags (*e.g.*, to select optimization levels) are specified. As any tester in the field is aware, the fact that the program worked with one compiler does not guarantee that it will work with the same compiler using a different set of flags, with another version of the same compiler, or with another compiler.

A compiler-based tool (or tools) could aid in compiler configuration testing. For example, it could try different compilers or, for a single compiler, different compile-time flags, and then report on configurations that lead to failures. As another example, in cases where the compiler has freedom in how a language feature is implemented, it could transform the program in different ways corresponding to different implementation decisions, thereby testing the sensitivity of the application to compiler implementation. Such a tool should help make the application more maintainable over time and across customer sites, and reduce time spent debugging compiler-related problems. We discuss specific related examples in Section 4.

3   The ROSE Compiler Infrastructure

The ROSE Project is a U.S. Department of Energy (DOE) project to develop an open-source compiler infrastructure for optimizing large-scale (on the order of a million lines or more) DOE applications [1, 2]. The ROSE framework enables tool builders who do not necessarily have a compiler background to build their own source-to-source optimizers. The current ROSE infrastructure can process C and C++ applications, and we are extending it to support Fortran90 as part of ongoing collaborations with Rice University. For C and C++, ROSE fully supports

122   D. Quinlan, S. Ur, and R. Vuduc

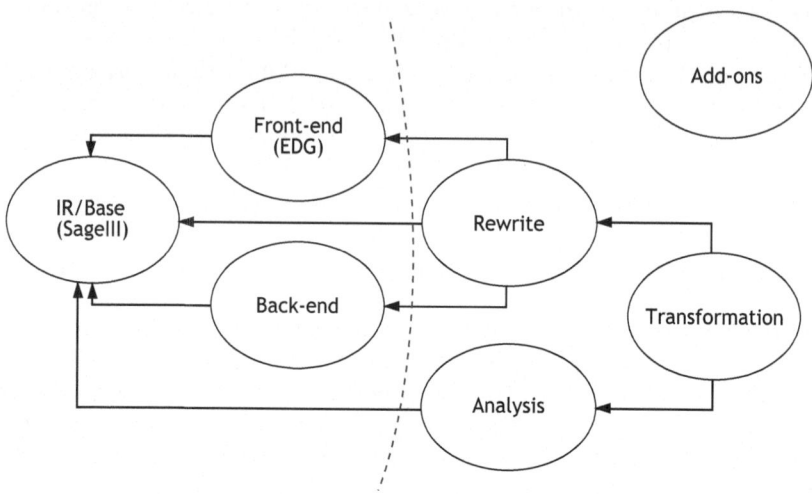

Fig. 1. Major components the ROSE compiler infrastructure, and their dependencies

all language features, and preserves all source information for use in analysis, and the rewrite system permits arbitrarily complex source-level transformations. Although research in the ROSE project emphasizes performance optimization, ROSE contains many of the components common to any compiler infrastructure, and is thus well-suited to addressing problems in testing and debugging.

The ROSE infrastructure contains several components to build source-to-source translators, shown as ovals in Figure 1. At the core of ROSE is the intermediate representation (IR) of the abstract syntax tree (AST) for C and C++ programs, SAGEIII, which is based on Sage II and Sage++ [12]. A complete C++ front-end is available that generates the SAGEIII AST. The AST preserves the high-level C++ language representation so that no information about the structure of the original application (including comments) is lost. A C++ back-end can be used to unparse the AST and generate C++ code. These three components (IR, front-end, and back-end), which all appear to the left of the dashed vertical line, compose the basic ROSE infrastructure.

The user builds a "mid-end" to analyze or transform the AST. ROSE assists mid-end construction by providing a number of mid-end components, shown to the right of the dashed vertical line in Figure 1, including a extensible traversal mechanism based on an attribute grammar (with fixed attribute evaluation for simplicity), AST queries, transformation operators to restructure the AST, and predefined optimizations. ROSE also provides support for library annotations whether they be contained in pragmas, comments, or separate annotation files. Finally, ROSE provides a number of additional packages ("add-ons") for visualizing the AST, building GUI interfaces to ROSE-based tools, etc. The dependencies among all the major components form a directed acyclic graph, shown by arrows in Figure 1. Thus, a tool built using ROSE can use just the subset of the ROSE infrastructure required.

3.1   Front-End

We use the Edison Design Group C and C++ front-end (EDG) [13] to parse
C and C++ programs. The EDG front-end generates an AST and performs a
full type evaluation of the C++ program. The EDG AST is represented as
a C data structure. Within the ROSE front-end, an initial phase translates
this C AST into a different object-oriented abstract syntax tree, SAGEIII, based
on Sage II and Sage++ [12]; additional phases in the ROSE front-end do further processing to handle template details, attach comments and C preprocessor
directives to AST nodes, and perform AST verification tests. The EDG work is
completely encapsulated as a binary which allows its general distribution (as a
binary library) with the rest of the ROSE source code. ROSE source including
EDG source is available to research groups who obtain the free EDG research
license. Alternative language specific front-ends are possible within the ROSE
front-end (e.g. Open64 front-end for Fortran90), abstracting the details of using
language specific front-ends (e.g. EDG front-end). SAGEIII is used uniformly as
the intermediate representation by the rest of the ROSE front-end (downstream
of the EDG front-end), the mid-end, and the back-end. Full template support
permits all templates to be instantiated, as required, in the AST. The AST
passed to the mid-end represents the program and all the header files included
by the program. The SAGEIII IR has approximately 240 types of IR nodes, as
required to fully represent the original structure of the application as an AST.

3.2   Mid-End

The mid-end permits the analysis and restructuring of the AST for performance
improving program transformations. Results of program analysis are accessible from AST nodes. The AST processing mechanism computes inherited and
synthesized attributes on the AST. An AST restructuring operation specifies
a location in the AST where code should be inserted, deleted, or replaced.
Transformation operators can be built using the AST processing mechanism in
combination with AST restructuring operations.

ROSE internally implements a number of forms of procedural and inter-
procedural analysis. Much of this work is in current development. ROSE currently includes support for dependence, call graph, and control flow analysis.

To support whole-program analysis, ROSE has additional mechanisms to store
analysis results persistently in a database (e.g., SQLite), to store ASTs in binary
files, and to merge multiple ASTs from the compilation of different source files
into a single AST (without losing project, file and directory structure).

3.3   Back-End

The back-end unparses the AST and generates C++ source code. Either all
included (header) files or only source files may be unparsed; this feature is important when transforming user-defined data types, for example, when adding
generated methods. Comments are attached to AST nodes (within the ROSE
front-end) and unparsed by the back-end. Full template handling is included
with any transformed templates output in the generated source code.

3.4   Features Relevant to Testing Tools

Of the major components, the following features are particularly relevant to the design and implementation of ROSE-based testing tools.

Full C++ information in AST. ROSE maintains an AST with all information required to rebuild the original source code, including:

- comments and C Preprocessor (CPP) directive information;
- pragmas;
- all C++ constructs; the SAGEIII IR consists of 240 different types of nodes;
- templates, including all required instantiations of templates;
- source file positions for every statement, expression, and name;
- full constant expression trees and their constant-folded representations;
- all macro information, including what macros were expanded and where;
- floating point constants represented as both values and strings.

Whole program analysis. Recent supporting work in ROSE permits the whole application (spanning hundreds of source files) to be merged into a single AST held in memory. File I/O, supporting the AST, permits the fast reading and writing of the AST to disk to support analysis tools. This work avoids the re-compilation of the application source to build the AST to support analysis or simple queries of the AST by ROSE-based tools.

GUI interface support for rapid tool development. ROSE supports QRose by Gabriel Coutinho at Imperial College, which implements a Qt interface for ROSE. QRose greatly simplifies the development of specialized analysis tools to select or present the AST or source code.

Robustness. ROSE has been tested on large-scale, production-quality, C++ scientific applications on the order of a millions lines of code. These applications make extensive use of C++ language and library features, including templates, the Standard Template Library (STL), and the Boost library, to name a few.

AST visualization tools. ROSE contains a number of graphical and non-graphical tools for visualizing the AST, which is useful when debugging ROSE-based tools or when trying to understand details of program structure.

Attribute-based tree traversals. ROSE has simple interfaces for traversing the AST and propagating context information through inherited and synthesized attributes. These interfaces are consistent among traversal and rewrite systems.

4   Examples of Testing Using ROSE

In this section, we show the interplay between compilers (specifically, ROSE) and testing through a series of specific concrete examples. The first three represent on-going research and development work on building tools for automated testing and debugging, while the fourth relates a number of anecdotal examples to motivate the need for configuration testing of compilers as discussed in Section 2.10.

4.1   Coverage Analysis and Testing of Multi-threaded Code

We are developing tools to address problems of code coverage and automated bug testing of multi-threaded for C and C++ applications, in collaboration with IBM, Haifa. In particular, we are connecting ROSE to the IBM ConTest tools developed originally for Java [7]. Our ROSE-based tool carries out three transformation tasks needed to interface to the ConTest tools:

1. Changing specific function names and their arguments
2. Instrumenting functions and conditions (blocks) for coverage
3. Instrumenting shared variables for coverage

In each case, the ROSE-based tool modifies the AST, and then outputs source code corresponding to the modified AST. Each source-to-source translation phase was supported by less than 50 lines of code and can be used to operate on applications of any size, including whole applications at once using the whole program analysis and transformation mechanisms within ROSE. Such transformations take seconds on hundreds of thousands of lines of code.

4.2   Automatic Checking of Measurement Units

An important source of errors in scientific applications are misuse of units of measurements, such as failing to convert a value in 'miles' to 'meters' before combining it with another value in 'meters' within some computation. OSPREY is a type analysis-based tool for soundly and automatically detecting possible measurement unit errors in C programs, developed by Zhendong Su and Lingxiao Jiang at the University of California, Davis [14]. We are collaborating with Su and Jiang to extend this work to C++ programs.

Prior work on OSPREY has shown that type inference mechanisms are an effective way to detect such errors. To be most effective, users need to introduce some lightweight annotations, in the form of user-defined type qualifiers, at a few places in the code to identify what variables have certain units. The type inference mechanism propagates this type information throughout the entire program to detect inconsistent usage. In ROSE, the type qualifiers can be expressed as comments or macros, since all comments and source file position information is preserved in the AST. Furthermore, the technique requires significant compiler infrastructure and program analysis support (inter-procedural control flow analysis and alias analysis).

4.3   Symbolic Performance Prediction

Another application demonstrating the use of ROSE is in the identification of performance bugs. In particular, we compute symbolic counts of program properties within loops and across function boundaries. The properties of interest include basic operation counts, numbers of memory accesses and function calls, and counts of global and point-to-point communication, among others. Our fully automated implementation uses basic traversals of the AST to gather raw symbolic terms, combined with calls to Maple, a popular commercial symbolic equation evaluation tool, to evaluate and to simplify sums and functions of these terms symbolically. Evaluated counts can then be inserted at various places in the AST as attached comments, so that the unparsed code includes comments indicating the symbolic counts. In connecting ROSE to Maple, we can thus make the performance of arbitrarily well-hidden high-level object-oriented abstractions transparent. Doing so permits the inspection of numerous aspects of program and subprogram complexity immediately obvious within general program documentation, desk-checking evaluation, or within the code review process.

4.4   Compiler-Configuration Testing

The main impact on the cost of a bug, which includes testing and debugging time, is the time between the time the bug was introduced and the time it was found [15]. As a program evolves through many versions, so does the compiler, and a future release may use different compiler or compiler options (*e.g.*, optimization level). In such cases, it is important to check the program under various options to detect if it is sensitive to specific compiler options or versions, and then either remove or carefully document such dependencies.

We encountered this particular problem recently in one of our projects when it stopped working for a client. It turned out that the client moved to a new compiler version that by default performed inlining, which changed location of objects in the heap, and caused our pointers to stop working. Automated "compiler-sensitivity testing" could have greatly simplified debugging. Understanding the root cause in the field was very difficult as it required copying the entire clients environment.

Another potential source of problems is a dependence of an application on a particular compiler's implementation of some language construct. There are a number of examples for C++ [16]:

1. Casts are sometimes required (explicit) and sometimes optional (implicit) with many compilers (C and C++ in particular). It is common for the rules for explicit casts to be relaxed to make it easier to compile older code. Across multiple compilers which casts are required to be explicit vs. implicit can result in non-portable code, and sometime a bug.
2. The order in which static variables are initialized (especially for static objects) is compiler-dependent, and notoriously different for each compiler. Since the language does not specify an order, applications should seek to reduce dependences on a particular order.
3. Infrequently used language features are often a problem, as the level of support for such features can vary greatly across compilers. These features include variadic macros, compound literals, and case ranges.
4. The compiler has some degree of freedom in choosing how fields of a structure or class are aligned, depending on how access-privilege specifies are placed. In particular,

```
class X {
   public: int x;
           int y;
};
```

is not the same as:

```
class X {
    public: int x;
    public: int y;
};
```

The first example forces the layout to be sequential in x and y, whereas the second example permits the compiler to reorder the field values. This freedom is not likely implemented in many compilers, but can be an issue in both large scientific and systems applications.

In all of these cases, it can be very difficult to track down the problem if an application somehow depends on a particular implementation. We are developing support within ROSE through bug pattern analysis tools that can help identify such implementation dependencies.

5   Related Work

We review related work in the general area of alternative open compiler infrastructures for C++, and place our current research and development in the context of existing static defect detection tools. We also include a brief discussion of a developer-centric, non-analysis based technique that extends aspect-oriented programming for testing. The following discussion emphasizes source-level tools, and we do not mention related work on tools that operate on binaries, or the important class of dynamic testing tools.

5.1   Open Compiler Infrastructures for C++

Although there are a number of available open-source compiler infrastructures for C, Fortran, and Java [17, 18], there are relatively few for C++, including g++ [19], OpenC++ [20], MPC++ [21], and Simplicissimus [22]. ELSA is a robust C++ front-end based on the Elkhound GLR parser generator [23], but the IR is best-suited to analysis (and not source-level transformation) tasks. ROSE could use ELSA instead of EDG with appropriate IR translations, either directly or through some external C++ AST format such as IPR/XPR [24]. Like several of these infrastructures, we are developing ROSE to handle realistic large-scale (million lines or more) applications in use throughout the DOE laboratories. We distinguish ROSE by its emphasis on ease-of-use in building compiler-based tools for users who do not necessarily have a formal compiler background.

5.2   Automatic, Static Defect Detection

There is a large and rapidly growing body of work on compile-time automatic software defect detection (*i.e.*, bug detection) systems. This research explores techniques that trade-off soundness (*i.e.*, finding all bugs), completeness (reporting no false positive errors), time and space complexity, and level of user interaction (*i.e.*, the degree to which user is required to provide manual annotations). Below, we provide just a sample of related projects, roughly grouped into four classes: bug pattern detectors, compiler-based program analyzers, model checkers, and formal verifiers based on automated theorem provers. (This categorization is somewhat artificial as many of the tools employ hybrid techniques.) Since no single class of techniques is ideal for all bugs of interest, our aim in ROSE is to provide an extensible, open infrastructure to support external research groups building equivalent tools for C++.

Bug pattern detectors. This class of tools uses a minimal, if any, amount of program analysis and user-supplied annotation. Bug pattern tools are particularly effective in finding errors in language feature and library API usage, but may also support diverse testing activities such as code reviews [11]. The classical example is the C LINT tool [25], which uses only lexical analysis of the source. Recent work on Splint (formerly, LClint) extends the LINT approach with lightweight annotations and basic program analysis for detecting potential security flaws [26]. More recently, Hovemeyer and Pugh have implemented the FINDBUGS framework for finding a variety of bug patterns in Java applications with basic program analysis, observing that many errors are due to a misunderstanding of language or API features [5]. This particular observation certainly applies to C++ applications, where many usage rules are well-documented [27], with some current progress toward automatic identification of STL usage errors (see STLlint, below). C++ in particular has additional challenges since the resolution of overloaded operators requires relatively complex type evaluation which can require more sophisticated compiler support.

Compiler-based static analyzers. Compared to bug pattern detectors, tools in this class use deeper analysis methods (*e.g.*, context-sensitive and flow-sensitive inter-procedural analysis for deadlock detection [28]) to improve analysis soundness while keeping the annotation burden small.

Type checkers constitute an important subclass, since a number of defect detection problems can be mapped to equivalent type analysis problems. The CQUAL tool uses constraint-based type inference to propagate user-specified type qualifiers to detect Y2K bugs, user- and kernel-space trust errors, correct usage of locks, and format string vulnerabilities, among others, in C programs [29]. CQUAL analyses are sound, but require some user annotation for best results. Work with Zhendong Su and Lingxiao Jiang are integrating these ideas into ROSE via OSPREY, a type qualifier-based system for checking the usage of scientific measurement units [14].

Researchers have used or extended classical program analysis in a variety of ways to create customized, lightweight static analyzers. This body of work includes meta-level compilation (MC[30] and the commercial version that became a basis for Coverity Inc.[31]), symbolic execution techniques as embodied in PREfix [32, 33] and STLlint [34], property simulation as in ESP [35], and reduction of program property checking to boolean satisfiability [36, 37]. Another interesting example in this class of tools is the highly regarded commercial tool, JTest, which combines static and dynamic (*e.g.*, coverage, test execution) analysis techniques [38]. The ROSE infrastructure supports the development of similar tools for C++ by providing an interface to a robust C++ front-end, and we (and others) are extending the available analysis in ROSE through direct implementation as well as interfacing to external analysis tools like OpenAnalysis [39].

Software model checkers. We consider, as members of this class, tools which explore the state-space of some model of the program. These tools include FeaVer for C programs [40], the Java PathFinder [41, 42], Bandera (and Bogor) for Java [43, 44], MOPS for C [45], SLAM for C [46], and BLAST for C [47]. These tools generally employ program analysis or theorem provers to extract a model, and then apply model checkers to verify the desired properties of the model. It is also possible to apply model checking in some sense directly to the source, as suggested in work on VeriSoft [48].

Although model checkers constitute powerful tools for detecting defects, they can be challenging to implement effectively for at least two reasons. First, their effectiveness is limited by how accurately the abstract models represent the source code. Secondly, they may require whole-program analysis. In ROSE, we have tried to address the first issue by maintaining an accurate representation of input program, including all high-level abstractions and their uses, thereby potentially enabling accurate mappings of the C++ source code to models. To address the second, we provide support for whole-program analysis, as outlined in Section 3.

Formal verifiers. Verification tools such as ESC [49], which are based on automated theorem provers, are extremely powerful but typically require users to provide more annotations than with the above approaches. ESC specifically was originally developed for Modula-3 but has been recently extended to apply to Java. ROSE preserves all comments and represents C/C++ pragmas directly, thus providing a way to express these annotations directly in the source.

5.3   Aspect-Oriented Testing

Aspect-oriented programming (AOP) permits the creation of generic instrumentation [50]. The central idea in AOP is that although the hierarchical modularity mechanisms of object-oriented languages are useful, they are unable to modularize all concerns of interest in complex systems. In the implementation of any complex system, there will be concerns that inherently cross-cut the natural modularity of the rest of the implementation. AOP provides language mechanisms that explicitly capture cross-cutting structures. This makes it possible to program cross-cutting concerns in a modular way, and achieve the usual benefits of improved modularity [51]. The most common use of AOP is to implement logging.

However, because AOP frameworks were designed with simplicity of mind, they do not permit context-dependent instrumentation. For example, one may use aspects to insert instrumentation at the beginning of all methods (possibly of a specific type), but cannot limit the instrumentation based on some attribute of the method. However, AOP has more recently also been shown to be useful in testing [52, 53].

6 Conclusions

The need for an open, extensible compiler infrastructure for testing is motivated simultaneously by (1) the desire to automate or semi-automate the many kinds of activities required for effective testing, (2) the fact that each activity has its own unique analysis and transformation requirements from a compiler infrastructure, and (3) that each testing team will require its own set of customized tools. Our goals in developing ROSE are to facilitate the development of all these kinds of existing and future testing tools by providing a robust, modular, complete, and easy-to-use foundational infrastructure. We are currently pursuing a number of different projects to support testing specifically, as discussed in Section 4. In both our current and future work, we are extending the analysis and transformation capabilities of ROSE, in part by interfacing ROSE with external tools to leverage their capabilities and also working with other research groups.

References

1. Dan Quinlan, Markus Schordan, Qing Yi, and Andreas Saebjornsen. Classification and utilization of abstractions for optimization. In *Proc. 1st International Symposium on Leveraging Applications of Formal Methods*, Paphos, Cyprus, October 2004.
2. Markus Schordan and Dan Quinlan. A source-to-source architecture for user-defined optimizations. In *Proc. Joint Modular Languages Conference*, 2003.
3. Qing Yi and Dan Quinlan. Applying loop optimizations to object-oriented abstractions through general classification of array semantics. In *Proc. Workshop on Languages and Compilers for Parallel Computing*, West Lafayette, Indiana, USA, September 2004.
4. Daniel Jackson and Martin Rinard. Software analysis: A roadmap. In *Proc. Conference on the Future of Software Engineering (International Conference on Software Engineering)*, pages 133–145, Limerick, Ireland, 2000.
5. David Hovemeyer and William Pugh. Finding bugs is easy. *SIGPLAN Notices (Proceedings of Onward! at OOPSLA 2004)*, December 2004.
6. Dawson Engler and Mandanlal Musuvathi. Static analysis versus software model checking for bug finding. In *Proc.International Conference on Verification, Model Checking, and Abstract Interpretation*, Venice, Italy, 2004.
7. Orit Edelstein, Eitan Farchi, Evgeny Goldin, Yarden Nir, Gil Ratsaby, and Shmuel Ur. Testing multithreaded Java programs. *IBM Systems Journal: Special Issue on Software Testing*, February 2002.
8. Shmuel Ur and A. Ziv. Off-the-shelf vs. custom made coverage models, which is the one for you? In *Proc. International Conference on Software Testing Analysis and Review*, May 1998.

9. Gregg Rothermel and Mary Jean Harrold. A safe, efficient regression test selection technique. *ACM Trans. Softw. Eng. Methodol.*, 6(2):173–210, 1997.

10. Gil Ratsaby, Baruch Sterin, and Shmuel Ur. Improvements in coverability analysis. In *FME*, pages 41–56, 2002.

11. Eitan Farchi and Bradley R. Harrington. Assisting the code review process using simple pattern recognition. In *Proc. IBM Verification Conference*, Haifa, Israel, November 2005.

12. Francois Bodin, Peter Beckman, Dennis Gannon, Jacob Gotwals, Srinivas Narayana, Suresh Srinivas, and Beata Winnicka. Sage++: An object-oriented toolkit and class library for building fortran and C++ restructuring tools. In *Proceedings. OONSKI '94*, Oregon, 1994.

13. Edison Design Group. EDG front-end. www.edg.com.

14. Lingxiao Jiang and Zhendong Su. Osprey: A practical type system for validating the correctness of measurement units in C programs, 2005. (*submitted*; wwwcsif.cs.ucdavis.edu/~jiangl/research.html).

15. NIST. The economic impacts of inadequate infrastructure for software testing. Technical Report Planning Report 02–3, National Institute of Standards and Technology, May 2002.

16. Bjarne Stroustrop. *The C++ programming language*. Addison-Wesley, 3rd edition, 2000.

17. M. S. Lam S. P. Amarasinghe, J. M. Anderson and C. W. Tseng. The suif compiler for scalable parallel machines. In *Proc. SIAM Conference on Parallel Processing for Scientific Computing*, Feb 1995.

18. G.-A. Silber and A. Darte. The Nestor library: A tool for implementing Fortran source to source transformations. *Lecture Notes in Computer Science*, LNCSD9(1593), 1999.

19. Free Software Foundation. GNU Compiler Collection, 2005. gcc.gnu.org.

20. Shigeru Chiba. Macro processing in object-oriented languages. In *TOOLS Pacific '98, Technology of Object-Oriented Languages and Systems*, 1998.

21. Yutaka Ishikawa, Atsushi Hori, Mitsuhisa Sato, Motohiko Matsuda, Jorg Nolte, Hiroshi Tezuka, Hiroki Konaka, Munenori Maeda, and Kazuto Kubota. Design and implementation of metalevel architecture in C++—MPC++ approach. In *Proc. Reflection '96 Conference*, April 1996.

22. Sibylle Schupp, Douglas Gregor, David Musser, and Shin-Ming Liu. Semantic and behavioral library transformations. *Information and Software Technology*, 44(13):797–810, April 2002.

23. Scott McPeak and George C. Necula. Elkhound: A fast, practical GLR parser generator. In *Proc. Conference on Compiler Construction*, Barcelona, Spain, April 2004.

24. Bjarne Stroustrop and Gabriel Dos Reis. Supporting SELL for high-performance computing. In *Proc. Workshop on Languages and Compilers for Parallel Computing*, Hawthorne, NY, USA, October 2005.

25. S. C. Johnson. Lint, a C program checker, April 1986.

26. David Evans and David Larochelle. Improving security using extensible lightweight static analysis. *IEEE Software*, pages 42–51, Jan 2002.

27. Scott Meyers. *Effective C++: 50 specific ways to improve your programs and design*. Addison-Wesley, 2nd edition, 1997.

28. Amy Williams, William Thies, and Michael D. Ernst. Static deadlock detection for Java libraries. In *Proceedings of the 19th European Conference on Object-Oriented Programming*, Glasgow, Scotland, July 2005.

29. Jeffrey S. Foster, Tachio Terauchi, and Alex Aiken. Flow-sensitive type qualifiers. In *Proc. ACM SIGPLAN Conference on Programming Language Design and Implementation*, pages 1–12, Berlin, Germany, June 2002.

30. Seth Hallem, Benjamin Chelf, Yichen Xie, and Dawson Engler. A system and language for building system-specific, static analyses. In *Proc. ACM SIGPLAN Conference on Programming Language Design and Implementation*, Berlin, Germany, June 2002.

31. Coverity Inc. Coverity source code security tool. www.coverity.com.

32. William R. Bush, Jonathan D. Pincus, and David J. Sielaff. A static analyzer for finding dynamic programming errors. *Software–Practice and Experience*, 30:775–802, 2000.

33. Sarfraz Khurshid, Corina Pasareanu, and Willem Visser. Generalized symbolic execution for model checking and testing. In *Proc. International Conference on Tools and Algorithms for Construction and Analysis of Systems*, Warsaw, Poland, April 2003.

34. Douglas Gregor and Sibylle Schupp. STLlint: Lifting static checking from languages to libraries. *Software: Practice and Experience*, 2005. (*to appear*).

35. Manuvir Das, Sorin Lerner, and Mark Seigle. ESP: Path-sensitive program verification in polynomial time. In *Proc. ACM SIGPLAN Conference on Programming Language Design and Implementation*, Berlin, Germany, June 2002.

36. Edmund Clarke, Daniel Kroening, and Flavio Lerda. A tool for checking ANSI C programs. In *Proc. International Conference on Tools and Algorithms for Construction and Analysis of Systems*, volume LNCS 2988, Barcelona, Spain, March 2004.

37. Yichen Xie and Alex Aiken. Scalable error detection using boolean satisfiability. In *Proc. Principles of Programming Languages*, Long Beach, CA, USA, January 2005.

38. Parasoft Corporation. Jtest, 2005. www.parasoft.com.

39. Michelle Mills Strout, John Mellor-Crummey, and Paul D. Hovland. Representation-independent program analysis. In *Proc. ACM SIGPLAN-SIGSOFT Workshop on Program Analysis for Software Tools and Engineering*, September 2005.

40. Gerard J. Holzmann and Margaret H. Smith. Automating software feature verification. *Bell Labs Technical Journal*, 5(2):72–87, 2000.

41. Willem Visser, Klaus Havelund, Guillame Brat, Seung-Joon Park, and Flavio Lerda. Model checking programs. *Automated Software Engineering Journal*, 10(2), April 2002.

42. Gerard J. Holzmann. The model checker SPIN. *IEEE Trans. on Software Engineering*, 23(5):279–295, May 1997.

43. James C. Corbett, Matthew B. Dwyer, John Hatcliff, Shawn Laubach, Corina S. Păsăreanu, and Robby. Bandera: Extracting finite-state models from Java source code. In *Proc.International Conference on Software Engineering*, pages 439–448, Limerick, Ireland, 2000.

44. Robby, Matthew B. Dwyer, and John Hatcliff. Bogor: An extensible and highly-modular model checking framework. In *Proc. Joint Meeting of the European Software Engineering Conference and ACM SIGSOFT Symposium on the Foundations of Software Engineering*, March 2003.

45. Hao Chen, Drew Dean, and David Wagner. Model checking one million lines of C code. In *Proc. Network and Distributed System Security Symposium*, San Diego, CA, USA, February 2004.

46. Thomas A. Ball and Sriram K. Rajamani. The SLAM project: Debugging system software via static analysis. In *Proc. Principles of Programming Languages*, January 2002.
47. Thomas A. Henzinger, Ranjit Jhala, Rupak Majumdar, and Gregoire Sutre. Software verification with BLAST. In *Proc. 10th SPIN Workshop on Model Checking Software*, volume LNCS 2648, pages 235–239. Springer-Verlag, 2003.
48. Patrice Godefroid. Software model checking: the VeriSoft approach. Technical Report ITD-03-44189G, Bell Labs, 2003.
49. David L. Detlefs, K. Rustan M. Leino, Greg Nelson, and James B. Saxe. Extended static checking. Technical Report SRC-159, Compaq Systems Research Center, December 18 1998.
50. Gregor Kiczales, John Lamping, Anurag Menhdhekar, Chris Maeda, Cristina Lopes, Jean-Marc Loingtier, and John Irwin. Aspect-oriented programming. In Mehmet Akşit and Satoshi Matsuoka, editors, *Proceedings European Conference on Object-Oriented Programming*, volume 1241, pages 220–242. Springer-Verlag, Berlin, Heidelberg, and New York, 1997.
51. Gregor Kiczales, Erik Hilsdale, Jim Hugunin, Mik Kersten, Jeffrey Palm, and William G. Griswold. An overview of AspectJ. *Lecture Notes in Computer Science*, 2072:327–355, 2001.
52. Daniel Hughes and Philip Greenwood. Aspect testing framework. In *Proceedings of the Formal Methods for Open Object-based Distributed Systems and Distributed Applications and Interoperable Systems Student Workshop*, Paris, France, November 2003.
53. Shady Copty and Shmuel Ur. Multi-threaded testing with AOP is easy, and it finds bugs! In *Euro-Par*, pages 740–749, 2005.

Effective Black-Box Testing with Genetic Algorithms

Mark Last[1,*] , Shay Eyal[1], and Abraham Kandel[2]

[1] Department of Information Systems Engineering, Ben-Gurion University of the Negev,
Beer-Sheva 84105, Israel
{mlast, shayey}@bgu.ac.il
[2] Department of Computer Science and Engineering, University of South Florida,
Tampa, FL 33620, USA
kandel@csee.usf.edu

Abstract. *Black-box (functional) test cases* are identified from functional requirements of the tested system, which is viewed as a mathematical function mapping its inputs onto its outputs. While the number of possible black-box tests for any non-trivial program is extremely large, the testers can run only a limited number of test cases under their resource limitations. An *effective* set of test cases is the one that has a high probability of detecting faults presenting in a computer program. In this paper, we introduce a new, computationally intelligent approach to automated generation of effective test cases based on a novel, Fuzzy-Based Age Extension of Genetic Algorithms (FAexGA). The basic idea is to eliminate "bad" test cases that are unlikely to expose any error, while increasing the number of "good" test cases that have a high probability of producing an erroneous output. The promising performance of the FAexGA-based approach is demonstrated on testing a complex Boolean expression.

Keywords: Black-box testing, Test prioritization, Computational intelligence, Genetic algorithms, Fuzzy logic.

1 Introduction and Motivation

Software quality remains a major problem for the software industry worldwide. Thus, a recent study by the National Institute of Standards & Technology [3] found that "the national annual costs of an inadequate infrastructure for software testing is estimated to range **from \$22.2 to \$59.5 billion**" (p. ES-3) which are about 0.6 percent of the US gross domestic product. This number does not include costs associated with catastrophic failures of mission-critical software systems such as the Patriot Missile Defense System malfunction in 1991 and the \$165 million Mars Polar Lander shutdown in 1999.

In today's software industry, the design of test cases is mostly based on the human expertise, while *test automation tools* are limited to execution of pre-planned tests only. Evaluation of test outcomes is also associated with a considerable effort by human testers based on often imperfect knowledge of the requirements specification. It is not surprising that this manual approach to software testing results in such heavy losses to the economy. In this paper, we demonstrate the potential use of genetic

---

* Corresponding author.

S. Ur, E. Bin, and Y. Wolfsthal (Eds.): Haifa Verification Conf. 2005, LNCS 3875, pp. 134 – 148, 2006.
© Springer-Verlag Berlin Heidelberg 2006

algorithms for *automated design* of effective test cases that have a high probability of detecting faults presenting in a computer program.

A system *fails* when it does not meet its specification [17]. The purpose of testing a system is to discover *faults* that cause the system to fail rather than proving the code correctness, which is often an impossible task [3]. In the software testing process, each *test case* has an identity and is associated with a set of inputs and a list of expected outputs [10]. *Functional (black-box) test cases* are based solely on functional requirements of the tested system, while *structural (white-box) tests* are based on the code itself. According to [10], black-box tests have the following two distinct advantages: they are independent of the software implementation and they can be developed in parallel with the implementation.

The number of combinatorial black-box tests for any non-trivial program is extremely large, since it is proportional to the number of possible combinations of all input values. On the other hand, testing resources are always limited, which means that the testers have to choose the tests carefully from the following two perspectives:

1. Generate *good* test cases. A *good* test case is one that has a high probability of detecting an as-yet undiscovered error [19]. Moreover, several test cases causing the same bug may show a pattern that might lead the programmer to the real cause of the bug [16].
2. Prioritize test cases according to a *rate of fault detection* – a measure of how quickly those test cases detect faults during the testing process [5].

As indicated by Jorgensen [10], the above requirements of test effectiveness suffer from a circularity: they presume we know all the faults in a program, which implies that we do not need the test cases at all! Since in reality testers do not know in advance the number and the location of bugs in the tested code, the practical approach to generation of black-box tests is to look for bugs where they *are expected* to be found, while trying to avoid redundant tests that test the same functionality more than once. Thus, the best-known black-box testing techniques include *boundary value testing* (presumes that errors tend to occur near the extreme values), *equivalence class testing* (requires to test only one element from each class), and *decision table-based testing* (based on a complete logical specification of functional requirements). Since all the systematic approaches to test generation cover only a tiny portion of possible software inputs, *random tests*, sometimes called "monkey tests", can discover new bugs in the tested system [16]. Even if a "monkey test" is smart enough to examine the results of each test case and evaluate its correctness, this is an extremely inefficient approach, since many random tests may be completely redundant, testing the same functionality and the same code over and over again. On the other hand, as shown in [10], it may take a huge number of random tests to test certain functionality of the software at least once.

The testers are interested to rank their test cases so that those with the highest priority, according to some criteria, are executed earlier than those with lower priority. This criterion may change according to some performance goal (e.g. code coverage, reliability confidence, specific code changes, etc.). The dominant criterion is the *rate of fault detection* as described earlier. An increased rate of fault detection can provide earlier feedback on the tested software and earlier evidence that quality goals have not been met yet [5]. Such information is particularly useful in *version-specific test case prioritization* while using *regression testing*. In the case of regression testing, the information gathered in previous runs of existing test cases might help to prioritize the test cases for subsequent runs. Obviously, such information is unavailable during initial testing.

Genetic Algorithms (GA) provide a general-purpose search methodology, which uses principles of the natural evolution [6]. In this paper, we introduce a new, computationally intelligent approach to improving the effectiveness of a given test set by eliminating "bad" test cases that are unlikely to expose any error, while increasing the number of "good" test cases that have a high probability of producing an erroneous output. The algorithm is started by randomly generating an initial test set, where each test case is modeled as a *chromosome* representing one possible set (vector) of input values. To apply a genetic algorithm to the test generation problem, the evaluation function should emulate the capability of a given test case to detect an error in a tested program. "Good" chromosomes (test cases) that discover the errors are given a positive reward, while a zero score is given to those that do not. This way the algorithm *prioritizes* a given set of test cases. At each iteration, the "population" of test cases is updated by mutating and recombining its most prospective chromosomes. Consequently, the percentage of fault-exposing test cases is expected to increase from one generation to the next. The performance of the basic genetic algorithm is enhanced by the novel Fuzzy-Based Age Extension of Genetic Algorithms (FAexGA) introduced by us in [12].

This paper is organized as follows. Section 2 presents an overview of the genetic algorithms methodology and its application to the test generation problem. The design of initial testing experiments and the obtained results are described in Sections 3 and 4 respectively. Section 5 concludes the paper.

2   Software Testing with Genetic Algorithms

2.1   Genetic Algorithms: An Overview

Genetic Algorithms (GAs) are general-purpose search algorithms, which use principles inspired by natural genetics to evolve solutions to problems. As one can guess, genetic algorithms are inspired by Darwin's theory about evolution [9]. They have been successfully applied to a large number of scientific and engineering problems, such as optimization, machine learning, automatic programming, transportation problems, adaptive control, etc. (see [13][14]).

GA starts off with population of randomly generated chromosomes, each representing a candidate solution to the concrete problem being solved, and advances towards better chromosomes by applying genetic operators based on the genetic processes occurring in nature. So far, GAs have had a great measure of success in search and optimization problems due to their robust ability to exploit the information accumulated about an initially unknown search space. Particularly GAs specialize in large, complex and poorly understood search spaces where classic tools are inappropriate, inefficient or time consuming [8].

As mentioned, the GA's basic idea is to maintain a population of chromosomes. This population evolves over time through a successive iteration process of competition and controlled variation. Each state of population is called generation. Associated with each chromosome at every generation is a *fitness* value, which indicates the quality of the solution, represented by the chromosome values. Based upon these fitness values, the selection of the chromosomes, which form the new generation, takes place. Like in nature, the new chromosomes are created using genetic operators such as *crossover* and *mutation*.

The fundamental mechanism consists of the following stages (as described in Figure 1):

1. Generate randomly the initial population.
2. Select the chromosomes with the best fitness values.
3. Recombine selected chromosomes using crossover and mutation operators.
4. Insert offsprings into the population.
5. If a stop criterion is satisfied, return the chromosome(s) with the best fitness. Otherwise, go to Step 2.

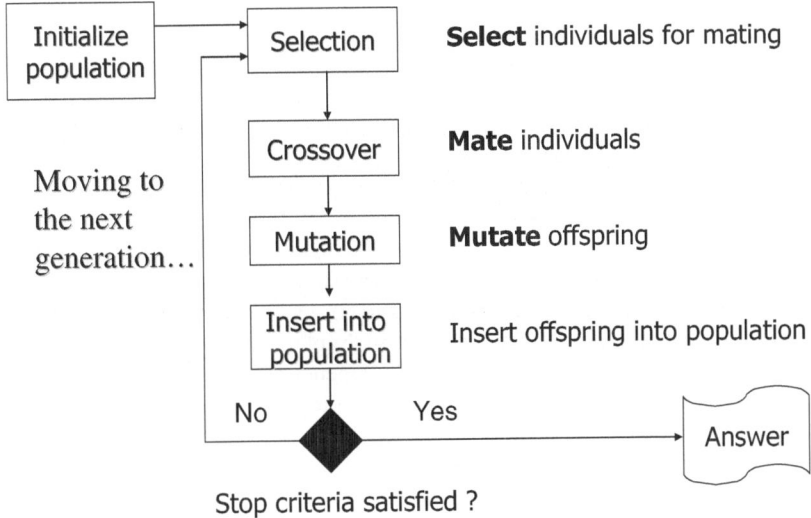

Fig. 1. Simple Genetic Algorithm Fundamental Mechanism

To apply genetic algorithm to a particular problem, such as test case generation, we need to determine the following elements [13]:

1. Genetic representation for potential solutions to the problem (e.g., test cases).
2. Method to create an initial population of potential solutions.
3. Evaluation function, which scores the solution quality (also called objective function).
4. Genetic operators that alter the composition of the off-springs.
5. Values of various parameters used by the genetic algorithm (population size, probabilities of applying genetic operators, etc.).

Each element is briefly discussed below.

Representation refers to the modeling of chromosomes into data structures. Once again terminology is inspired by the biological terms, though the entities genetic algorithm refers to are much simpler than the real biological ones. *Chromosome* typically refers to a candidate data solution to a problem, often encoded as a bit string. Each element of the chromosome is called *allele*. In other words, chromosome is a sequence of alleles. For example, consider 1-dimension binary representation: each allele is 0 or 1, and a chromosome is a specific sequence of 0 and 1's. According to [14], a *proper representation* for a given problem should cover the whole search space, avoid infeasible solutions and redundant data, and require a minimal computational effort for evaluating the objective function.

Initialization. This genetic operator creates an initial population of chromosomes, at the beginning of the genetic algorithm execution. Usually initialization is random, though it may be biased with a pre-defined initialization function, as suggested in [18].

The *selection* operator is used to choose chromosomes from a population for mating. This mechanism defines how these chromosomes will be selected, and how many offsprings each will create. The expectation is that, like in the natural process, chromosomes with higher fitness will produce better offsprings. Therefore, selecting such chromosomes at higher probability will eventually produce better population at each iteration of the algorithm. Selection has to be balanced: too strong selection means that best chromosomes will take over the population reducing its diversity needed for exploration; too weak selection will result in a slow evolution. Classic selection methods are *Roulette-Wheel, Rank based, Tournament, Uniform*, and *Elitism* [6][9].

The *crossover* operator is practically a method for sharing information between two chromosomes: it defines the procedure for generating an offspring from two parents. The crossover operator is considered the most important feature in the GA, especially where building blocks (i.e. schemas) exchange is necessary [1]. The crossover is data type specific, meaning that its implementation is strongly related to the chosen representation. It is also problem-dependent, since it should create only feasible offsprings. One of the most common crossover operators is *Single-point crossover*: a single crossover position is chosen at random and the elements of the two parents before and after the crossover position are exchanged to form two offsprings.

The *mutation* operator alters one or more values of the allele in the chromosome in order to increase the structural variability. This operator is the major instrument of the genetic algorithm to protect the population against pre-mature convergence to any particular area of the entire search space [14]. Unlike crossover, mutation works with only one chromosome at a time. In most cases, mutation takes place right after the crossover, so it practically works on the offsprings, which resembles the natural process. The most common mutation methods are:

1. Bit-flip mutation: given chromosome, every bit value changes with a mutation probability.
2. Uniform mutation: choose one bit randomly and change its value.

Evaluation function, also called objective function, rates the candidate solutions quality. This is the only single measure of how good a single chromosome is compared to the rest of the population. So given a specific chromosome, the evaluation function returns its score. The score value is not necessary the fitness value, which the genetic algorithm works with, as mentioned earlier. The fitness value is typically obtained by a transformation function called *scaling*. It is important to note that the evaluation process itself has been found to be very 'expensive' due to the time and resources it consumes [1]. Therefore it is worthwhile to simplify the evaluation function as much as possible.

The basic genetic algorithm has a set of parameters, which define its operation and behavior. Within the basic genetic algorithm, the parameter settings are fixed along the run [9]. The list of common GA parameters is given below:

1. *Population size (N)* – this parameter defines the size of the population, which may be critical in many applications: If N is too small, GA may converge quickly, whereas if it is too large the GA may waste computational resources. Regarding evaluation, if the number of fitness evaluations is not a concern, larger population sizes improve the GA's ability to solve complex problems [1]. However, in natural environments the population size changes according to growth rate, and tends to stabilize around an appropriate value [2].

2. *Chromosome length (L)* – defines the number of allele within each chromosome. This number is influenced by the chosen representation and the problem being issued.

3. *Number of generations (Ngen)* – defines the number of generations the algorithm will run. It is frequently used as a stopping criterion.

4. *Crossover probability (Pc)* – the probability of crossover between two parents. The crossover probability has another trade-off: if Pc is too low, then the sharing of information between high fitness chromosomes will not take place, hence reducing their capability to produce better offsprings. On the other hand, the crossover may also bring an offspring with lower fitness, and therefore if the Pc value is too high there is a chance we may get trapped in a local optimum.

5. *Mutation probability (Pm)* – the probability of mutation in a given chromosome. As mentioned earlier, this method element helps to prevent the population from falling into local extremes, but a too high value of Pm will slow down the convergence of the algorithm.

Finding robust methods for determining the universally best GA parameter settings is probably impossible, since the optimal values are problem-dependent and the GA parameters interact with each other in a complex way [2]. Furthermore, to attain an optimal exploitation/exploration balance, the parameter values may need adjustment in the course of running the algorithm. The GAVaPS is a Genetic Algorithm with Varying Population Size where each individual has a lifetime, which is measured in generations [1]. An individual remains in the population during its lifetime. Since parents are selected randomly, the selective pressure is assured by the fact that genomes representing better solutions have longer lifetimes. In [12], we have introduced a fuzzy-based extension of the GAVaPS algorithm, called FAexGA, which has outperformed the simple GA and the GAVaPS on a set of benchmark problems. The FAexGA algorithm is briefly described in the next sub-section.

2.2 Fuzzy-Based Age Extension of Genetic Algorithms (FAexGA)

Our general principle is that crossover probability Pc should vary according to age intervals during the lifetime: for both young and old individuals the crossover probability is naturally low, while there is a certain age interval, where this probability is high. The concepts of "young", old", and "middle-aged" are modeled as *linguistic variables* using the concepts of fuzzy set theory [11]. Therefore, very young offsprings would be less crossovered thus enabling exploration capability. On the other hand, old offsprings being less crossovered and eventually dying out would help avoiding a local optimum (premature convergence). The middle-age offsprings being crossovered most frequently would enable the exploitation aspect. As shown in [12], this approach is able to enhance the exploration and exploitation capabilities of the algorithm, while reducing its rate of premature convergence.

The general structure of the FAexGA algorithm is described in [12]. It is similar to the GAVaPS [1], with the following difference: in GAVaPS, the recombination of two parents takes place by crossover with a fixed crossover probability; in the fuzzy-based age extension of GA, the crossover probability Pc is determined by a Fuzzy Logic Controller (FLC). The FLC state variables include the age and the lifetime of the chromosomes to be crossovered (parents) along with the average lifetime of the current population.

The *fuzzification interface* of the Fuzzy Logic Controller defines for each parent the truth values of being {*Young, Middle-age, Old*}, which are the linguistic labels of the *Age* inputs. These values determine the grade of membership for each rule premise in the FLC rule base. This computation takes into account the age of only one parent at a time, and relies on the triangular membership functions like the ones shown in Fig. 2. The relative age mentioned on the x-axis in Fig. 2 refers to the ratio between chromosome's age and the average lifetime in the current population. Inspired by nature, the concept of person being *Young, Middle-age* or *Old* is relative to the average lifetime in the population; for example, if the average lifetime of a human population is about 77 years (76 for male and 78 for female), then a person whose age is between 35-55 is still considered as middle-age.

In our software testing case study (see the next section), we have experimented with three different settings of *Young upper limit* and *Old lower limit* parameters. These

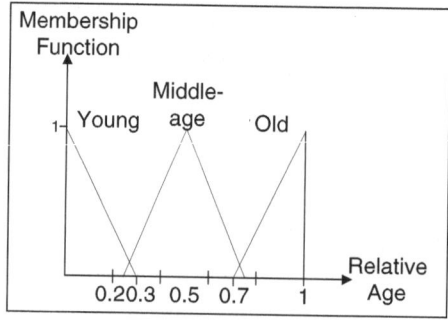

Fig. 2. Fuzzy Logic Controller

Table 1. Fuzzy-Based Age Extension of GA Tested Configurations

Parameter	FAexGA#1	FAexGA#2	FAexGA#3
FLC – "Young" upper limit	30%	25%	20%
FLC – "Old" lower limit	70%	75%	80%

Table 2. Fuzzy Rule Base

Parent I / Parent II	"Young"	"Middle-age"	"Old"
"Young"	Low	Medium	Low
"Middle-age"	Medium	High	Medium
"Old"	Low	Medium	Low

three configurations, denoted as **FAexGA#1, FAexGA#2**, and **FAexGA#3** respectively, are described in Table 1.

The fuzzy rule base we used in our experiments is presented in Table 2. Each cell defines a single fuzzy rule. For example, the upper left cell refers to the following rule: "**If** Parent I is *Young* and Parent II is *Young* **Then** crossover probability is *Low*".

The inference method used is MAX-MIN (see [11]): each rule is assigned a value equal to the minimum value of its conditions. Afterwards, the inference engine assigns each linguistic output a single value, which is the maximum value of its relative fired rules Thus, by the end of the inference process, the crossover probability will have degrees of truth for being *Low, Middle* and *High*.

The *Defuzzification interface* computes a crisp value for the controlled parameter *crossover probability* based on the values of its linguistics labels {*Low, Medium, High*}, which are the outputs of the rule base, and the triangular membership functions shown in Fig. 3. The settings of the linguistic labels were chosen according to previous studies. The defuzzification method used is *center of gravity (COG)* [11]: the crisp value of the output variable is computed by finding the center of gravity under the membership function for the fuzzy variable.

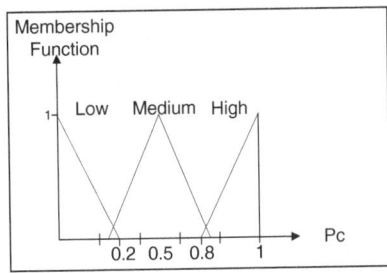

Fig. 3. The *Crossover Probability* Linguistic Variable

2.3   Generating Effective Test Cases with Genetic Algorithms

The problem of test set generation is a typical optimization problem in a combinatorial search space: we are looking for a minimal set of test cases that are most likely to reveal faults presenting in a given program. A genetic algorithm can be applied to the problem of generating an effective set of test cases as follows:

- *Representation.* Each test case can be represented as a vector of binary or continuous values related to the inputs of the tested software.
- *Initialization.* An initial test set can be generated randomly in the space of possible input values.
- *Genetic operators.* Selection, crossover, and mutation operators can be adapted to representation of test cases for a specific program.
- *Evaluation function.* Following the approach of [5], test cases can be evaluated by their *fault-exposing-potential* using buggy (mutated) versions of the original program.

3   Design of Experiments

Our case study is aimed at generating an effective set of test cases for a Boolean expression composed of 100 Boolean attributes and three logical operators: AND, OR, and NOT (called *correct expression*). This expression was generated randomly by using an external application. To define a simple evaluation function for each test case, we have generated an *erroneous expression* – the same Boolean expression as the correct one, except for a single error injected in the correct expression by randomly selecting an OR operator and switching it to the ANDone (or vice versa). Such errors are extremely hard to detect and backtrack in complex software programs, since they usually affect the outputs of only a small portion of combinatorial test cases (2^{100} test cases for the expression we have defined).

Table 3. Experimental Settings

#	Parameter	Fuzzy-based Age Extension of GA	GAVaPS	SimpleGA
	N = Population size	100	100	100
	L = chromosome length	100	100	100
	$Ngen$ = total number of generations	200	200	200
	Pc = crossover probability	Adaptive	0.9	0.9
	Pm = mutation probability	0.01	0.01	0.01
	Selection method	Random	Random	Roulette Wheel
	Crossover method	1-point	1-point	1-point
	Mutation method	Flip	Flip	Flip
	1-elitism	-	-	True
	Lifetime calculation strategy	Bi-Linear	-	-
	Reproduction ratio	0.3	0.3	-
	$MinLT$ = Minimum lifetime (number of generations)	1	1	-
	$MaxLT$ = Maximum lifetime (number of generations)	7	7	-

The chromosomes are 1-dimesional binary strings of 100-bit length. The value of the evaluation function F is calculated as follows:

$$F(T) = \begin{cases} 1, if\ Eval\_Correct(T) \neq Eval\_Erroneous(T) \\ 0, respectively \end{cases}$$

Where T is a 100-bit 1-dimesional binary chromosome representing a single test case and *Eval_Correct (T)* / *Eval_Erroneous (T)* are the binary results of applying chromosome T to the correct / erroneous expression respectively.

The experimental GAs' settings are stated in Table 3.

4 Summary of Results

Table 4 presents a summarized comparison for each performance measure obtained with each algorithm. The results of each algorithm were averaged over 20 runs.

Table 4. Comparative Evaluation of Genetic Algorithms

Measure\Algorithm	SimpleGA	GAVaPS	FAexGA#1	FAexGA#2	FAexGA#3
% Runs with solution	70%	60%	95%	70%	75%
% Solutions in the final population	1.267%	81.519%	99.934%	98.857%	87.183%
Unuse Factor	65.571%	62.269%	70.658%	68.714%	46.267%

The first measure (% Runs with solution) shows that the first configuration of the Fuzzy-Based Age Extension of Genetic Algorithms (FAexGA#1) has reached a solution in a significantly higher percentage of runs compared to all other algorithms. Solution means a test case that detects the injected error by returning different values when applied to the correct expression and the erroneous one. Surprisingly, GAVaPS reached a solution in fewer runs than the SimpleGA. The rest of the measures demonstrate a clear advantage of the FAexGA configurations: the number of solutions found in the SimpleGA final population is very small (about 1-2 solutions). GAVaPS found much more solutions than the SimpleGA while all FAexGA configurations found even more. FAexGA#1 is the most *effective* configuration, since it found the highest number of distinct solutions in its final population.

Table 4 also shows that all genetic algorithms prove to be considerably more *efficient* for this problem than the conventional test generation methods. According to [10], a program with logical variables, like a Boolean expression, should be tested by the *strong equivalence class* approach. Since each binary input in the tested expression has two equivalence classes (0 and 1), this would mean enumeration of $2^{100} \approx 1.27 \cdot 10^{30}$ distinct tests, which is a completely impossible task. Another alternative for testing such an expression is *random testing*, but the probability of one random test case to reveal a single operator change in a 100-term Boolean expression may be as low as $0.5^{99} \approx 1.58 \cdot 10^{-30}$, which is practically zero.

In addition, the FAexGA#1 unuse factor is the highest one. Unuse Factor was measured as $u = 1 - g / g_{max}$ where g is the number of generations until the first solution (i.e., test case that finds an error) appeared and g_{max} is the maximum number of generations [2]. The SimpleGA unuse factor is better than GAVaPS and the third configuration of the Fuzzy-Based Age Extension of Genetic Algorithms. Overall, the first configuration of the Fuzzy-Based Age Extension of Genetic Algorithms (FAexGA#1) has reached the best results with respect to all three performance measures.

Figure 4 shows the average percentage of solutions in each generation. As mentioned earlier, the SimpleGA found a very small number of solutions, while the first and the second configuration of FAexGA found solutions at quite a linear rate.

Figure 5 shows the population average genotypical diversity in runs where solution was found. It is calculated by comparing the inner structure of the chromosomes [7]. The SimpleGA diversity as well as the GAVaPS diversity remains high most of the generations because the number of solutions found is relatively low. The Fuzzy-Based Age Extension of GA diversities are significantly lower (especially #1 and #2) and get even lower in the final generations due to the fact that a high percentage of the population are good solutions.

Figure 6 describes the population average genotypic diversity in runs where solution was *not* found. All algorithms diversity is relatively high due to the fact that exploration occurs, but without a success.

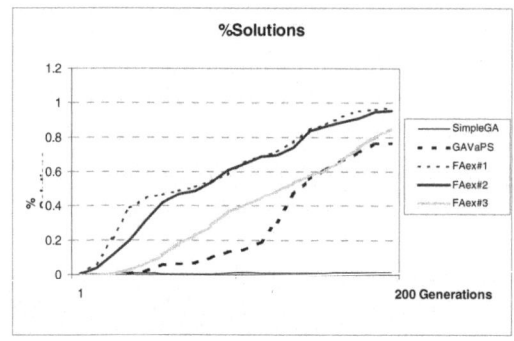

Fig. 4. Average Percentage of Solutions in the Population

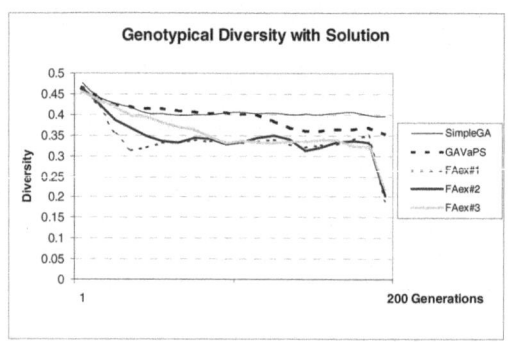

Fig. 5. Genotypical Diversity of Each Algorithm Solutions

Figures 7-11 illustrate for each algorithm the differences between the diversities in runs where solutions were found to those where solutions were not found (*G. Average* stands for runs with solutions and *B. Average* for runs without solutions). Since SimpleGA and GAVaPS found a smaller number of solutions, there are no differences between their diversities. On the contrary, all three configurations of the Fuzzy-Based

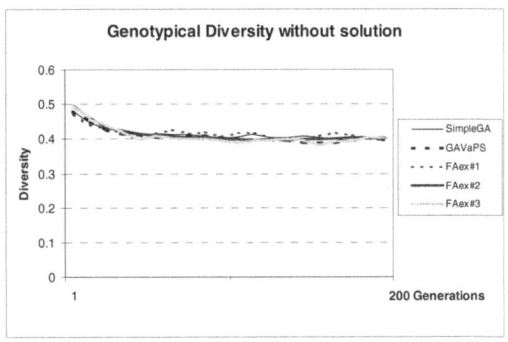

Fig. 6. Genotypical Diversity of Each Algorithm Non-Solutions

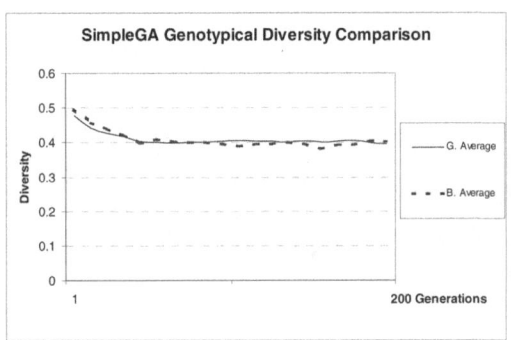

Fig. 7. SimpleGA Genotype Diversities

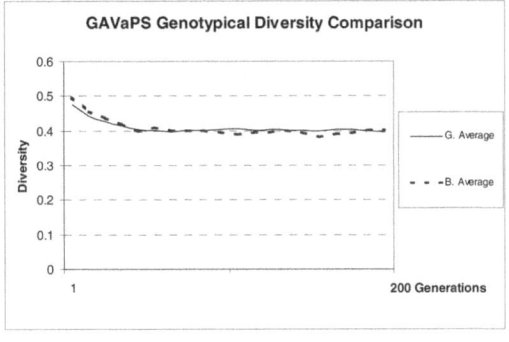

Fig. 8. GAVaPS Genotype Diversities

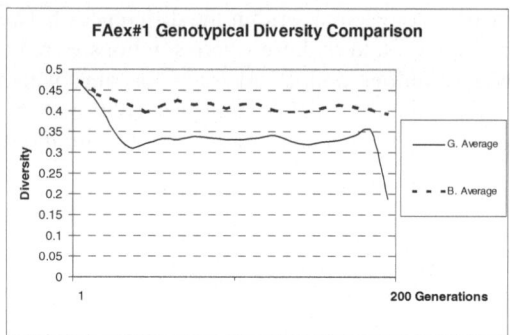

Fig. 9. FAexGA#1 Genotype Diversities

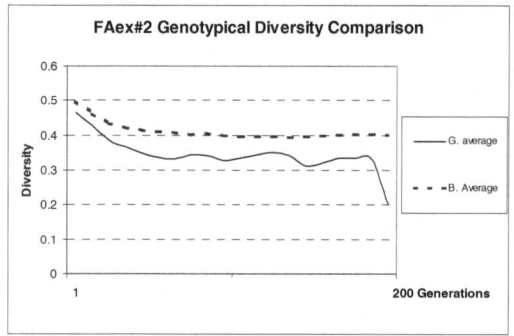

Fig. 10. FAexGA#2 Genotype Diversities

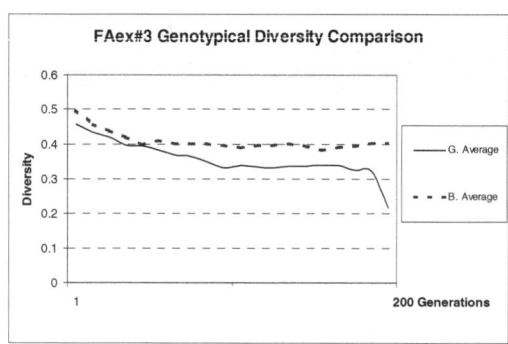

Fig. 11. FAexGA#3 Genotype Diversities

Age Extension of Genetic Algorithms present significant differences between runs where solutions were found to those where solutions were not found. Therefore, it can be concluded that FAexGA maintains the diversity in a suitable manner: it enables exploitation while still retaining some level of diversity for potential exploration.

5 Conclusions

In this paper, we have introduced a new, GA-based approach to automated generation of effective black-box test cases. From the case study, we can conclude that the Fuzzy-Based Age Extension of Genetic Algorithm (FAexGA) is much more efficient for this problem than the two other evaluated algorithms (SimpleGA and GAVaPS). First, FAexGA has a much higher probability to find an error in the tested software. Second, the error would be found much faster, which should result in saving a lot of resources for the testing team. In addition, the number of distinct solutions produced by FAexGA is significantly higher, which may be useful for investigation and identification of the error itself by the software programmers. Future research includes application of the proposed methodology to testing of real programs, comparison of generated test sets to the tests designed by professional testers, and development of more sophisticated, possibly continuous evaluation functions for the evolved test cases.

Acknowledgments. This work was partially supported by the National Institute for Systems Test and Productivity at University of South Florida under the USA Space and Naval Warfare Systems Command Grant No. N00039-01-1-2248 and by the Fulbright Foundation that has awarded Prof. Kandel the Fulbright Senior Specialists Grant at Ben-Gurion University of the Negev in November-December 2005.

References

[1] Arabas, J., Michalewicz, Z., Mulawka, J.: GAVaPS – a Genetic Algorithm with varying population size, In Proc. of the first IEEE Conference on Evolutionary Computation, (1994) 73-78.

[2] Deb, K., Agrawal, S.: Understanding interactions among genetic algorithm parameters, In: W. Banzhaf, C. Reeves (eds.), Foundations of Genetic Algorithm 5, Morgan Kaufmann, San Francisco, CA, (1998) 265-286.

[3] DeMillo, R.A. & Offlut, A.J.: Constraint-Based Automatic Test Data Generation, IEEE Transactions on Software Engineering 17, 9 (1991) 900-910.

[4] Eibon, A. E., Hinterding, R., Michalewicz, Z.: Parameter Control in Evolutionary Algorithm, IEEE Transactions on Evolutionary Computation, (1999) 124-141.

[5] Elbaum, S., Malishevsky, A. G., Rothermel, G.: Prioritizing Test Cases for Regression Testing, in Proc. of ISSTA '00 (2000). 102-112.

[6] Goldberg, D.E.: Genetic Algorithms in Search, Optimization and Machine Learning, Addison-Wesley, Reading, MA, 1989.

[7] Herrera F., and Lozano, M.: Adaptation of genetic algorithm parameters based on fuzzy logic controllers. In Herrera F. and Verdegay J. (eds) Genetic Algorithms and Soft Computing, Physica Verlag, (1996) 95-125.

[8] Herrera F., and Magdalena, L.: Genetic Fuzzy Systems: A Tutorial. Tatra Mt. Math. Publ. (Slovakia), 13, (1997) 93-121

[9] Holland, J. H.: Genetic Algorithms, Scientific American, 267(1) (1992) 44-150.

[10] Jorgensen, P. C.: Software Testing: A Craftsman's Approach. Second Edition, CRC Press, 2002.

[11] Klir G. J., and Yuan, B.: Fuzzy Sets and Fuzzy Logic: Theory and Applications, Prentice-Hall Inc., 1995.

[12] Last M., and Eyal, S.: A Fuzzy-Based Lifetime Extension of Genetic Algorithms, Fuzzy Sets and Systems, Vol. 149, Issue 1, January 2005, 131-147.

[13] Michalewicz, Z.: Genetic Algorithms + Data Structures = Evolution Programs, Verlag, Heidelberg, Berlin, Third Revised and Extended Edition, 1999.

[14] Mitchell, M.: An Introduction to Genetic Algorithms, MIT Press, 1996.

[15] National Institute of Standards & Technology. "The Economic Impacts of Inadequate Infrastructure for Software Testing". Planning Report 02-3 (May 2002).

[16] Patton, R.: Software Testing, SAMS, 2000.

[17] Pfleeger, S. L.: Software Engineering: Theory and Practice. 2nd Edition, Prentice-Hall, 2001.

[18] Reeves, C. R.: Using Genetic Algorithms with Small Populations, Proc. Of the Fifth Int. Conf. On Genetic Algorithms and Their Applications, Morgan Kaufmann, (1993) 92-99.

[19] Schroeder P. J., and Korel, B.: Black-Box Test Reduction Using Input-Output Analysis. In Proc. of ISSTA '00 (2000). 173-177.

Optimal Algorithmic Debugging and Reduced Coverage Using Search in Structured Domains

Yosi Ben-Asher[1], Igor Breger[1], Eitan Farchi[2], and Ilia Gordon[1]

[1] Comp. Sci. Dep. Haifa University, Haifa, Israel
[2] I.B.M. Research Center, Haifa, Israel

Abstract. Traditional code based coverage criteria for industrial programs are rarely met in practice due to the large size of the coverage list. In addition, debugging industrial programs is hard due to the large search space. A new tool, REDBUG, is introduced. REDBUG is based on an optimal search in structured domain technology. REDBUG supports a reduced coverage criterion rendering the coverage of industrial programs practical. In addition, by using an optimal search algorithm, REDBUG reduces the number of steps required to locate a bug. REDBUG also combines testing and debugging into one process.

1 Introduction

One of the problems of adequately testing industrial programs by meeting coverage criteria is that the number of coverage tasks is too large. For example, covering all the define-use relations of a given industrial program might prove impractical, and indeed such coverage criteria are rarely met in practice.

Given a program based coverage criterion [11] defined on some directed graph determined by the program under test P, we propose a corresponding, intuitively appealing, reduced coverage criterion. The directed graph could be the program's static call graph, the program's control flow graph, the program's define-use graph, etc. The coverage criterion can be any one of the standard criteria defined on such graphs such as: statement coverage, branch coverage, multicondition coverage, or define-use coverage ([4], [5], [10], [11]). Reduced coverage requires meeting a much smaller coverage task list and is thus more practical than the standard coverage criteria. Further research is required to determine the confidence level achieved by this new set of reduced coverage criteria.

Fixing some code based coverage C, a process for obtaining a C reduced coverage criterion and a process, called algorithmic debugging, for assisting the programmer in debugging the implementation ([6], [9], [8], [7], [3]) are described. Then a method for combining both processes in one tool, called Reduced Coverage and Algorithmic Debugging, REDBUG, is described. Both processes leverage an optimal search algorithm defined on the relevant program directed graph ([2], [1]).

In large industrial programs debugging is a tedious and time consuming task due to the large size of the search space. We address this problem by applying

S. Ur, E. Bin, and Y. Wolfsthal (Eds.): Haifa Verification Conf. 2005, LNCS 3875, pp. 149–157, 2006.
© Springer-Verlag Berlin Heidelberg 2006

an optimal search algorithm to a run-time directed graph representing the program P. By doing this, we improve REDBUG algorithmic debugging compared to traditional algorithmic debugging and reduce the number of queries required to locate the fault.

This paper is organized as follows. First, search in structured domains is introduced. Next, reduced coverage and optimal algorithmic debugging is defined. Finally, a detailed example is given.

2   Search in Structured Domains

For the purpose of defining the concept of reduced coverage criteria and the optimal algorithmic debugging process, we first explain the general notion of an optimal search algorithm in structured domains. Searching in structured domains is a staged process whose goal is to locate a 'buggy' element. The current stage of the search process is modeled by a set α. At the current stage, the 'buggy' element can be any element in α. The possible queries at this stage α are modeled by a set of sets $\{\beta_1, \ldots, \beta_k\}$, such that $\beta_i \subset \alpha$. Each query β_i either directs the search to β_i in the case of a positive answer ($'yes'$) to the query, or directs the search to $\alpha \backslash \beta_i$ in the case of a negative answer ($'no'$). This process continues until $|\alpha| = 1$ and a buggy element is located. A search algorithm, denoted Q_D, is a decision tree whose nodes indicate which query should be used at each stage of the search. An optimal search algorithm minimizes the number of queries required to locate any buggy element.

Restricted types of structured search domains, such as trees or directed graphs, have been studied in [2]. In [2], it was shown that an optimal search algorithm for trees, Q_D, can be computed in $O(|D|^4)$ where D are the tree's nodes. In the case of a tree, the current stage of the search process is modeled by a sub-tree. The set of queries possible at each stage of the search is modeled by all the sub-trees of this tree, and an available sub-tree is queried at each stage of the search. If the queried sub-tree does not contain the buggy node, the search continues with the complement tree. An example of an optimal search algorithm for a tree is given in figure 1. The arrows point to the

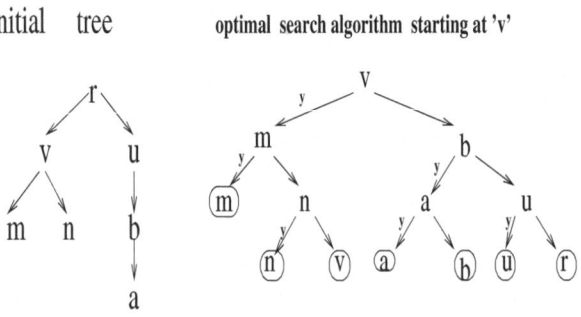

Fig. 1. Searching in a tree

next query. This search algorithm takes three queries in the worst case. Any other search which does not start at node $'v'$ takes at least four queries in the worst case[1].

3   Reduced Coverage and Optimal Algorithmic Debugging

Coverage and algorithmic debugging are next defined.

Coverage - Coverage of a program P, requires that the user will find a sequence of inputs $I_1, \ldots, I_k$, until the execution of P on those inputs, $P(I_1), \ldots, P(I_k)$, satisfies a condition. For example, the condition might require that all statements, branches or define-use relations of the program have been executed. The goal of the coverage process is
  - to find an input I_j that exposes a fault in P or
  - to give evidence that increases the confidence that P has no fault if the coverage condition is met and no failing run, $P(I_j)$, occurred.

Algorithmic debugging - Algorithmic debugging is a machine guided search taken by the user to locate a fault in the run $P(I_j)$ once the run $P(I_j)$ fails. The algorithm tells the user where to place the first breakpoint and query variable's value. Then, based on some user feedback, the algorithm determines the program location of the next breakpoint to be placed by the user, and so forth. This search can be fully automated when a database of appropriate pre-post conditions is available.

We are now ready to define reduced coverage. We are given a program P, a directed graph, G_P, defined by P and a code based coverage criterion C. We are further given that each component, or node, u_i of the graph G_P corresponds to a subset of P's statements. We use the notion of an optimal search of G_P to define a reduced coverage criterion. We say that P is reduced covered by a set of inputs I if

  - $u_1, \ldots, u_k$ correspond to an optimal sequence of queries applied by the optimal search algorithm when G_P is treated as a structured search domain and the "buggy" element is not found.
  - the coverage criterion C is obtained on $u_1, \ldots, u_k$ by I.

As the number of nodes required to optimally query a directed graph is small compared to the number of nodes in the graph, the number of coverage tasks to cover, such as code branches to cover, is greatly reduced rendering the task of obtaining a reduced coverage criterion practical.

---

[1] Intutivaly, $'v'$ separates the tree in the "middle". Consider choosing $'u'$ instead of $'v'$ as $'u'$ also, intuitively, separates the tree in the "middle". An adversary would chose to answer $'no'$. Next, you best chose $'v'$. This time an adversary would answer $'yes'$ resulting in two more queries, e.g., $'m'$ answered by $'no'$ and then $'n'$ answered by $'no'$. Overall, four queries were used which is worst than the number of queries that result in the worst case when choosing $'v'$.

If the reduced coverage is met, then it is as if an adversary chose the worst components for us to cover (in terms of the way these components are related to each other in G_P). Next, each of the chosen components are covered according to the coverage criterion C. As a result our confidence level that the program P does not have a fault increases.

Applying our method, the optimal search algorithm is given a directed graph G_P determined by the program and output a list of components to be covered $u_1, \ldots, u_k$. The programmer attempts to cover each component, u_i, according to the coverage criterion C. If a failure occurs the algorithm debugging stage is invoked guiding the programmer in the debugging of the failure. The algorithmic debugging phase applies the same optimal search algorithm to a run-time directed graph determined by the program P, such as the dynamic program call graph. In this way the user search for the bug is guided. As an optimal search is used to guide the algorithmic debugging stage, the number of breakpoints to set an inspect by the user is greatly reduced rendering the debugging task of large industrial components simpler.

In REDBUG, reduced coverage and the new algorithmic debugging process were implemented for the static and dynamic call graph of the program. The rest of the paper concentrates on the details of the implementation.

4   Detailed Example

We demonstrate how REDBUG is used to obtain reduced coverage and debug program failures. REDBUG is currently designed to work with call coverage and contains the following components:

1. An instrumentation module that can generate the static call graph C_P of a given program P. The resulting static graph C_P is converted to a tree by selecting a spanning tree of C_P. For a given input I of P, this module can also produce the dynamic call graph $C_{P(I)}$. Note that for call coverage, $C_{P(I)}$ is in fact a tree, as each call to a function has a single caller. Thus, we refer to C_P as the static tree of P and to $C_{P(I)}$ as the dynamic tree of $P(I)$.
2. The search module, which computes the optimal search algorithm A_{C_P} for the static tree of P and $A_{C_{P(I)}}$ for the dynamic tree of P. This module is interactive, and based on the answer for the last query (fed to it by the user) it prompt the next node to query in C_P or $C_{P(I)}$.
3. Finally, the debugger is used to determine if a given call to a function $f(...)$ is buggy or not. We use the debugger breakpoint mechanism to locate a specific call to a function in $C_{P(I)}$. Once a suitable breakpoint is reached, we use the debugger to check the values of $f(...)$'s variables and see if their value is as expected. In the following example, we inserted a check for the validity of a pre-post condition in every function that is queried. We prompt $'yes'$ if the pre-post condition of a given call is $'false'$, meaning that failure is in that call and $'no'$ otherwise.

We have chosen a simple program[2] that computes the value of roman numbers. For the input "MIX" the program should compute 1009, for "MXI" the program should compute 1011, etc. Many combinations are not allowed, e.g., $'MXM','LL','IIIVII'$ for which an error message should be printed.

The program implements the grammar of roman numbers using the following routines:

- thousands() - converts each 'M' to +1000
- fivehundreds() - converts 'CM' to +900 and each 'D' to +500
- hundreds() - converts 'CD' to +400 and each 'C' to +100
- fifties() - converts 'XC' to +90 and each 'L' to +50
- tens() - converts 'XL' to +40 and each 'X' to +10
- fives() - converts 'V' to +5
- ones() - converts 'IX' to +9, 'IV' to +4 and each 'I' to +1
- match(), next_token() and error() - process the input and report errors

A simple pre-post condition function $prePost(int\ before, int\ after, int\ max\_inc)$ is used to check if a given call to a function contains a bug. This is done to leverage the coverage process. When during the coverage process a test fails we immediately know which function is failing and the debugging phase is simplified. Obviously, the proposed notion of query depends on the ability of the programmer to insert such tests at suitable places in the code. Due to the way functions to be covered are chosen in the reduced coverage process , the list of functions to cover is usually small. As a result, the task of inserting pre-post conditions for the list of functions to cover becomes easier.

First instrumentation is applied to generate the static call tree of the program. The search module is then applied on the resulting tree. Figure 2 is a screen shot of the the search module showing: the static tree, and the first query in an optimal search of the call tree. Thus, the first function to cover is $thousands()$. We can see that we need to cover only four functions out of the total of eleven functions.

Next we select the inputs to cover $thousands()$, these inputs contain three roman numbers: XXVII for which $thousands()$ outputs the correct value of 27, $MMXLVI$ for which $thousands()$ outputs the wrong value of 46 (instead of 2046) and $MMXXXIV$ for which $thousands()$ outputs the correct value 2034.

In order to locate and correct the bug, we move to the algorithmic debugging phase, generate the dynamic tree and apply the search tool on the dynamic tree (starting from the first call to $thousands()$). The dynamic tree contains 35 nodes and as is depicted in figure 3 can be searched in at most six queries. Figure 3 describes the situation after three queries on the dynamic tree. At this stage, we pressed 'yes' to the first query on $thousands()$, 'no' to the second query on $thousands()$ (marked at the bottom of the window in figure 3), 'yes' to $fivehundreds()$, and 'yes' to $fifties()$. We are now advised to query $fives()$, call number 4 and see if it is erroneous or not (see figure 3).

---

[2] The program was written by Terry R. McConnell.

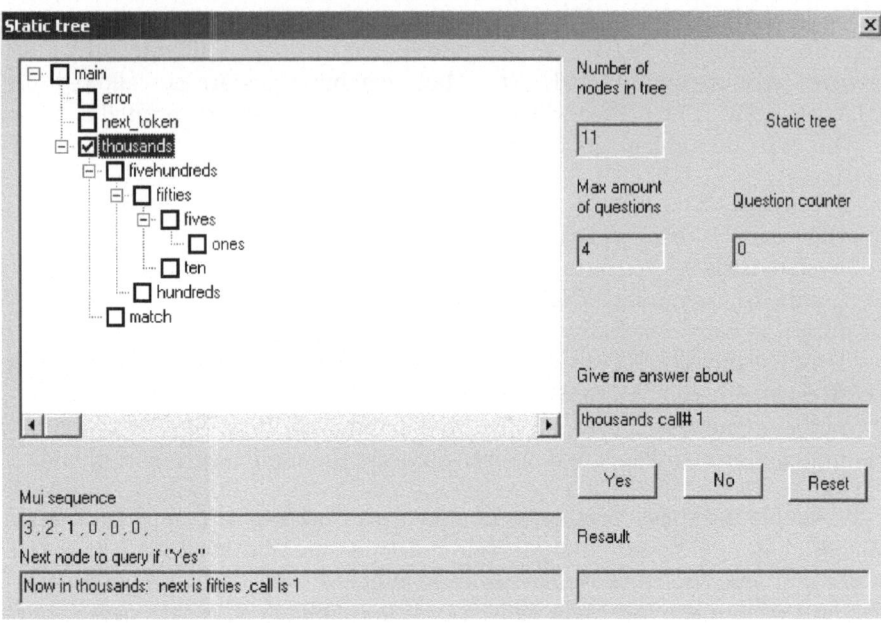

Fig. 2. Search tree for *Roman.c*

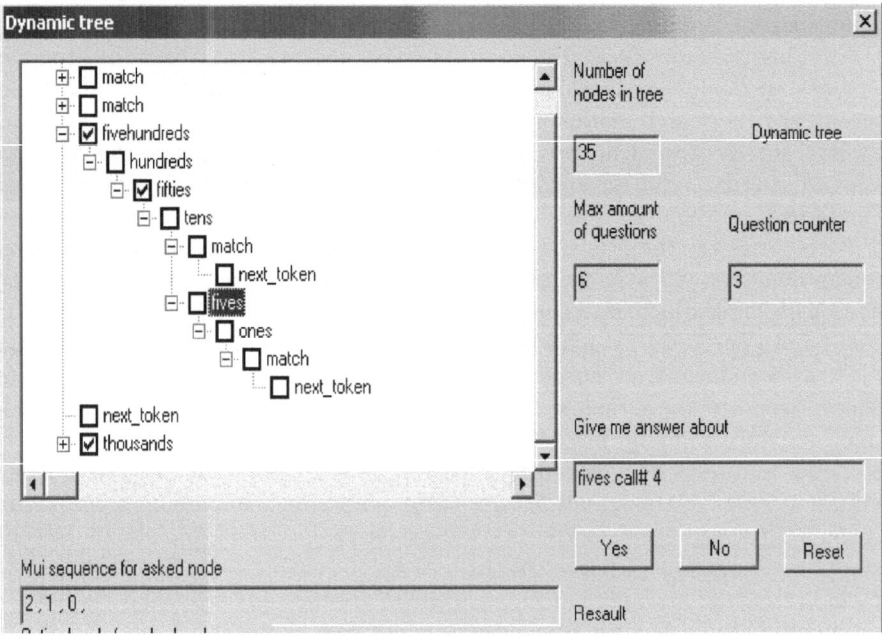

Fig. 3. Next query is *fives*() call no. 4

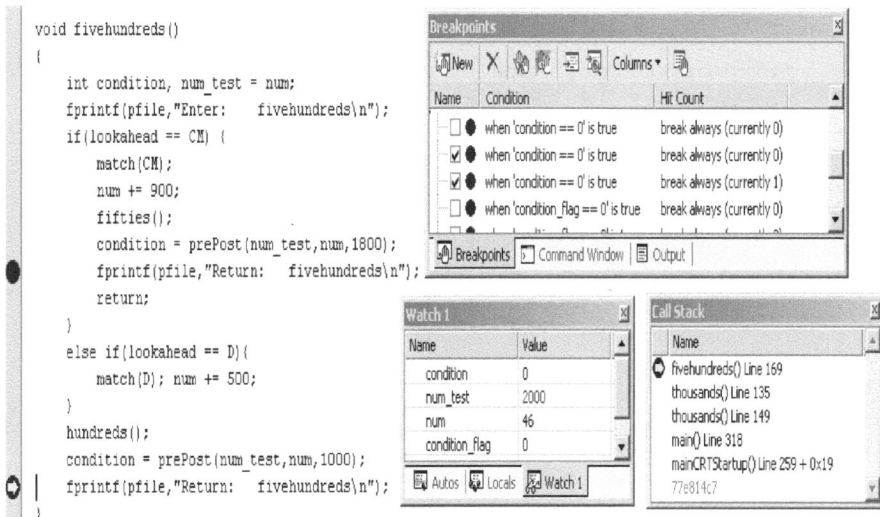

Fig. 4. Using the debugger to evaluate a query

We evaluate each query to a specific call using the debugger. First we insert a call to $prePost()$ before the function returns and insert a breakpoint after the calls to $prePost()$. By examining the result of the pre-post conditions and the value of other variables we are able to decide if the current call executed correctly or not. For example, in figure 4 we can check the return value of $prePost()$ and other variables using the $Watch$ window. In figure 4 we have two active breakpoints, and the debugger stops with a false $condition$ from $prePost(num\_test, num, 1000)$, so that this call is buggy. The debugger is also used to reach a specific call , e.g., $fives()$, call number four, of the dynamic tree. This is done using a global counter in the code which is incremented every time $fives()$ is called. The breakpoint is set to stop only if the value of this counter is the desired one (four in the case of $fives()$, call number four).

We use the debugger to evaluate call number four, of $fives()$ and find out that the pre-post condition is true. Consequently we prompt $'no'$ to the search tool. As can be seen from the tree in figure 3, the next query will be on $match()$ for which we also prompt $'no'$ (after checking the program state using the debugger). Thus, the search process ends by locating the bug in $tens()$. We use the debugger, one last time to examine the pre-post condition and try to locate the bug. Figure 5 depicts this stage, the call to $prePost(num\_test, num, 80)$ return false, as initially the value accumulated so far is $num\_test = 2000$, yet the value returned by $tens()$ which should have been between num_test and $num\_test + 80$ has been reduced to $num = 40$. As we know that the call to $fives()$ is correct, we find the bug in statement $num = 40$ that should have been $num + = 40$.

After fixing that bug, we continue in the reduced coverage process and as no further bugs are found the process ends. This process obtained reduced call coverage using three inputs and used six tests to locate the failure.

```
void tens()
  {
      int condition, num_test = num, count;
      fprintf(pfile,"Enter:     tens\n");
      if(lookahead == XL) {
          match(XL);
          num = 40; /* ERROR!!! must be "num += 40;" */
          fives();
          condition = prePost(num_test,num,40);
          fprintf(pfile,"Return:    tens\n");
          return;
      }
      count = 0;
      while(lookahead == X){
          if(count++ >= 3)error("Too many Xs\n");
          num += 10;       match(X);
      }
      fives();
      condition = prePost(num_test,num,40);
      fprintf(pfile,"Return:    tens\n");
  }
```

Fig. 5. Using the debugger to locate the bug in *tens*()

5   Conclusion

A tool called REDBUG that combines software testing and algorithmic debugging is introduced. The tools uses the same underlying technology of optimal search in structured domains to support the testing and debugging phases.

Traditional code based coverage criteria have large coverage task list rendering them impractical. The notion of an optimal search in structure domain is used to define a new set of reduced coverage criteria corresponding to the traditional coverage criteria rendering the coverage of industrial programs practical. If a reduced coverage criteria is met, then it is as if an adversary chose the most complex components for the user to cover. More research is required to determine the confidence level these new reduced coverage criteria provide.

In large industrial components the problem of locating a failure requires many hours of setting breakpoints and observing of program behavior. By using an optimal search algorithm, REDBUG is able to reduce the number of breakpoints a programmer should set in-order to find the bug. REDBUG will be especially useful in setting were there are the call stack is big or when there are a lot of communication layers resulting in large search domains.

Further research will implement REDBUG on other program defined graphs. It would be interesting to see how much saving is obtained by the new algorithmic debugging method on a set of real life industrial examples. Finally, the issue of the confidence level that the new set of reduced coverage criteria introduce should be addressed.

References

1. Y. Ben-Asher and E. Farchi. Compact representations of search in complex domains. *International Game Theory Review*, 7(2):171–188, 1997.
2. Y. Ben-Asher, E. Farchi, and I. Newman. Optimal search in trees. In *8'th Annual ACM-SIAM Symposium on Discrete Algorithms (SODA97), New Orleans*, 1997.
3. Jong Deok Choi and Andreas Zeller. Isolating failure inducing thread schedules. *International Symposium on Software Testing and Analysis*, 2002.
4. L. A. Clarke. comparison of data-flow path selection criteria. *IEEE Transaction on Software Engineering*, 1985.
5. L. A. Clarke. An investigation of data flow path selection criteria. *Work Shop On Software Testing, Banff, Canada*, 1986.
6. Peter Fritzson et. al. Generalized algorithmic debugging and testing. *ACM Letters on Programming Languages and testing*, 1:303–322, 1992.
7. Fritzson Peter Kamkar Mariam and Shahmehri Nahid. Interprocedural dynamic slicing applied to interprocedural data flow testing. In *Proceedings of the Conference on Software Maintenance*, 1993.
8. Nahid Shahmehri Mariam Kamkar and Peter Fritzson. Interprocedural dynamic slicing and its application to generalized algorithm debugging. In *Proceedings of the International Conference on Programming Language*, 1992.
9. Mikhail Auguston Peter Fritzson and Nahid Shahmehri. Using assertions in declarative and operational models for automated debugging. *Journal of Systems and Software*, 25(3):223–232, June 1994.
10. Elaine J. Weyuker. Axiomatizing software test data adequacy. *IEEE Transaction on Software Engineering*, SE-12(12), December 1986.
11. Elaine J. Weyuker. The evaluation of program-based software test data adequacy criteria. *Communications of the ACM*, 31(6), June 1988.

Benchmarking and Testing OSD for Correctness and Compliance

Dalit Naor, Petra Reshef, Ohad Rodeh, Allon Shafrir,
Adam Wolman, and Eitan Yaffe

IBM Haifa Research Laboratory, University Campus, Carmel Mountains, Haifa, 31905
{dalit, petra, orodeh, shafrir, wolman, eitany}@il.ibm.com

Abstract. Developers often describe testing as being tedious and boring. This work challenges this notion; we describe tools and methodologies crafted to test object-based storage devices (OSDs) for correctness and compliance with the T10 OSD standard. A special consideration is given to test the security model of an OSD implementation. Additionally, some work was carried out on building OSD benchmarks. This work can be a basis for a general-purpose benchmark suite for OSDs in the future, as more OSD implementations emerge. Originally designed to test performance, it was surprisingly useful for discovering unexpected peculiar behaviors and special type of bugs that are otherwise not considered bugs.

The tool described here has been used to verify object-disks built by Seagate and IBM research.

1   Introduction

Developers often find testing tedious and boring. This work attempts to challenge this image. We describe a tool set created in order to test object-storage-devices (OSDs) for correctness and compliance with the T10 OSD standard [16,13], which proved to be a difficult and challenging part of our overall development effort. Aside from correctness, performance is also an important quality of an implementation. Initial work was carried out on benchmarking OSDs.

Generally speaking, the T10 standard specifies an object-disk that exports a two-level object-system through a SCSI based protocol. The OSD contains a set of partitions each containing a set of objects. Objects and partitions support a set of attributes. A credential based security architecture with symmetric keys provides protection.

Our group, at IBM research, built an OSD together with a test framework; a description of our object-store related activities can be found in [6]. For testing there were several choices ranging from white-box to black-box testing. Black-box testing was preferred and chosen to be the primary methodology so as to make the testing infrastructure independent of the OSD implementation. However, in addition to the black box testing, we developed some limited capabilities based on gray-box techniques to test and debug our own OSD. Building upon knowledge of the internals of the target implementation, this could improve coverage considerably.

S. Ur, E. Bin, and Y. Wolfsthal (Eds.): Haifa Verification Conf. 2005, LNCS 3875, pp. 158–176, 2006.

We aimed to built a small and light tool set that would achieve a good a coverage as we could get. A `tester` program was written to accept scripts containing OSD commands. The `tester` sends the commands to the target OSD and then checks the replies. This creates a certain workload on the target.

This kind of testing falls under the sampling category. Out of the possible scenarios of commands reaching the target only a sample is tested. Passing the tests provides a limited guarantee of compliance and correctness. The crux of the matter is to identify and test the subset of the scenarios that provide the best coverage. To address the sampling issue and increase coverage, a `generator` program was written to generate scripts with special characteristics. In this paper we did not pursue functional or structural approaches to the testing problem.

The choice of black-box testing proved a fortunate choice later on. During 2005 we were part of a larger IBM research group that built an experimental object-based file-system. This system was demonstrated in Storage Networking World in the spring of 2005 [5]. The file-system was specified to work with any compliant OSD. To demonstrate this we worked with Seagate^{TM} who provided their own OSDs. Our group was commissioned to test correctness and compliance of our own target as well as Seagate's. This was a necessary prerequisite before wider testing within the file-system could be carried out.

The main contribution of this paper is to report on our practices and experience in testing object stores that are compliant with the new OSD T10 standard. Object storage is an emerging storage technology in its infancy, expected to gain momentum in the near future. So far, very few OSD implementations have been reported, and even fewer are compliant with the OSD standard. Hence, we believe that our work regarding the validation of such implementation and conformance with the standard will be relevant and valuable to the community at large. Our tools include a language which allows to specify OSD commands and their expected behavior (based on the standard), as well as a program (called `tester`) to verify whether a given implementation complies with the expected behavior. These are infrastructural tools that can be used to develop "test-vectors" for the OSD T10 standard. We also argue and demonstrate in this paper that although close in spirit to a file system, T10 compliant OSDs have unique characteristics that distinct their testing and validation from one of a traditional file system. One of the most notable differences is the OSD security model and its validation. We also note the interesting connection between performance testing and bug hunting.

In this paper we describe the tools we developed. We emphasize how these tools were tailored to address the specific difficulties in testing an OSD, both for compliance and correctness, by providing specific examples. The rest of the paper is organized as follows. Section 2 describes the T10 specification and walks through some of the difficulties it poses for testing. Section 3 describes the testing infrastructure. Section 4 describes the techniques used to locate bugs. An important aspect of the OSD T10 protocol that requires special testing tools and techniques is the OSD security model. Section 5 describes the mechanism developed to test the security aspects of OSDs. Section 6 talks about the benchmarks devised to measure performance. Section 7 describes related work and Section 8 summarizes.

2   The OSD Specification

2.1   T10 Overview

An object-disk contains a set of partitions each containing a set of objects. Objects and partitions are identified by unsigned 64-bit integers. Objects can be created, deleted, written into, read from, and truncated. An additional operation that was needed for the experimental object-based file-system and will be standardized in the near future is `clear`. Clearing means erasing an area in an object from a start offset to an end offset. Partitions can be created and deleted. The list of partitions can be read off the OSD with a list operation. A list operation is much like a `readdir` in a file-system; a cursor traverses the set of partition-ids in the OSD. Similarly, the set of objects in a partition can be read by performing a list on the partition.

Partitions and objects have attributes that can be read and written. A single OSD command can carry, aside from other things, a list of attributes to read and a list of attributes to write. There are compulsory attributes and user-defined attributes. Compulsory attributes are, for example, object size and length. User-defined attributes are defined by the user outside the standard. There are no size limitations on such attributes. For brevity considerations, user-defined attributes are not addressed here.

A special root object maintains attributes that pertain to the entire OSD. For example, the used-capacity attribute of the root counts how much space on disk is currently allocated.

An important aspect of a T10 compliant OSD is its security enforcement capabilities. The T10 standard defines a capability-based security protocol based on shared symmetric-keys. The protocol allows a compliant OSD to enforce access-control on a per-object basis according to a specified security policy. The enforcement uses cryptographic means. The protocol allows capability revocations and key refresh. The standard protocol defines three different methods to perform the validation, depending on the underlying infrastructure for securing the network. In this work we only consider one of the methods, the CAPKEY method.

2.2   Difficulties

The specification poses several serious difficulties to a testing tool. Three examples are highlighted in this section: testing atomicity and isolation guarantees, testing parallelism, and verifying quotas.

Atomicity and isolation guarantees. The OSD standard specifies only weak atomicity and isolation guarantees between commands. For example, a write command does not necessarily have to be completed atomically and some of its data may be lost (but not the meta-data). Also, if two write commands into overlapping ranges are issued simultaneously, the resulting object data may be interleaved. The weak guarantees are traded for higher performance - since stricter guarantees require the implementation to perform more locking and serialization, thus hurting the performance of typical commands. However, this also creates non-determinism. For example, assume two writes W_1, W_2 to the same extent in an object are sent. The result is specified to be some mix of the data from W_1 and W_2. The mix might be limited by the actual atomicity provided by the OSD specific implementation, thus it is implementation dependent.

Parallelism. Commands sent in parallel to an OSD may be executed in any order. For example, if a command $C_1 =$ create-object(o) is sent concurrent with $C_2 =$ delete-object(o) then there are two possible scenarios.

1. The OSD performs C_1 and then C_2. The object is created and then deleted. Both commands return with success.
2. The OSD performs C_2 and then C_1. The object is deleted and then created. The delete fails because the object did not exist initially. C_1 returns with success; C_2 returns with an `object-does-not-exist` error. Object o remains allocated on the OSD.

In general, the non-determinism that results from executing multiple commands on the OSD concurrently poses a big challenge on verification.

Quotas pose another kind of problem. They are specified as being fuzzy. For example, consider an object with a certain quota limit. The target *must* signal an out-of-quota condition upon the next write into the object if the object-data exceeds the quota limit. However, it *may* signal, at its own discretion, an out-of-quota even if the written data is less than the quota but 'close' to it by a certain confidence-margin. The upshot is that it is not possible to write a simple test to check for quota enforcement.

Another issue is that object, partition, and LUN[1] used-capacity are not completely specified. For this discussion we shall focus solely on objects. An object's used capacity is defined to reflect the amount of space the object takes up on disk, including meta-data. However, an implementation has freedom in its usage of space. For example, in one implementation one byte of live data may consume 512 bytes of space, whereas in another implementation it may consume 8192 bytes (one 4K page for its meta-data an one 4K page for data). Since various sizes are legal a single one-size-fits all test is impossible to devise.

3   Infrastructure

3.1   The Components

Our OSD code, including the testing infrastructure, is structured as follows (see Figure 1):

---

[1] LUN denotes a *Logical Unit* of storage.

- tester: a relatively simple program that reads scripts of OSD commands, sends them to the target, and verifies their return values and codes.
- iSCSI OSD initiator: an addition to the Linux kernel of T10-specific iSCSI extensions. Specifically: bidirectional commands, extended CDBs[2], and the T10 command formats[10].
- iSCSI target: a software iSCSI target.
- Front-End (FE): module on the target that decodes T10 commands.
- Reference implementation (simulator): a simple as possible OSD[11].
- Real implementation (OC): an optimized OSD.

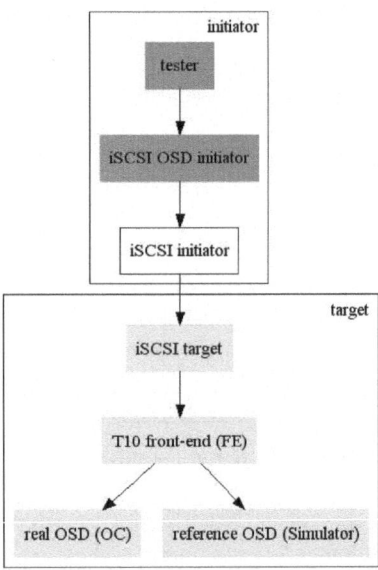

Fig. 1. OSD code structure: the set of components

The engineering principle followed while building the components was to leverage a small, well tested, module to test and verify a larger module. The simpler the module is the more confidence we had in it. We aimed to keep testing-modules simple and with low line counts.

3.2 Tester and Script Language

The design point was to build a simple tester program. There was interest in minimizing the amount of state kept on the tester side. Minimizing state would simplify the tester and improve its reliability[3].

---

[2] The CDB is the command descriptor block of a SCSI command.
[3] The total line-count for the tester is about 8000 lines of C code.

A script language was tailored specifically for the OSD T10 standard. The commands fall into two categories. The first category contains simple T10 commands such as `create`, `delete`, `read`, `write`, and `list`. Each command can be accompanied with the expected return code and values. The second category contains composite commands such as the device snapshot command. This command performs a snapshot of the entire contents of the target or of a specific object. This ability is used to compare the contents of the target with other targets such as the reference implementation target or a different target implementation.

The commands are grouped into blocks which can be defined recursively. The types of blocks are:

- Sequential block: The `tester` waits for each command to complete before submitting the next command in the block.
- Parallel block: Commands in the block are submitted to the target concurrently. It is the responsibility of the script writer to ensure valid scripts (e.g. creating a partition must terminate successfully before creating an object within it if the object creation is expected to complete successfully). The order in which the commands are actually submitted to the target is not defined.[4]

A simple example is:

```
create oid=o1;
write oid=o1 ofs=4K len=4K;
delete oid=o1;
```

This script create and object, writes 4K of data into it, and then deletes it. A more complex script is:

```
par {
  create oid=o1;
  create oid=o2;
}

par {
  write oid=o1 ofs=20K len=4K;
  write oid=o2 ofs=8K len=4K;
}
```

This script creates two objects concurrently and then writes into them concurrently.

---

[4] For example, a set of commands that are submitted to the iSCSI layer simultaneously may be sent to the target in any order, since the iSCSI layer might change the order in which it sends the commands.

One can use the `seq` operator to create sequential blocks:

```
create oid=o2;

par {
  seq { create oid=o1; write oid=o1;
        delete oid=o1 }
  clear oid=o2 ofs=30K len=512;
}
```

The `tester` execution phases are:

- Parse script, building a DAG (directed acyclic graph) representing command dependencies.
- submit commands according to the specified order, handling target responses for each command and verifying the result.

3.3   Reference Implementation

A reference-implementation, or *simulator*, was built. The simulator was the simplest OSD implementation we could write. It uses the file system to store data; an object is implemented as a file and a partition is implemented as a directory. The simulator core is implemented with 10000 lines of C code. Incoming commands are executed sequentially; no concurrency is supported. Having a reference implementation turned out to be extremely useful. Due to simplicity and non-concurrent nature it had very few bugs but still preserved the OSD semantics. Thus, it could be used as a reference point for debugging our 'real' implementation, and for testing other implementations. It was recently released for public use on IBM's AlphaWorks[11].

3.4   Script Generator

A *script-generator* automatically creates scripts that are fed into the `tester`. The generator accepts parameters such as error percentage, create-object percentage, delete-object percentage, etc. It attempts to create problematic scenarios such as multiple commands occurring concurrently on the same object. The expected return code for each command is computed and added to the script. The tester, upon execution, verifies the correctness of the return codes received from the target. Generated scripts are intended to be deterministic; only such scripts have verifiable results. However, non-deterministic scripts are also useful. The generator can create such scripts with the intent of stability testing; the target should not crash.

Although its primary use was to increase testing coverage via automatic generation of interesting tests, the *script-generator* turned out to also be useful as a debugging tool in the following manner. Hunting for a bug that was initially identified with a very long and automatically generated script (with, say, 10000 lines) required the generation of many shorter scripts; instead of manually generating the short scripts, one could use the generator's input arguments wisely to produce scripts that narrow the search space.

3.5   Gray-Box Testing

For gray box testing *crash command, configurations*, and *harness-mode*, were added. These are specific to our own implementation; they are not generic for all OSDs. The target OSD was to be modified as little as possible to allow gray-box testing. This would ensure that the tests were measuring something close to the real target behavior.

The crash command is a special command outside of the standard command set. It causes the OSD to fail and recover; it is used for testing recovery. We believe it should be added to the standard in order to enable automated crash-recovery testing.

A configuration file contains settings for internal configuration variables, such as, the number of pages in cache, the number of threads running concurrently, and the size of the s-node cache. Running the same set of tests with different configuration files increases the coverage. It also provides a faster way to expose corner-case bugs by using small configurations.

Harness-mode is a deterministic method of running the OSD target. The internal thread package and IO scheduling is switched to a special deterministic mode. This mode attempts to expose corner cases and race-conditions by slowing down or speeding up threads and IOs. The `tester` program is linked directly with the target thereby removing the networking subsystem. This creates a *harness* program. The harness can read a script file and execute it. The full battery of tests is run against the harness. If a bug is found it is can be reproduced deterministically.

A regression suite composed of many scripts is used in testing. The regression contains short hand-crafted scripts that test simple scenarios and large 5,000-10,000 line tests created with the script-generator.

4   Techniques

This section describes a number of techniques employed by the testing suite to test and verify non-trivial properties of an OSD implementation. The techniques were proved to be extremely useful for identifying bugs in the system and as debugging tools. The techniques that were developed to test the security aspects of an implementation are discussed separately in Section 5.

4.1   Verifying Object Data

Data written to objects requires verification. Is the data on-disk equivalent to the data written into it by the user? A two pronged approach was taken. A light-weight verification method of *self-certifying* the data was employed for all reads and writes, and a heavy-weight method of *snapshots* was used occasionally.

The light-weight method consisted of writing *self-certifying data* to disk and verifying it when reading the data back. For writes that are 256 bytes or more, the `tester` writes 256-byte aligned data into objects. At the beginning of a 256byte chunk a header containing the object-id and offset is written. When reading data from the OSD the `tester` checks these headers and verifies them. Since the data is self-certifying the `tester` does not need to remember which object areas have been written. A complication arises with *holes*. A hole is an area in an object that has not been written into. When a hole is read from disk the OSD returns an array of zeros. This creates a problem for the tester because it cannot distinguish between the case where the area is supposed to be a hole, and the case where the user wrote into the hole but the target "lost" the data.

Snapshots are the heavy-weight method. A snapshot of an object disk is an operation performed by the tester. The tester reads the whole object-system tree off the OSD and records it. In order to verify that a snapshot is correct the tester compares it against a snapshot taken from the simulator. Technically, the object-system tree is read by requesting the list of the partitions and then the list of the objects in each partition. All the data, including attributes, from the root, partitions, and objects, is read using read and get-attribute commands.

Using snapshots helps avoiding maintaining a lot of state at the tester. The tester does not need to keep track of the state of the target. It just needs to compare its state against another OSD. Theoretically, if we had n different implementations we could compare them all against each other. In practice, the regression suite runs against three different implementations: harness, simulator, and real-target. The snapshots are compared against each other.

We should note that the problem of verifying object data is very similar to verifying file data, and therefore the two abovementioned methods are similar to standard practices in testing of file system, expect for the treatment of holes.

4.2 Crash Recovery

Recovery is a difficult feature to verify because it exhibits an inherently non-deterministic behavior. For each command that was executing at the time of the failure, the recovered OSD state may show that it had not been started, it was partially completed, or totally finished.

To cope with that problem, we allow for checking the consistency of only a partial section of the system. For example, in the script three objects, `o1`, `o2`,`o3`, are created and written into. After the write to `o1` completes the OSD is

```
par {
  seq { create oid=o1;
        write oid=o1 ofs=4K len=512;
        crash }
  seq { create oid=o2;
        write oid=o2 ofs=20K len=90K
        }
  seq { create oid=o3;
        write oid=o3 ofs=8K len=8K }
}

snap_obj oid=o1;
```

instructed to crash and recover. Finally, the snapshot from object o1 is taken. It is later compared against the snapshot from the reference implementation. The state of object o2 and o3 is unknown; they may contain all the data written into them, some of it, or none. They may not exist at all.

5 Testing of Security Mechanisms

The T10 OSD standard specifies mechanisms for providing protection. In this paper we refer to these as the *Security Mechanisms* (corresponding to [16, Section 4.10] on *Policy management* and *Security*). This section discusses the techniques used to test the security mechanisms implemented by an OSD target, specifically the CAPKEY security method.

Most of the tested security mechanisms have to do with validation of commands. Testing the implementation for correct validation is extremely important: a minor inconsistency of an OSD target with the specification may violate all protection guarantees.

5.1 Validation of an OSD Command

The OSD standard uses a credential based security architecture. Each OSD command carries an additional set of *security fields* to be used by the security mechanisms. We refer to this set of fields as a *credential*; this is a simplification, for the sake of brevity, of the credential definition as specified in the OSD standard.

Every incoming command to the OSD target requires the following flow of operations to be executed by the target:

1. Identify the secret key used to authenticate the command. This stage requires access to previously saved keys.
2. Using the secret key, authenticate the contents of all command fields (including the security fields).
3. Test that the credential content is applicable to the object being accessed. This stage requires access to previously saved attributes of the object.
4. Test that all actions performed by this command are allowed by the credential.

The T10 OSD standard specifies the exact content for each security field in the command. A target permits the execution of a command if all security fields adhere to this specification.

We say that a credential is *good* if (1) it is valid and (2) its appropriate fields permit the command that 'carries' it. We say that a credential is *bad* if either it is invalid (*e.g.* has bad format or is expired) or if it does not permit the operations requested by the command that 'carries' it. *Valid Commands* are commands carrying good credentials, whereas commands carrying bad credentials are considered *invalid*.

5.2 Testing Approach

For a given command and a given target state, there may be many possible good credentials and many possible bad credentials. A perfect testing suite should test:

For every OSD command:
  For every possible target state:

1. Send the command with all possible good credentials.
2. Send the command with all possible bad credentials.

The general problem of increasing coverage, whether command coverage or target state coverage, is addressed in Section 3. There we describe how a clever combination of the `tester` together with the `script-generator` can yield increased coverage. Section 5.5 describes the integration of special *security state* parameters into the `tester` and `generator` for the purpose of testing the security mechanisms. Considering the regression suite described in Section 3 as the basis, we now focus on the problem of testing security mechanisms for a given command in a given target state.

The number of possibilities per command is too large to be fully covered. A random sampling approach is therefore used for this problem as well.

5.3 Generation of a Single Credential

For a given command with given state parameters, a single *good credential* is generated using a constraint-satisfaction approach[4,1] as follows:

1. Build a set of constraints which the credential fields should satisfy.
   - The constraints precisely define what values would make a good credential as specified by the standard.
   - A single constraint may assert that some field must have a specific value (*e.g.* the object-id in the credential must be same as the object-id field in the command). Alternatively, it may assert that a certain mask of bits should be set (*e.g.* for a WRITE command, the WRITE permission bit must be set).
   - A single field may have several constraints, aggregated either by an AND relation or an OR relation. For example, 'the access tag field should be identical to the access tag attribute' OR 'it may be zero' (in which case is isn't tested).
2. Fill all fields with values that satisfy the constraints.
   - For a field that has several options, one option is randomly chosen and satisfied.
   - Each field is first filled with a 'minimal' value (*e.g.* the minimum permission bits required for the command). Then it may be randomly modified within a range that is considered 'don't-care' by the constraint (*e.g.* adding permission bits beyond the required ones).

Generating a *bad credential* starts by generating a good credential as described above. We then randomly select a single field in the credential and randomly 'ruin' its content so that it no longer satisfies its constraints.

How is this generation technique integrated with the existing tester? The `tester` controls whether a good or a bad credential is generated for a given command, and it expects commands with bad credentials to fail (that is, if they are completed successfully by the target it it is considered an error). Commands with good credentials are expected to behave as if security mechanisms do not exist.

When generating a bad credential, our tester generates one invalid field at a time. This is justified since almost all causes for rejecting a command are based on a single field. The standard does however specify that some fields should be validated before others. Since we only generate one invalid field at a time, this specification is not tested in our scheme.

Randomness is implemented as pseudo-randomness. This allows the tester to control the seed being used for the pseudo-random generation, enabling reproducing any encountered bug.

We now describe how to generate multiple random credentials for each command. This is required in order to cover as many rules involved with validating a command.

5.4   Generation of Many Credentials

For a given command, the tester has several modes for generating credentials:

- A **deterministic mode**, where minimal credentials are generated.
- A **normal mode**, where each command is sent once with a random good credential. This mode allows testing many good credentials while activating the regression suite for other purposes.
- A **security-testing** mode described below.

The **security-testing** mode sends each command $2N$ times, N times carrying a random good credential and N times carrying a random bad credential. This allows testing N different combinations of a bad credential and N different combinations of a good credential. However, since OSD commands are not idempotent, this should be done carefully due to the following difficulties:

1. Each script command may depend on successful completion of previous commands. Hence, it is desirable to generate a command (whether with a good or a bad credential) only after all previous commands completed successfully.
2. On the other hand, some commands cannot be executed after they were already executed once (e.g. creating an object). Hence, a command should not be re-sent after it was already sent with a good credential.

A script is executed in the **security-testing** mode in two stages, while each command is sent multiple times. The underlying assumption is that every script ends by 'cleaning up' all modifications it made, thus restoring the target to its old state.

First Phase: Each command in the script is sent N times with N randomly generated *bad* credentials, expecting N rejections. The command is sent once again, this time with a *good* credential, expected to succeed.

Second Phase: The script is executed $N-1$ time in the normal mode (carrying a random good credential for each command), thus completing $N - 1$ more good credentials for each command.

This technique proved to be very practical, mainly when used with long automatically-generated scripts. Its main contribution was in testing multiple bad paths. For example, a command accessing a non-existing object using an invalid credential or a command reading past object length while also accessing attributes that are not permitted by the credential.

5.5   Generating Security-States at the Target

The OSD security model defines a non-trivial mechanism for *management of secret keys.* It involves interaction between the OSD target and a stateful Security manager. To test this mechanism, the testing infrastructure should enable the generation of scenarios such as:

- Bad synchronization of key values between the security manager and the target.
- A client uses an old credential that was calculated with a key that is no longer valid.

To generate such scenarios, we let the *tester* act as a security manager and maintain a *local-state* of key values shared with the target. In addition to the regular set_key command we introduced two special script commands: set_key_local, set_key_target used to simulate scenarios where the key is updated only at one side. These extensions are within the black-box testing paradigm and do not require any modification of the target. These two commands proved very handy: by replacing normal set_key commands with these special commands, scripts with good key management scenarios can be easily transformed into scripts that simulate bad key management scenarios.

Another security mechanism requiring a state-aware tester is *revocation.* The OSD security model offers a per-object revocation mechanism via a special object attribute called the *policy/access tag.* By modifying this attribute, a policy manager may revoke all existing credentials allowing access to this object. Our tester was extended to support this mechanism by keeping the state of the policy/access tags for selected objects; this allowed to verify an implementation of the revocation mechanism. By introducing a simplified keys-state and the revocation-attribute in the script-generator we enabled generation of many revocation scenarios as well as many key-management scenarios.

6   Benchmarks

Once an OSD is built a natural question to ask is what is its performance. Or rather, how well does it perform? Benchmarking is a complex issue. There are many benchmarks for file-systems and block disks; both close cousins to object-disks. However, we argue that these benchmarks alone are not adequate for measuring OSD performance, and that OSD benchmarking may need more specific tools.

Block-disk benchmarks contain only read and write commands while OSDs support a much richer command set. In the file-system world NFS-servers and NAS-boxes are the closest to object-disk. However, NFS benchmarks such as Spec-SFS [14] contain a lot of directory operations that OSDs do not support. Furthermore, the workload on a NAS-box is quite different than the expected workload for an object-disk. Here are a couple of comparison points:

1. In Spec-SFS the NFS lookup operation takes up 27% of the workload. An OSD does not support an operation similar to lookup.
2. Some architectures place the RAID function above OSDs. This means that OSDs will contain file-parts and see read-modify-write RAID transactions. This bears very little similarity to NFS style workloads.
3. An OSD supports a rich set of operations that are unique to OSD's and do not translate directly to file-system operations. A good example is an OSD Collection. OSDs supports an 'add to collection' operation. One possible use for this function is to add objects to an *in-danger* list that counts objects that may need to be recovered in the event of file-system failure. This operation affects the workload and it is not clear what weight it should be given in a benchmark.

The OSD workload is dependent on the file-system architecture it is used with. Therefore, if one sets out to build a Spec-SFS like benchmark for OSDs there are a lot of question marks around the choice of operation weights. As OSD-based file-systems are in their infancy, we expect the 'right' choice of operation weights to converge as the field matures and more OSDs emerge. As part of our OSD code, we built an initial framework for OSD benchmarks. We expect these benchmarks to be used as a tool to evaluate strengths/weaknesses of a specific OSD implementation, and to be used to design an OSD application on top of it. In the future, there will undoubtedly be a need to develop a 'common criteria' with which standard OSDs can be compared and evaluates, very much like file systems implementations.

Although the benchmarks were originally designed to measure performance, we discovered that they can be very useful in locating unexpected peculiar behaviors. Such behaviors turned out to be an indication of special types of bugs which may not be classified as such. Below we provide a few examples of benchmark results that helped identify a problem, a weakness or a bug in our system. Currently, this is the main use of the benchmark tool in our system. We believe that the linkage between performance testing and bug hunting is an interesting observation that may benefit the testing community at large.

6.1   OSD-Benchmarks Suite

We developed a skeleton for a benchmark suite which we believe can be extended and tuned in the future. Our suite is composed of two types: synthetic and spec-SFS like. All benchmarks are written as client executables, using the OSD initiator asynchronous API. Currently, they measure throughput and latency on the entire I/O path, but other statistics information can be gathered as needed.

- The `synthetic benchmarks` are built to test specific hand-made scenarios. They are useful for isolating a particular property of the system such as locking, caching or fragmentation and analyzing it. Currently the synthetic benchmarks consider the case of many small objects, or alternatively a single large object (these are common file-system scenarios, similar to approach taken in [12]). Basic measurements include throughput and latency of Read, allocating-Write and non-allocating (re-)Write commands as a function of the I/O size (ranging from 4K to 64K).

- The `Spec-sfs-like benchmarks` create, as pre-test stage, a large number of objects, and select a small part of it as working set. It then chooses a command from an underlying distribution, randomly picks an object from the working set, selects arguments for the command from a given distribution and initiates the command. Statistics are gathered on a per-command-type basis.

6.2   Benchmark Examples

Parameter tuning. The example in Figure 2 depicts throughput performance (in MBytes/sec) of our OSD implementation that was obtained from one of the synthetic benchmarks for all-cache-hit Reads, allocating-Writes and non allocating-Write commands as a function of the I/O size. Our goal was to reach the maximum raw TCP performance over a 1Gb/sec network. In general, the throughput grows as a function of the I/O size. However, Figure 2 shows irregular behavior for Writes (but not for Reads) when I/O size is 32K. This called for a closer analysis of a Write command, which requires multiple Requests-to-transfer (R2T) messages of various lengths. These R2T parameters need to be tuned to eliminate the observed irregularity.

Target behavior. The next example is taken from the Spec-sfs like benchmarks. It considers only Read and Write commands (all other weights are set to zero), with uniform size of 64K. Latency statistics for Reads are depicted in Figure 3. When Reads are all cache-hits, latency per command is distributed uniformally around 5-20 msec. However, in the example below, a bimodal behavior is observed with two very distinct means, indicating a mixture of cache-hit Reads as well as cache-misses.

Identifying bugs. The third example shows how we traced a bug in the Linux SCSI system using the benchmarks framework. As we ran longer benchmarks and plotted maximum command latency we observed that there are always a small (statistically negligible) number of commands whose latency is substantially larger than the tail of the distribution. This indicated starvation in the system.

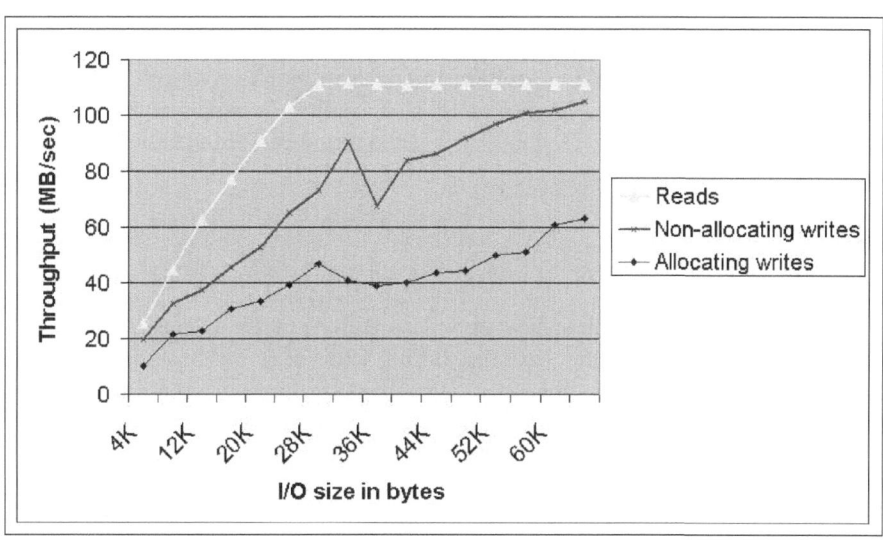

Fig. 2. Throughput performance for all-cache-hits reads and writes (in MBytes/sec as a function of the I/O size); irregularity is observed for writes at 32K

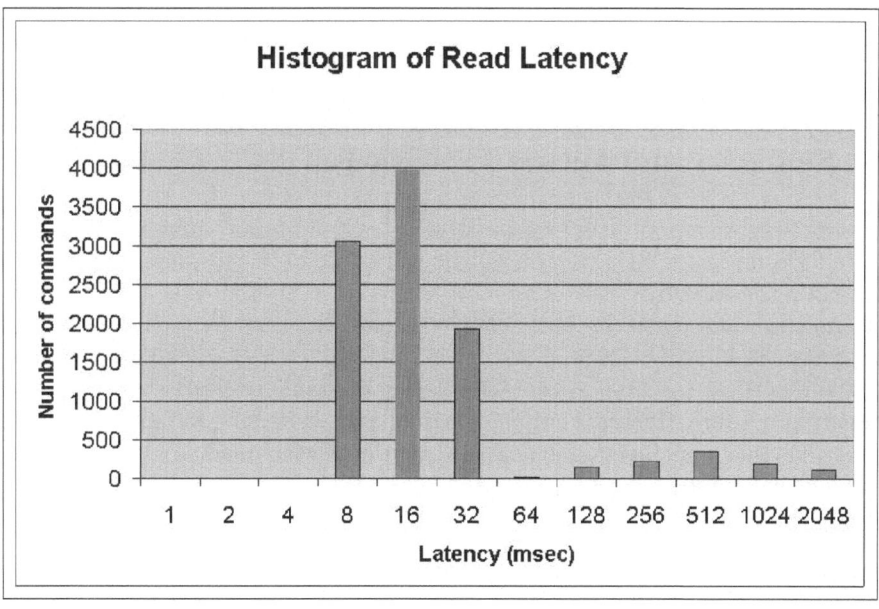

Fig. 3. Latency of Read commands (in msecs); A bimodal distribution is observed due to a mixture of cache hits and cache-misses

Indeed, by looking closely we found that in the Linux SCSI implementation[8], SCSI commands are submitted in LIFO instead of FIFO order without avoiding starvation (via a timeout mechanism for example)[5]. Since our benchmarks are designed not to leave the target idle, the LIFO behavior caused the said starvation. As a result, we patched the Linux kernel to support the appropriate ordering of OSD commands.

7   Related Work

A model-checking approach can be very powerful for file-system checking, as was shown in [17]. Proof systems can also be used to verify an implementation [2]. These approaches are different from the sampling approach that was taken in this paper.

File-system debugging using comparison is used in NFS-tee [15]. NFS-tee allows testing an NFSv3 server against a reference implementation by situating a proxy in-front of the two systems. A workload is executed against the two systems and their responses are compared; a mismatch normally means a bug. This approach is close to ours, however, NFS-tee does not combine any of the snapshot or gray-box techniques we employ.

There are many file-system testing and compliance suites, in fact, too many to attempt to list here (among the popular ones are [7,9]). However, most suites do not check file-system recovery. Furthermore, we haven't found a test-system that took the path described here. Other I/O benchmarking are also applicable to OSD benchmarking[3].

8   Summary and Future Extensions

In this paper we report on our extensive effort of building a comprehensive testing suite for compliant OSDs, and our initial work on building a common criteria for evaluating them. Object stores are new, and to this day there are only a handful of implementations. As the technology emerges, the need for such tools will be apparent. To our knowledge, our work is the first attempt to address this need.

We report on the unique characterization of standard OSDs that made the testing procedure different and challenging, and show how we addressed these issues. Further work is required as more experience with building OSDs is gained. Among these are:

- Improve testing coverage by enhancing the script-generation to address non-determinism beyond what it currently supports.
- Extend the script language to define broader recursive scripts, thus exploiting more complicated patterns of parallelism.

---

[5] This behavior is documented in the Linux code in scsi_lib.c. Commands are placed at the head of the queue to support the scsi_device_quiesce function. Apparently this has not been a problem in most systems since they do not overload the scsi midlayer.

- Testing the other T10 security method (CMDRSP and ALLDATA).
- Testing advanced functionalities in the OSD T10 standard, *e.g.* Collections.
- Enrich the benchmarks with real use-case data.

Some related issues that are more fundamental to testing and require further work include:

- Compute the efficiency of our testing tools (e.g. #bugs per Kloc) and develop a quantitative coverage criteria.
- Contrast the sampling approach for testing OSDs with other functional and structural approaches to testing.
- Investigate the applicability of these tools to other domains.

Acknowledgments

Efforts in IBM Haifa ObjectStone team related to testing our object store implementation have been going on almost from day one. Hence, many people in the team contributed ideas and work to this mission throughout the years. Special thanks to Guy Laden who wrote the first script generator, Grisha Chockler who wrote the first version of the tester and Itai Segall who conceived the benchmarks suite. Avishay Traeger's input during the writeup of this paper was useful. Thanks to our colleagues from the IBM Haifa verification team: Roy Emek and Michael Veksler, and to the anonymous referees for their useful insights related to the general area of verification. Finally, thanks to Stuart Brodsky, Sami Iren and Dan Messinger from Seagate who experienced with our tester and were part of the design of testing the CAPKEY security method.

References

1. A. Aharon, D. Goodman, M. Levinger, Y. Lichtenstein, Y. Malka, C. Metzger, M. Molcho, and G. Shurek. Test program generation for functional verification of powerpc processors in ibm. In *DAC '95: Proceedings of the 32nd ACM/IEEE conference on Design automation*, pages 279–285, New York, NY, USA, 1995. ACM Press.
2. K. Arkoudas, K. Zee, V. Kuncak, and M. Rinard. Verifying a File System Implementation. In *Proceedings of the Sixth International Conference on Formal Engineering Methods (ICFEM 2004)*, 2004.
3. P. M. Chen and D. A. Patterson. A new approach to i/o performance evaluation: self-scaling i/o benchmarks, predicted i/o performance. *ACM Transactions on Computer Systems (TOCS)*, 12(4), november 1994.
4. R. A. DeMillo and A. J. Offutt. Constraint-based test data generation. *IEEE Transactions on Software Engineering*, 17(9), september 1991.
5. *A Demonstration of an OSD-based File System, Storage Networking World Conference, Spring 2005*, April 2005.
6. M. Factor, K. Meth, D. Naor, O. Rodeh, and J. Satran. Object storage: The future building block for storage systems. a position paper. In *Proceedings of the 2nd International IEEE Symposium on Mass Storage Systems and Technologies, Sardinia Italy.*, pages 119–123, June 2005.

7. *IOZone Filesystem Benchmark.* http://www.iozone.org/.
8. *The Linux 2.6.10 SCSI Mid-layer Implementation, scsi_do_req API.*
9. NetApp.   *The PostMark Benchmark.*   http://www.netapp.com/tech_library/ 3022.html.
10. *A T10 iSCSI OSD Initiator.* http://sourceforge.net/projects/osd-initiator.
11. *IBM Object Storage Device Simulator for Linux.* Released on IBM's AlphaWorks, http://www.alphaworks.ibm.com/tech/osdsim.
12. M. I. Seltzer, K. A. Smith, H. Balakrishnan, J. Chang, S. McMains, and V. N. Padmanabhan. File system logging versus clustering: A performance comparison. In *USENIX Winter*, pages 249–264, 1995.
13. SNIA - Storage Networking Industry Association. *OSD: Object Based Storage Devices Technical Work Group.* http://www.snia.org/tech_activities/workgroups/osd/.
14. Standard Performance Evaluation Corporation.   *SPEC SFS97_R1 V3.0 Benchmarks*, August 2004. http://www.spec.org/sfs97r1.
15. Y.-L. Tan, T. Wong, J. D. Strunk, and G. R. Ganger.   Comparison-based File Server Verification. In *USENIX 2005 Annual Technical Conference*, April 2005.
16. R. O. Weber. *SCSI Object-Based Storage Device Commands (OSD), Document Number: ANSI/INCITS 400-2004.* InterNational Committee for Information Technology Standards (formerly NCITS), December 2004. http://www.t10.org/drafts.htm.
17. J. Yang, P. Twohey, D. R. Engler, and M. Musuvathi. Using Model Checking to Find Serious File System Errors. In *OSDI*, pages 273–288, 2004.

A Kernel-Based Communication Fault Injector for Dependability Testing of Distributed Systems*

Roberto Jung Drebes, Gabriela Jacques-Silva,
Joana Matos Fonseca da Trindade, and Taisy Silva Weber

Instituto de Informática – Universidade Federal do Rio Grande do Sul,
Caixa Postal 15064 – 90501-970 – Porto Alegre, RS, Brazil
{drebes, gjsilva, jmftrindade, taisy}@inf.ufrgs.br

Abstract. Software-implemented fault injection is a powerful strategy to test fault-tolerant protocols in distributed environments. In this paper, we present ComFIRM, a communication fault injection tool we developed which minimizes the probe effect on the tested protocols. ComFIRM explores the possibility to insert code directly inside the Linux kernel in the lowest level of the protocol stack through the load of modules. The tool injects faults directly into the message exchange subsystem, allowing the definition of test scenarios from a wide fault model that can affect messages being sent and/or received. Additionally, the tool is demonstrated in an experiment which applies the fault injector to evaluate the behavior of a group membership service under communication faults.

1 Introduction

Specification errors, project errors and inevitable hardware faults can lead to fatal consequences during the operation of distributed systems. Developers of mission critical applications guarantee their dependable behavior using techniques of fault tolerance. Examples of such techniques include detection and recovery from errors, and error masking through replicas. The implementation of these techniques should be strictly validated: it should be assured that recovery mechanisms are capable of masking the occurrence of faults, and that unmasked faults lead the system through a known, expected, fault process.

Experimental validation is the main goal of fault injection. A fault injection test experiment lies in the introduction of faults from a given scenario into a system under test, the target, to observe how it behaves under the presence of such faults. Many software-implemented fault injectors exist for different platforms. When dealing with communication faults, most of them, however, are specific to some proprietary technology. For example, EFA [1], runs exclusively on the A/ROSE distributed operating system, while SFI [2] and DOCTOR [3] are restricted to the HARTS distributed system. ORCHESTRA [4], was originally

---

* Partially supported by HP Brazil R&D and CNPq Project # 472084/2003-8.

S. Ur, E. Bin, and Y. Wolfsthal (Eds.): Haifa Verification Conf. 2005, LNCS 3875, pp. 177–190, 2006.

developed for the Mach operating system and later ported to Solaris. When tools are not locked to a specific architecture, they may be designed to work with a specific runtime environment, such as FIONA [5] or the network noise simulator for Contest [6], which only support Java applications.

ComFIRM (*COMmunication Fault Injection through operating system Resources Modification*) is a tool for experimental validation, built exploring the extensibility of Linux through the load of kernel modules. This allows it to minimize the probe effect in the kernel itself. The tool aims to inject communication faults to validate fault tolerance aspects of network protocols and distributed systems.

Like ORCHESTRA, the tool provides many options to inject communication faults and allows specifying the test experiments through scripting. However, different than ORCHESTRA, ComFIRM was conceived exploring specific characteristics of the open source Linux kernel, which is available to many different architectures. Even if both ORCHESTRA and ComFIRM are script based, a fundamental difference exists: the former uses a high-level interpreted language, while the latter uses a special bytecode oriented language, which gives ComFIRM more power to specify test conditions and actions, and a lower intrusiveness in the target and supporting systems.

This paper emphasizes ComFIRM's fault model, design and use. Section 2 introduces fault injection concepts. Section 3 describes the tool's components and its mechanisms. Section 4 presents design and implementation issues of ComFIRM. Section 5 demonstrates the application of the tool in a practical experiment using the JGroups middleware as a target. Final considerations and future perspectives are presented in the closing section.

2   Experimental Fault Tolerance Validation Using Fault Injection

A software fault injector is a code that emulates hardware faults corrupting the running state of the target system. Software fault injection modifies the state of the target, forcing it to behave as if hardware faults occur. During a test run, the injector basically interrupts or changes, in a controlled way, the execution of the target, emulating faults through the insertion of errors in its different components. For example, the contents of registers, memory positions, code area or flags can be changed.

Network protocols and distributed systems assume the role of the target when considering communication environments. The injection of faults is an important experimental tool for their dependability analysis and design validation: the developer can obtain dependability measures such as fault propagation, fault latency, fault coverage and performance penalty under faults. Fault injection can also be used to evaluate the failure process of non fault-tolerant systems when faults are present. Also, and perhaps more important in distributed systems, targets can be tested when they operate under abnormal timing conditions.

The validation of protocols and distributed systems, however, meets inherent space and timing constraints. While fault injection activities can be used to test the targets under special timing conditions by delaying messages, for instance, the intrinsic fault injector activities impose an extra load on the target which can bring unintentional timing alterations to task execution time. Thus, when developing an injector, special care should be taken to limit the injector intrusiveness on the target system to minimize the probe effect on the dependability measure.

A complete fault injection campaign includes the following phases. First, a fault scenario for the test experiment should be specified. This includes the decision of location, type and time of injected faults. Next, the effective injection of faults should take place. During this phase, experiment data must be collected, and as a final step the data must be analyzed to obtain the dependability measures of the target protocol. For this paper, our major concerns are the creation of scenarios, control of location, type and time of fault injection and the effective injection of faults.

2.1 Building Kernel-Based Fault Injectors

Software fault injectors can disturb the target system and mask its timing characteristics. This happens because the fault injector usually alters the code of the target inserting procedures that interrupt its normal processing. When dealing with the communication subsystem of an operating system, those procedures can be executed as user programs or as components of the kernel. The closer they are to the hardware, the more efficient those procedures tend to be. One alternative is placing the fault injector at operating system level, which is the approach used by ComFIRM.

The exact location inside the operating system depends on how it is structured. The fault injector can operate, for example, through system libraries, system calls or other executable resources of the system (like interruption or exception routines, device drivers or modules). Whenever the target protocol requests services from these operating system resources, the fault injector can act and change the execution flow to inject faults.

If a fault injector is located inside the operating system, the intrusiveness, or probe effect, in the target protocol code is greatly minimized, since no context switches are necessary between the target and injector, beyond those originally existing between user and kernel contexts. Even if the target protocol is not disturbed or altered directly, however, the fault injector could disturb the operating system integrity: the intrusiveness moves from the target protocol to the operating system. An efficient kernel-based fault injection tool, then, must restrict the way it alters the original kernel, both in terms of timing and code alteration.

3 ComFIRM: A Communication Fault Injection Tool

ComFIRM intends to address the main issues regarding the injection of software-implemented communication faults, that is, validating fault-tolerant mechanisms of network protocols and distributed systems. Like ORCHESTRA, the fault types it supports are: receive and send omission, link and node crash, timing and byzantine faults. These classes represent a typical fault model for distributed systems, as detailed in Section 3.1.

Currently, ComFIRM explores the Netfilter architecture [7] available in versions 2.4 and 2.6 of the Linux kernel [8]. ComFIRM can be applied to any device running recent versions of the Linux operating system. Since it uses high-level features of Linux, it is architecture independent, and so can run both in servers and workstations as well as embedded devices. Beyond that, our choice of Linux as platform was influenced by its free software nature. This makes a large amount of Linux information available, such as user manuals, kernel documentation, newsgroup archives, books and, of course, source code.

Two basic guidelines were followed while developing the tool. First, the injector should allow the configuration of the experiment during runtime. Also, the description of fault scenarios and fault activation should be possible using simple rules, which combined could be used to describe more complex scenarios.

Runtime configuration is done through virtual files created in the proc file system. Four files are used: writing to ComFIRM_Control, general commands can be passed to the tool, like activation, deactivation, and log detail level; reading from ComFIRM_Log, the sequence of fault injector events can be obtained; finally, ComFIRM_RX_Rules and ComFIRM_TX_Rules receive the fault injection rules that are used for reception and transmission, respectively.

3.1   Fault Model

Usual fault models for distributed systems, as the one suggested by Cristian [9], can be implemented using communication faults. A node crashes when it stops to send or receive messages. A node is omitting responses when it does not send or receive some of the messages. It is said to have timing faults when it delays or advances messages. Byzantine behavior happens when the values (contents) of messages are altered, messages are duplicated, or if a node sends contradictory messages to different nodes.

For any given experiment, a fault scenario is described from the chosen fault model. Considering a communication fault injector, the scenario depends on the way the tool handles communication messages, that is, how messages are selected and manipulated [10].

A scenario must describe what message the injector must select to manipulate, be it by deterministic or probabilistic means. It is not enough to operate randomly on messages, unless the distribution of faults follows some known model.

With ComFIRM, messages can be selected by the following criteria: message contents, message flow or counters and flags. Content based selection considers some pattern in the message, as confirmation messages or connection close requests. When selection is based on message flow, contents are ignored, and all messages which match a rule are considered as a single group. It is then possible, for example, to manipulate a percentage of the messages, or even one in each n messages. Selection by external elements considers conditions based on timers, user variables and measures of some physical parameter. It is the combination of the types of selection that gives the tool great flexibility in describing scenarios.

After determining which messages will be selected, it must be specified how to manipulate them. Three classes of manipulation can be identified: message internal actions, actions on the message itself, or unrelated to the messages. The first class defines actions that change some message field. The second, defines the most common actions, as losing messages (where the selected message is simply discarded), delaying messages (late delivery) and duplication. The last class includes actions that manipulate counters, timers, variables and user level warnings. Again, the possibility to combine several actions allows the definition of sophisticated fault scenarios for a test experiment.

The injection of timing faults can test the target under conditions that may trigger timing bugs, deadlocks or race conditions. Developers usually assumes that communication is "instantaneous", or even when that is not the case, they are unable to practically test their systems under degraded network performance in a controlled manner. With ComFIRM, not only the developer can test his applications with late message delivery – applied deterministically through message content examination, and not statistically as usually provided by network emulation tools such as NIST Net [11] and Dummynet [12] – but he can also alter message contents, forcing explicit message retransmissions which may result in timing situations.

3.2   Selection and Manipulation of Messages

ComFIRM's operation is performed using a bytecode language developed to describe an experiment through operation codes. The choice of a bytecode language to represent the fault scenarios is related to the lower timing intrusiveness its compact and efficient representation imposes on the experiments. Since the tool executes inside the operating system kernel, the interpretation of a higher level language would need to either be done in kernel space, which presents many challenges, or in user space, which would require frequent context switches between these modes to run the fault injection scripts, resulting on a significant performance penalty.

Using this representation, rulesets are composed of individual rules and can be independently configured for both the transmission and the reception flows. A rule is composed of a set of operations, or instructions, which determine either selection conditions or manipulation actions, arranged in a nested *if-then* structure. Instructions in a rule are evaluated in sequence until a test condition returns false, or an action specifying that a message should be dropped, duplicated or delayed is reached. If a test is evaluated false, processing is resumed in the next rule. If a message was discarded, duplicated or delayed, no further rules are processed for that packet. Figure 1 presents the relationship between instructions, rules, and rulesets. When specifying rulesets, each rule should be prepended by its rule length, and a null (zero length) rule specifies the end of the set.

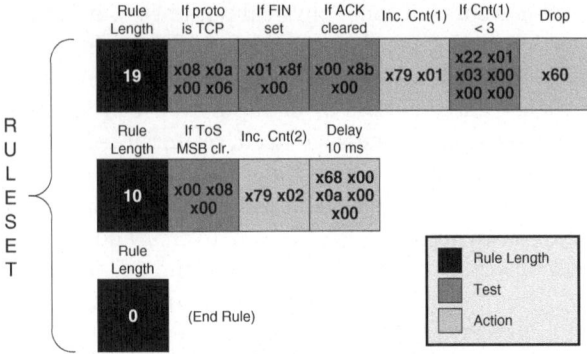

Fig. 1. ComFIRM example rules

Table 1. ComFIRM selection instructions

x00	Test bit	x20	Test counter
x08	Test byte	x28	Test random byte
x10	Test word	x30	Test timer
x18	Test double word	x38	Test flag

Table 2. ComFIRM selection instruction modifiers

x00	Test if equal	x00	Test if cleared
x01	Test if greater	x01	Test if set
x02	Test if less		

The injector can select a message by its contents, through message flows, or based on external elements, like timers, counters and flags. Selection instructions are given in Table 1. It is possible to test the contents on a bit, byte, word (16 bits) or double word (32 bits) level. The modifiers presented in Table 2 should be applied to these operations, to determine relations. The bit and flag testing operations (instruction codes 0x00 and 0x38) can only determine if a bit is cleared or set (0x00 and 0x01 modifiers). The remaining operations can test if a value is equal, greater or less than the operand (0x00, 0x01 and 0x02 modifiers, respectively).

Bit testing can be used to verify specific protocol control flags. Testing of bytes, words and double words is interesting when testing other protocol fields, like addresses, commands, sequence numbers and offsets. For selection based on message flow, the injector can test a counter value or it can randomly select messages. ComFIRM provides up to 256 general use 32 bit counters, which should be manually incremented or decremented, through explicit manipulation operations. For random selection, the injector can obtain a kernel supplied pseudo-random value. For selection of messages based on external elements, timers and flags can be used. Timer resolution when running on the x86 architecture is that of the

Table 3. ComFIRM manipulation instructions

x40	Modify bit	x70	Duplicate message
x48	Modify byte	x78	Modify counter
x50	Modify word	x80	Modify timer
x58	Modify double word	x88	Modify flag
x60	Drop message	x90	Dump message
x68	Delay message		to log

Table 4. ComFIRM manipulation instruction modifiers

x00	Assign value	x00	Clear
x01	Increment	x01	Set
x02	Decrement	x02	Complement

Linux kernel, that is, 1 millisecond in current versions. Just like the counters, 32 single bit flags are available to the test operator. With all these selection operations, precise selection conditions can be specified.

Selection instructions are passive and by itself could be used for monitoring, but are not enough for active fault injection. Manipulation operations, presented in Table 3, can be combined to the previous operations for defining complex fault scenarios. ComFIRM supports manipulating bits, bytes, words (16 bits) and double words (32 bits) from the message contents, acting over the external elements: timers, flags and counters; as well as delaying, dropping and duplicating messages. An extra instruction can dump the contents of a message to the fault injector event log.

Similar to the selection instructions, there are modifiers which can be applied to manipulation instructions, as shown in Table 4. For bit and flag related operations (0x40 and 0x88), their operands can be cleared, set or complemented. These actions can be used to emulate *stuck-at* or byzantine faults. The values of bytes, words, double words, timers and counters, with the appropriate modifiers, can be assigned, incremented or decremented. Operations which drop, duplicate, delay and dump messages do not support any modifier.

3.3   Fault Manifestation

It should be noted that when changing contents of a message through manipulation, all error detection and/or correction codes that the protocol may implement, become invalid. If the goal is to test the detection mechanisms of transport protocols, this approach is suitable. If it is to test higher-level protocols, we should observe which supporting protocols are used. If a supporting protocol like UDP, which provides unreliable datagram service, is used, manipulation of message contents makes sense and the injected faults can reach higher-level target protocols. However, if lower protocols that handle errors, such as TCP, are used as the basis for multicast or membership target protocols, it will be the supporting protocol itself that will react and recover from errors. The target protocol on the upper layers will be unaware of the injected faults.

While TCP is extensively used in network applications, it is stream oriented and has FIFO semantics, what, in addition to not supporting multicast, makes it ill-suited for messaging in group communication systems [13]. UDP packet delivery, however, while unreliable, is commonly used as the basis for the development of communication protocols where the required reliability is implemented on upper layers, such as group communication systems. ComFIRM can directly alter the contents of UDP and TCP messages, but as explained above TCP alterations are masked by the protocol itself. Regarding UDP, even when its checksum is employed, the injector can make the receiving application ignore the verification code by writing a value of zero in its location. This is interpreted as if the sender did not enable the checksum mechanism. The destination, when receiving a message like this, cannot tell whether it was the fault injector or the sender who set this value. In this way, the alteration in the packet contents generated by ComFIRM are not discarded and can reach the receiving end.

3.4   Creating a Fault Scenario

Rulesets described in Section 3.2 should map to a specific instance of the supported fault model. These instances are the fault scenarios of a fault injection campaign. For instance, discarding messages emulates crash and omission faults. Delaying messages emulates timing faults and when the contents of the messages are modified, or messages are duplicated, byzantine faults can be emulated.

To illustrate how rules can be created to test implementations of network protocols and applications, here we present how ComFIRM can be applied to test specific conditions or events.

For example, a test operator may be interested in watching the behavior of a TCP implementation when requests to close a connection are lost. To do that, a rule must be specified that, for each message: (i) checks if its transport level protocol is TCP, value 6, (ii) verifies if it contains a request to close a connection (FIN bit set and ACK bit cleared) and (iii) drops the message. This rule would discard every close connection request. The operator may add a counter to the rule, to discard, for instance, only the first 3 requests processed. Such a rule would, after matching request to close a connection, increment a counter and check if its value is less than '3', dropping the message only if so.

Another rule may be set to test how a target system behaves when IP packets with low-precedence Type of Service (ToS) bits (in the range 0-3) are delivered with a higher delay. For that, this rule would check the priority bits and, if they are found to have a low value, introduce and artificial delay of, say, 10 ms. This is done testing the most significant bit of the ToS field, bit #8, and adding a delay of 10 ms. Additionally, the rule can increment a counter to monitor how many times this action was performed. The increment should happen before the delay instruction is executed, since that instruction causes instruction evaluation to stop.

The representation as bytecode of the ruleset containing the two rules described above can be seen in Figure 1.

4   Implementation Details

The current version of ComFIRM is a major rewrite of an original architecture which was very intrusive to the Linux kernel. It required modifications in key functions of the network subsystem and the timer handler, since Netfilter hooks were not available. Wrappers were used in the packet sending and receiving functions and the periodic timer handler had to be adapted as well. This modifications had to be made in the original kernel source files, and so recompilation of the full kernel was necessary to instrument it. Also, the tool had to be constantly updated whenever a new kernel version was released. While this was functional, it was not a clean solution. The changed kernel could behave in an unanticipated way, out of its original specification, invalidating the results of any tests performed over it. The rewrite of ComFIRM using the Netfilter framework, significantly minimizes this probe effect.

Netfilter [7] is a kernel framework for packet manipulation that defines specific points, or hooks, in the protocol stack where callback functions can be registered. When packets reach these points, they are passed to the functions which have full access to their contents. The functions can read the packets and decide if they should continue the traversal of the protocol stack or not. Netfilter supports callback functions for the IPv4, IPv6 and DECnet protocol families. Hooks are available on many locations: where local packets arrive and leave the host and where packets are forwarded, both before and after routing.

ComFIRM is implemented as a Netfilter module for the IPv4 protocol family. For incoming packets, it registers a callback function in the NF_IP_PRE_ROUTING hook, which processes packets identified as IP before routing occurs. For outgoing packets, a function is registered at the NF_IP_POST_ROUTING hook, where all processing of IP packets is already done and they are just about to be sent to the transmission media. These callbacks parse the fault injection rules, repeatedly for every packet sent and/or received. ComFIRM's evaluation of a rule can return to the protocol stack how the packet should be handled. The Netfilter supported actions used by ComFIRM are NF_ACCEPT, NF_DROP and NF_STOLEN. NF_ACCEPT is used when a selection rule does not match the packet, that is, it should be processed normally, or when a packet is matched but the only action rules are content, flag or counter manipulation. NF_DROP, as the name implies, is used when an action indicates that a packet must be dropped. NF_STOLEN is a special case of drop, where the packet does not continue the traversal of the protocol stack as well, but its allocated resources are not freed. This is used for delayed messages, which are queued for later processing inside the periodic timer handler. For message duplication, the message is sent or delivered during rule interpretation and when the full rule evaluation is finished a NF_ACCEPT guarantees that the sending or delivery will be performed again.

The other core component of ComFIRM is the processing of the virtual files created on the proc filesystem, which are used for inputting rules, control commands and reading the fault injection event log. To each of these virtual files, functions are associated to the open(), close(), read() and write() manipulation primitives. This gives the test operator access to the ComFIRM functions and structures, which are internal to the kernel, using standard system utilities.

The controlling virtual files and the callback functions are dynamic resources which are registered on module load and unregistered on module removal giving ComFIRM the characteristics of a *plugin* that can be easily added or removed from a running system, without reboot.

5   ComFIRM Demonstration

Here, we present the application of ComFIRM to a target system to observe how the communication faults it creates get manifested on the upper layers of a distributed application. For our testbed, we chose JGroups [14], an open source toolkit which provides reliable multicast communication for applications developed in Java.

JGroups is flexible, in that it allows its protocol stack to be adapted to different development requirements and/or network characteristics, through the inclusion of specific protocol layers. Also, the toolkit provides a group membership service (GMS), making it possible to create group views, consisting of nodes which can be spread across multiple LANs or WANs. Members can join or leave such groups, and a detection mechanism may notify each member whenever others join, leave, or even detect node crashes. Detected crashed members can then be removed from the group view, maintaining a consistent state. These features turn JGroups into an attractive alternative to the development of distributed systems and applications.

For this example, we started with a very simple protocol stack consisting of a multicast protocol at its base. Next, a loss-less transmission layer was added, through the addition of a NAKACK (negative acknowledgement) protocol. To provide membership change notifications (i.e. node join, leave or crash detection) the group membership service was added, by including the GMS protocol. This set of layers became the starting point for our experiments.

The behavior of the GMS layer was tested under three different scenarios: (i) without a fault detection mechanism, (iii) with a fault detection mechanism based on a heartbeat protocol (FD) and (iii) with a fault detection mechanism based on sockets (FD_SOCK). The modularity of JGroups protocol stack makes it easy to modify the scenarios, since it is only necessary to alter the stack configuration. It is not even necessary to recompile the target application.

On top of our target, JGroups, we developed an application which uses its services by joining a specific group and notifying the user when the membership changes. These view alterations are provided by the GMS. Using the fault injector under the JGroups middleware and our application above it, it is then possible to observe fault propagation from the network to the application. Three

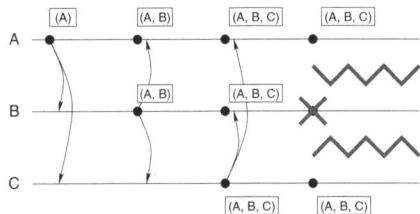

Fig. 2. JGroups' GMS used without a fault detector

instances of our JGroups' application were started, in three different machines: A, B and C. Each machine joins the group at a different time, and multicast messages notify group joins and leaves. ComFIRM was applied in host B to break connectivity of this machine with the other members of the group (A and C), that is, emulating a link crash.

The first scenario is illustrated in Figure 2. As expected, after B gets out of reach (when ComFIRM is activated), members A and C still consider that the machine is part of the group. Since the protocol stack uses a negative acknowledgement protocol – in which it is assumed by default that all messages reach all destinations, and only a transmission fault generates a retransmission request – members A and C do not receive any message from B and assume that no error has occurred. The lack of fault detection mechanisms causes an inconsistent group view.

Next, the application was configured to use JGroups' heartbeat based fault detection mechanism (FD). In this case, each member sends an *are-you-alive* message periodically to its neighbor. If a response is not received within a time interval (3 seconds by default), and for a fixed number of tries (2 by default), the member is suspected, and after a second round, excluded from the group view. Figure 3 corresponds to this situation. At left, it is shown a common case where the machine B simply fails (process failure). Since no responses are sent to the other machines' inquiries, they remove machine B from the group view. At right, ComFIRM is applied to break connectivity between the machines (link failure). While the application is still running, it cannot be reached by the other members, which update the view. This is not instantaneous. After the injector is activated, it takes a variable time until the other nodes acknowledge that the machine is unreachable, which depends on the time interval and number of tries described above, being typically in the order of 12 seconds ($3\times2\times2$).

Finally, the third scenario represents the situation in which a fault detector based on a socket mechanism (FD_SOCK) is used. The protocol establishes TCP connections between each member of the group and its neighbor. The advantage of this method is that no *are-you-alive* messages need to be sent, and so traffic is reduced. When a process is terminated, the socket is closed by the virtual machine and operating system. This causes TCP FIN requests to be sent, as illustrated in Figure 4. At left, it is shown that when these requests reach machines A and C, and the sockets are locally closed, they update their

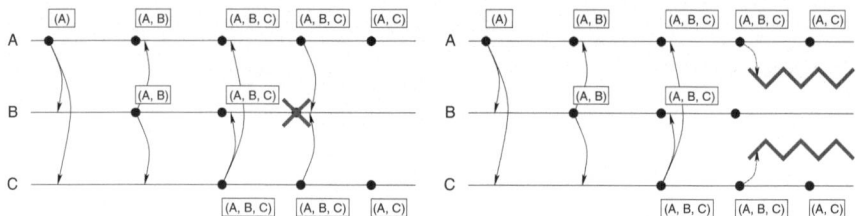

Fig. 3. JGroups' heartbeat based fault detector

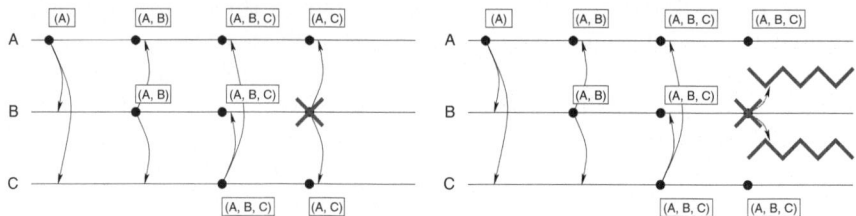

Fig. 4. JGroups' socket based fault detector

views to exclude B. At right, however, ComFIRM does not let the requests reach A and C, who keep their TCP connections in the established state, maintaining an inconsistent view. Since no data is sent through these TCP connections, only the TCP *keep alive* mechanism may detect the fault, but this mechanism is seldom used, and if so, it defines that at least 120 minutes should have elapsed to send the *keep alive* messages which could trigger the link crash detection. This demonstrates the unsuitability of the FD_SOCK protocol to detect router and connectivity faults.

This example does not aim to provide a complete example of a dependability evaluation of the JGroups' group membership service, but exemplify how Com-FIRM can be applied to such kind of evaluation. Link crash can be emulated through many ways, but ComFIRM's instructions allow the selection of specific messages, which helps in a more precise setting of the experiments. The advantage of inspecting and selecting messages deterministically, instead of randomly selecting messages, is that it is possible to test the fault tolerance mechanisms in a shorter time, since it is not necessary to wait for the random selection of a message which would trigger the mechanisms. Random selection, however, is also supported by ComFIRM (through the 'Test random byte' instruction, which can be used to build more complex random-based selection scenarios). By experimenting with the targets early in the development process, new bugs can be found and a greater confidence can be had on the correct operation of the fault tolerance mechanisms of distributed systems and protocols, building more reliable systems.

6   Concluding Remarks

Fault injection is a powerful technique to evaluate how protocols and distributed systems behave when faults occur. A fault injection tool allows the designer to measure the efficiency of detection, correction and error recovery mechanisms of a system before it is put into effective operation.

Several approaches to implement fault injectors are available. ComFIRM is implemented as a Linux kernel module using the Netfilter framework for communication message processing. This gives the tool full access to both the incoming and outgoing message flows, in a very clean and non-intrusive way. The bytecode instructions supported by ComFIRM allow messages to be inspected and selected in a deterministic or statistical way, and provides many actions to be performed with packets, which mimic the behavior of real faults: message drop and duplication, late delivery and content manipulation. The tool is fully operational and can be easily added to a running Linux system to perform experiments with distributed systems, as shown in Section 5.

In the future, ComFIRM may be adapted to work with other protocol families, like IPv6. Since this is an emerging infrastructure standard, many new protocols and applications are being developed on top of it. Their testing and validation are necessary and so a fault injection tool which supports IPv6 can be valuable. For such adaptation, ComFIRM only needs to be registered in the IPv6 Netfilter hooks, which already exist. Optimally, new instructions can be implemented which support larger variable sizes (which hold 128 bit addresses, for instance), or even which permits that the contents read from a given offset in a packet can be used as a parameter to the next instructions. This would help to process the variable size message headers concatenated through the *next header* field used in IPv6.

References

1. Echtle, K., Leu, M.: The EFA fault injector for fault-tolerant distributed system testing. In: Proc. of the IEEE Workshop on Fault-Tolerant Parallel and Distributed Systems. (1992) 28–35
2. Rosenberg, H.A., Shin, K.G.: Software fault injection and its application in distributed systems. In: Proc. of the 23rd Intl. Symposium on Fault Tolerant Computing. (1993) 208–217
3. Han, S., Shin, K.G., Rosenberg, H.: DOCTOR: An integrateD sOftware fault injeCTiOn enviRonment for distributed real-time systems. In: Proc. of the Intl. Computer Performance and Dependability Symposium. (1995) 204–213
4. Dawson, S., Jahanian, F., Mitton, T., Tung, T.L.: Testing of fault-tolerant and real-time distributed systems via protocol fault injection. In: Proc. of the 26th Intl. Symposium on Fault Tolerant Computing. (1996) 404–414
5. Jacques-Silva, G., Drebes, R.J., Gerchman, J., Weber, T.S.: FIONA: A fault injector for dependability evaluation of Java-based network applications. In: Proc. of the 3rd IEEE Intl. Symposium on Network Computing and Applications. (2004) 303–308

6. Farchi, E., Krasny, Y., Nir, Y.: Automatic simulation of network problems in UDP-based Java programs. In: Proc. of IEEE International Parallel & Distributed Processing Symposium 2004, Santa Fe, New Mexico (2004)
7. Russell, R., Welte, H.: Linux netfilter hacking HOWTO (2002) Available at: <http://www.netfilter.org/documentation/>.
8. Beck, M.: Linux Kernel Internals. 2nd edn. Addison Wesley (1998)
9. Cristian, F.: Understanding fault-tolerant distributed systems. Communications of the ACM **34** (1991) 56–78
10. Dawson, S., Jahanian, F.: Probing and fault injection of protocol implementations. Technical Report CSE-TR-217-94, University of Michigan (1994)
11. Carson, M., Santay, D.: NIST Net: a Linux-based network emulation tool. SIGCOMM Computer Communication Review **33** (2003) 111–126
12. Rizzo, L.: Dummynet: a simple approach to the evaluation of network protocols. SIGCOMM Computer Communication Review **27** (1997) 31–41
13. Wiesmann, M., Défago, X., Schiper, A.: Group communication based on standard interfaces. In: Proc. of the 2nd IEEE Intl. Symposium on Network Computing and Applications, Cambridge, MA (2003) 140–147
14. Ban, B.: JavaGroups - Group communication patterns in Java. Technical report, Department of Computer Science, Cornell University (1998)

Detecting Potential Deadlocks with Static Analysis and Run-Time Monitoring*

Rahul Agarwal, Liqiang Wang, and Scott D. Stoller

Computer Science Dept., SUNY at Stony Brook, Stony Brook, NY 11794-4400
{ragarwal, liqiang, stoller}@cs.sunysb.edu
http://www.cs.sunysb.edu/~{ragarwal, liqiang, stoller}

Abstract. Concurrent programs are notorious for containing errors that are difficult to reproduce and diagnose. A common kind of concurrency error is deadlock, which occurs when a set of threads is blocked each trying to acquire a lock held by another thread in that set. Static and dynamic (run-time) analysis techniques exist to detect deadlocks.

Havelund's GoodLock algorithm detects potential deadlocks at run-time. However, it detects only potential deadlocks involving exactly two threads. This paper presents a generalized version of the GoodLock algorithm that detects potential deadlocks involving any number of threads. Run-time checking may miss errors in unexecuted code. On the positive side, run-time checking generally produces fewer false alarms than static analysis.

This paper explores the use of static analysis to automatically reduce the overhead of run-time checking. We extend our type system, Extended Parameterized Atomic Java (EPAJ), which ensures absence of races and atomicity violations, with Boyapati *et al.*'s deadlock types. We give an algorithm that infers deadlock types for a given program and an algorithm that determines, based on the result of type inference, which run-time checks can safely be omitted. The new type system, called Deadlock-Free EPAJ (DEPAJ), has the added benefit of giving stronger atomicity guarantees than previous atomicity type systems.

1 Introduction

Concurrent programs are notorious for containing errors that are difficult to reproduce and diagnose at run-time. Some common kind of programming errors include deadlocks, data races and atomicity violations. A *deadlock* occurs when each thread is blocked trying to acquire a lock held by another thread. A *data race* occurs when two threads concurrently access a shared variable and at least one of the accesses is a write. Atomicity is a common higher-level correctness requirement that expresses non-interference between concurrently executed methods. A method is *atomic* if every execution of the program is equivalent to an execution in which that method is executed without being interleaved with other concurrently executed methods. This paper focuses on detecting deadlocks.

* This work was supported in part by NSF under Grant CCR-0205376 and CNS-0509230 and ONR under Grants N00014-02-1-0363 and N00014-04-1-0722.

S. Ur, E. Bin, and Y. Wolfsthal (Eds.): Haifa Verification Conf. 2005, LNCS 3875, pp. 191–207, 2006.

The GoodLock algorithm [Hav00] detects potential deadlocks at run-time. However, it detects only potential deadlocks involving two threads, *i.e.*, each of those threads is blocked trying to acquire a lock held by the other thread. This paper presents a generalized version of GoodLock algorithm, that detects potential deadlocks involving any number of threads in other executions of the program.

Static analysis can also detect potential deadlocks. Boyapati, Lee and Rinard [BLR02] introduce a type system that ensures Java programs are deadlock-free. That type system extends Boyapati and Rinard's Parameterized Race Free Java (PRFJ) type system [BR01], which ensures Java programs are race-free.

Run-time checking and static analysis are both useful. Static analysis can guarantee that all executions of a program are deadlock-free; run-time checking cannot. However, due to limitations of the type system, some deadlock-free programs are not typable; the resulting warnings from the typechecker are called *false alarms*, and they may be difficult to diagnose. On the other hand, run-time checking generally produces fewer false alarms than static analysis; this is a significant practical advantage, since diagnosing all of the warnings from static analysis of large codebases may be expensive.

This paper extends our type system, Extended Parameterized Atomic Java (EPAJ) [SAWS05], which ensures absence of races and atomicity violations, with the deadlock types described in [BLR02]. The new type system, called Deadlock-Free EPAJ (DEPAJ), ensures absence of deadlocks due to locks and, as an added benefit, gives stronger atomicity guarantees than EPAJ and Atomic Java [FQ03], which do not consider deadlocks and hence may classify a method as atomic even if it could deadlock in the middle—something that cannot happen if the method executes without interruption by other threads.

The type systems and run-time analysis algorithms considered in this paper only attempt to detect potential deadlocks caused by locks. They do not consider wait/notify or other forms of condition synchronization and hence do not detect deadlocks due to them.

Manually annotating code with the necessary type annotations can be a significant burden, especially for legacy code. Type inference reduces the annotation burden by automatically determining types for some or all parts of the program. This paper presents a type inference algorithm for [BLR02]'s basic deadlock types.

Static analysis can be used to decrease the overhead of run-time checking, in the following way. First, our type inference algorithm infers deadlock types for the program. Run-time deadlock detection is then focused on fragments of code which were not typable. The user can inspect the run-time warnings, which are more likely to indicate real errors and can provide more detailed and specific diagnostic information; then, if desired, the user can inspect warnings from the typechecker. The goal is to reduce the overhead of run-time checking to a level where it can be used unobtrusively throughout the testing process, or even in deployed systems, instead of only during a limited period of testing focused on concurrency errors.

The rest of the paper is organized as follows. Sections 2, 3, 4 and 5 describe run-time detection of potential deadlocks, deadlock types, type inference for deadlock types, and our type system DEPAJ respectively. Section 6 presents our techniques for focused run-time detection of potential deadlocks. Section 7 presents our experiments, Section 8 discusses related work.

2   Run-Time Detection of Potential Deadlocks

The GoodLock algorithm [Hav00] detects potential deadlocks at run-time. It records a run-time lock tree for each thread. The run-time lock tree for a thread represents the nested pattern in which locks are acquired by the thread. Each node of the run-time lock tree is labeled with a lock and represents the thread acquiring that lock. There is an edge from a node n_1 to a node n_2 if n_1 represents the most recently acquired lock that the thread holds when it acquires the lock associated with n_2. At each instant, each run-time lock tree has one node designated as the *current node*; the path from the root of the tree to that node represents the nested acquires of locks held by that thread at that instant. If a thread re-acquires a lock that it already holds, its run-time lock tree does not contain a node representing the re-acquire. When a thread acquires a lock that it does not already hold, if there is already a child of the current node labeled with that lock, that child becomes the current node, otherwise a new child labeled with that lock is created and becomes the current node. At the end of the execution of the program, if there exist threads t_1 and t_2 and locks l_1 and l_2 such that t_1 acquires l_2 while holding l_1, and t_2 acquires l_1 while holding l_2, then a warning of potential deadlock is issued, unless there is a common lock, called a gate lock, that is held by both threads when they acquire l_1 and l_2; the gate lock prevents the acquires of l_1 and l_2 from being interleaved in a way that leads to deadlock. The worst-case time complexity of the algorithm is $O(|T|^3 \times |Thread|^2)$, where $|T|$ is the size of the largest run-time lock tree, and $Thread$ is the set of threads. However, this algorithm only detects potential deadlocks caused by interleaving of lock acquires in two threads.

We present a generalized version of the GoodLock algorithm that detects potential deadlocks involving any number of threads. In particular, it checks whether there exist distinct threads $t_0, \ldots, t_{m-1}$ and locks $l_0, \ldots, l_{m-1}$ such that, for all $i = 0..m-1$, t_i holds lock l_i while acquiring lock $l_{i+1 \ mod \ m}$. Note that we always ignore a thread re-acquiring a lock it already holds, so a thread acquiring $l_{i+1 \ mod \ m}$ while holding l_i implies $l_{i+1 \ mod \ m}$ and l_i are different locks. In the absence of other constraints on the schedule (*e.g.*, due to gate locks or start-join synchronization), such acquires can be interleaved in a way that leads to deadlock. We call this the Potential for Deadlock from Locks Ignoring GateLocks (PDL-IGL) condition.

The algorithm constructs a run-time lock tree for each thread during execution, as described above. At the end of the execution, it constructs a run-time lock graph, which is a directed graph $G = (V, E)$, where V contains all the nodes of all the run-time lock trees, and the set E of directed edges contains (1) *tree edges:* the directed (from parent to child) edges in each of the run-time lock trees, and (2) *inter edges:* bidirectional edges between nodes that are labeled with the same lock and that are in different run-time lock trees.

For a run-time lock graph G, a *valid path* is a path that does not contain consecutive inter edges and such that nodes from each lock tree appear as at most one consecutive subsequence in the path. Similarly, a *valid cycle* is a cycle that does not contain consecutive inter edges and nodes from each thread appear as at most one consecutive subsequence in the cycle.

As an example, Figure 1 shows the run-time lock graph for the illustrative program in Figure 2. The graph in Figure 1 contains several cycles including the following three, where $1i^{Tj}$ denotes the node for lock $1i$ in the run-time lock tree for thread j: $13^{T1} \rightarrow 13^{T2} \rightarrow 13^{T4} \rightarrow 13^{T1}$, $11^{T1} \rightarrow 12^{T1} \rightarrow 12^{T2} \rightarrow 13^{T2} \rightarrow 13^{T1} \rightarrow 14^{T1} \rightarrow 14^{T3} \rightarrow 11^{T3} \rightarrow 11^{T1}$, and $13^{T1} \rightarrow 14^{T1} \rightarrow 14^{T4} \rightarrow 13^{T4} \rightarrow 13^{T1}$.

The first cycle is not valid because it contains two or more consecutive inter edges. The second cycle is not valid because nodes from thread T1 appear in more than one subsequence. The third cycle is valid and hence indicates a potential deadlock. Specifically, it indicates that the program in Figure 2 can deadlock if thread 1 acquires lock 13 and waits for lock 14 and thread 4 acquires lock 14 and waits for lock 13.

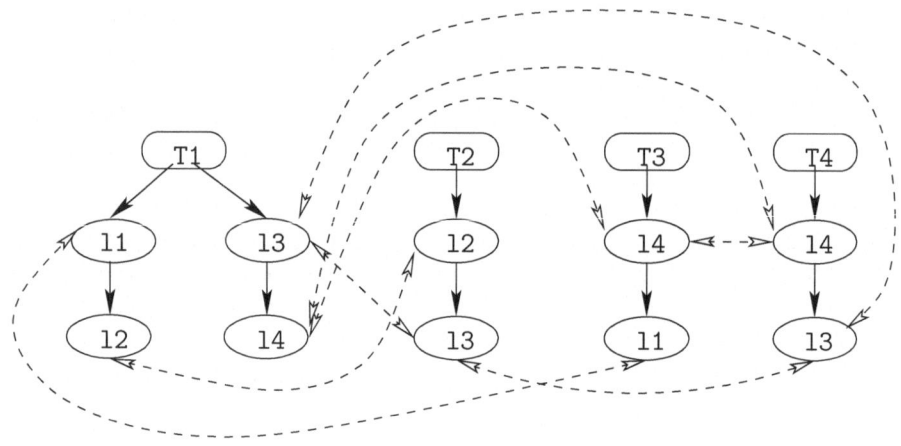

Fig. 1. Run-time lock graph

Now we show that PDL-IGL holds iff the run-time lock graph G contains a valid cycle. Suppose there exist distinct threads $t_0, \ldots, t_{m-1}$ and locks $l_0, \ldots, l_{m-1}$ such that for all $i = 0..m-1$, t_i holds lock l_i while acquiring lock $l_{i+1 \bmod m}$. Let n_i and n_i' denote the nodes in T_i corresponding to the acquire of l_i and the acquire of $l_{i+1 \bmod m}$ nested within it, respectively. Since thread t_i acquires lock l_i and waits for lock $l_{i+1 \bmod m}$, there is a path from n_i to n_i' in run-time lock tree T_i for t_i (because, n_i' is nested below n_i). Note that this path is made of tree edges. The locks l_i and $l_{i+1 \bmod m}$ are distinct, so this path contains at least one tree edge. Also, there

```
Thread 1:
  sync(11) {
    sync(12) {
    }
  }
  sync(13) {
    sync(14) {
    }
  }
```

```
Thread 2:
  sync(12) {
    sync(13) {
    }
  }
```

```
Thread 3:
  sync(14) {
    sync(11) {
    }
  }
```

```
Thread 4:
  sync(14) {
    sync(13) {
    }
  }
```

Fig. 2. Synchronization behavior of 4 threads. `sync` abbreviates `synchronized`.

is an inter edge from n_i' in run-time lock tree T_i to $n_{i+1 \; mod \; m}$ in run-time lock tree $T_{i+1 \; mod \; m}$ in G (by construction). These tree edges and inter edges together form a valid cycle.

Next, we show that existence of a valid cycle C in G implies that the PDL-IGL condition holds. The cycle involves nodes from more than one lock tree, because nodes of a single tree cannot be involved in a cycle. Suppose, C had nodes n_i and n_i' in run-time lock tree T_i for thread t_i, $i \in 0..m-1$ (without loss of generality, we can just consider the beginning and end nodes in the consecutive subsequence from the same thread). Also, nodes n_i' and $n_{i+1 \; mod \; m}$ are labelled with the same lock (they are consecutive nodes from different lock trees and this is only possible through an inter edge which connects two similar labeled locks). Thus, existence of C implies there exist distinct threads $t_0, \ldots, t_{m-1}$ and locks $l_0, \ldots, l_{m-1}$ (node n_i corresponds to lock l_i and node n_i' corresponds to lock $l_{i+1 \; mod \; m}$) such that, for all $i = 0..m-1$, t_i holds lock l_i while acquiring lock $l_{i+1 \; mod \; m}$. Hence, the PDL-IGL condition holds.

Our algorithm to detect existence of a valid cycle traverses all valid paths starting from the root of each lock tree in G using a modified depth-first search (DFS) algorithm, called DFS-ValidCycle, which differs from standard DFS in two ways. First, it traverses only valid paths, because it extends the current path (on the search stack) only with edges satisfying both criteria for validity. Second, a node all of whose neighbors have been explored may be explored multiple times (along incoming inter edges); this is necessary because the set of threads with some lock-tree nodes on the stack might be different on different visits, so the set of valid paths that can be explored by continuing the search from that node is different. The algorithm terminates when a valid cycle is found or all valid paths have been explored. Pseudo-code for the algorithm appears in [AWS05].

To see that the algorithm traverses every valid path, consider a valid path P that begins at a node n in a lock tree T. Extending P by prepending the edges on a path from the root of T to n produces a valid path that is explored by the algorithm when DFS-ValidCycle is started from the root of T. Note that a cycle involving P will be detected, because we check in the algorithm whether n' is anywhere on the stack (not just on the bottom).

To show the worst-case complexity of the algorithm, we consider the number of valid paths in the run-time lock graph. Let $S(k)$ be the number of valid paths in k lock trees $T_1, \ldots, T_k$, assuming the path visits those lock trees in that order. Then $S(k) = S(k-1) + N_k \times N_{k-1}$, where N_k and N_{k-1} are the number of nodes in lock trees T_k and T_{k-1} respectively, because for each node n in T_{k-1}, the valid paths ending at n can be extended in N_k different ways. Thus, the total number of valid paths is $O(|V|^{|Thread|})$, where $|V|$ is the total number of nodes in the graph, and $|Thread|$ is the total number of threads. There are $|Thread|!$ permutations of $T_1, \ldots, T_k$, and each step of extension or backtracking takes constant time, so the overall worst-case complexity of this algorithm is $O(|V|^{|Thread|} \times |Thread|!)$.

The algorithm can be optimized by observing that many valid paths share a common suffix. Define an ordering on edge types: tree-edge $\geq$ inter-edge. This reflects the fact that in the definition of validity, a tree edge implies fewer restrictions on the next edge in the path. For each node n, n.visits is a set of pairs `<ts,et>`, where ts is a set of threads, and et is an edge type. The meaning of `<ts, et>` $\in$ n.visits is that n has been visited along an edge with type et with a stack containing nodes from the lock trees of the threads in ts. If we start the modified DFS at every node n, we do not need to explore a node n' if n'.visits contains a pair `<ts1,et>` such that the current stack contains all nodes from the lock trees of the threads in $ts1$ and n' is being visited along an edge with type less than or equal to et. If those conditions hold, then no valid cycles are reachable by continuing the search from n'. This is because there is no valid path from n' back to n that avoids the lock trees on the stack, because if there were, the search would have detected the cycle (containing n and n') and terminated during the visit that added that tuple to n'.visits. Pseudo-code for the optimized algorithm appears in [AWS05].

The worst-case complexity of the optimized algorithm is $O(2^{|Thread|} \times |V|^3)$, It is easy to see that each node can have $O(2^{|Thread|})$ items in its visits set. Hence, each node can be explored $O(2^{|Thread|})$ times and during each visit it may need to visit its out-edges. There are at most $|V|$ out-edges from each node. Since we repeat the algorithm for each node, the overall worst-case complexity of the algorithm is $O(2^{|Thread|} \times |V|^3)$.

If the number of threads is a constant, then the algorithm is polynomial in the number of nodes in the run-time lock graph.

However, the algorithm does not consider gate locks and therefore produces false alarms whenever some common lock acquired by at least two threads prevents deadlocks. To eliminate these false alarms, we extend the algorithm to check whether there exist distinct $t_0 \ldots t_{m-1}$ and locks $l_0 \ldots l_{m-1}$ such that for all $i = 0..m-1$, t_i holds lock l_i while acquiring lock $l_{i+1 \bmod m}$ and there do not exist t_i, t_j, and l such that t_i and t_j hold l when acquiring l_i and l_j, respectively. (Such a lock l is called a *gate lock* for the cycle). We call this the Potential for Deadlocks from Locks (PDL) condition.

To check the PDL condition, we modify the algorithm to backtrack (instead of halting) when a valid cycle is encountered, so the algorithm explores all valid cycles, and we check for every valid cycle generated whether there is a gate lock, *i.e.*, whether no two nodes in different run-time lock trees have ancestors labeled with the same lock. This can be done in $O(|V|^2 \times |Lock|)$ time for each valid cycle , where $|Lock|$ is the number of locks. If a valid cycle without a gate lock is found, potential for deadlock is reported.

3   Deadlock Types

Boyapati, Lee and Rinard [BLR02] introduce a static type system that ensures Java programs are deadlock-free. The deadlock types express a partial order among the locks. The typing rules ensure that whenever a thread holds multiple locks, the thread acquires the locks in descending order. This ensures absence of cyclic waiting and therefore implies absence of deadlocks.

The rest of this section briefly describes Parameterized Race Free Java (PRFJ) [BR01], and [BLR02]'s deadlock types. In PRFJ, as in its predecessor Race Free Java [FF00], types indicate the synchronization discipline (also called "protection mechanism" or "owner") used to co-ordinate accesses to each object. To allow different instances of a class to use different protection mechanisms, each class is parameterized by formal owner parameters, which may be instantiated with other formal owner parameters, final expressions (*i.e.*, expressions whose value does not change) representing locks, or special owners (described below).

A final expression used as an owner specifies a lock that must be held when the object is accessed. There are four special owners: thisThread, self, readonly and unique. readonly indicates that the object is readonly and cannot be updated. unique means that there is a unique reference to the object. thisThread means that the object is thread-local (*i.e.*, unshared). self means that the object is protected by its own lock. The owner of an object is said to *guard* all of its fields.

Method declarations may have a accesses clause that contains a set of final expressions; the owners of these expressions are locks, those locks must be held when the method is invoked.[1]

Deadlock types associate a lock level with each lock. The typing rules ensure that if a thread acquires a lock l_2 (which the thread does not already hold) while holding a lock l_1, then l_2's level is less than l_1's level; in other words, locks are acquired in descending order. Lock levels and the partial order on them are defined by statements of the form LockLevel l_1 = new; l_2 < l_1. In PRFJ, only locks on objects with owner self can be acquired (acquiring locks on other objects is not useful for showing race-freedom), so lock levels are associated only with objects with owner self.

In this paper, we focus on [BLR02]'s basic deadlock types, in which all instances of a class are associated with the same lock level. An extension that supports polymorphism in lock levels, i.e., that allows classes to be parameterized with formal lock level parameters, is presented in [BLR02], but in our experience, this extra flexibility is rarely useful, and it makes type inference much more difficult.

---

[1] For simplicity, we ignore the distinction between owners and root owners in this overview.

```
class Account<self:l1> {
  int balance;
  Vector<self:l2> v = new Vector<self:l2>();
  Locklevel l2 < l1;

  int deposit(int x, int tid) locks l1, l2 {
    synchronized(this) {
      this.balance = this.balance + x;
      v.addElement(new Integer<readonly>(tid));
    }
  }
}

class Vector<self:l2> {

....

  synchronized void ensureCapacity(int minCapacity) locks l2, this {...}
  synchronized void addElement(Object<f> obj) locks l2 {
    if (elementCount == elementData.length)
      ensureCapacity(elementCount + 1);
    modCount++;
    elementData[elementCount++] = obj;
  }
}

Account<self:l1> a1 = new Account<self:l1>;
fork(a1){a1.deposit(100,1);}
fork(a1) {a1.deposit(100,2);}
```

Fig. 3. An example program with race-free types and deadlock types

In the deadlock type system, each method m is annotated with a locks clause that contains a set of lock levels. These lock levels are the maxima amongst the levels of locks that may be acquired when m is executed. To ensure that a program is free of deadlocks, the typing rule for method calls ensures that the caller only holds locks that are of a higher level than the levels in the called method's locks clause. A locks clause may also contain a lock l, which indicates that the thread invoking the method may hold a lock on object l. The typing rule for synchronized expression checks that the lock being acquired is l or has a lower level than l. This allows typing of programs in which, for example, a synchronized method of a class calls a synchronized method of the same class on the same object.

The program in Figure 3 illustrates race-free types and deadlock types. It shows a class Account whose owner is self with lock level l1. It has an instance field v of class type Vector with owner self and lock level l2. The main thread spawns two threads, each of which invoke the deposit method on account a1. The deposit method acquires a lock on this followed by a lock on its field v when the addElement method of v is invoked. This is consistent with the declared lock level ordering 12 < 11, since the lock on v is acquired after the lock on a1. The lock levels specified in the locks clause of deposit method satisfy the method invoke rule as it is called with no locks held (corresponding to lock level infinity which is greater than both 11 and 12). The addElement method of v is called with current lock level (the lock level of the most recently acquired lock but not yet released) 11 which is greater than 12 specified in the locks clause. Hence, the call to addElement also typechecks. When the addElement method calls the ensureCapacity method of v, the current lock level 12 is not greater than 12, rather it is the same. However, the ensureCapacity method also contains a lock, viz., this (which has lock level 12), in the locks clause which is held at the call site. The program typechecks because the typing rule for method calls allows the current lock level to be the same as the level of the lock l in the locks clause if the thread invoking the method already holds the lock on l. Reacquiring the lock on this in ensureCapacity method also typechecks, because the typing rule for synchronized expressions checks whether the newly acquired lock is the same as specified in the locks clause.

4   Static Type Inference for Deadlock Types

The following section presents a type inference algorithm for [BLR02]'s basic deadlock type system. The algorithm assumes that race-free types are already known. Type inference for race-free types is NP-hard, but in practice, race-free types can often be obtained using a SAT solver [FF04] or type discovery [AS04, ASS04].

The algorithm works as follows:

1. Each field, method parameter and local variable with owner self is initially assigned a distinct lock level. The levels are initially unordered. For each method, equality constraints among lock levels are generated based on assignment statements and method invocations. This is necessary for programs to typecheck as left and right hand side of an assignment must have the same type (modulo subtyping), and the type now includes the lock level when the owner is self. Similarly, for each call site, each argument to the called method must have the same lock level as the corresponding parameter. The constraints can then be solved using the standard Union-Find algorithm.

2. A static lock graph G_L is constructed that captures the locking pattern of the program. A synchronized statement is redundant if the final expression corresponding to the lock acquire appears nested below a lock acquire of the same final expression by a synchronized statement in the same method or if the final expression is the same as the rootowner of an expression in the accesses clause of the method containing this synchronized statement. For each synchronized statement in the program that is not redundant (including the implicit synchronized statement enclosing the body of each synchronized method), the graph contains a corresponding node, called a

lock node . For each method m in the program, the graph contains a corresponding node n_m, called a *method node*. There is an edge from a lock node n_1 to a lock node n_2 if the synchronized block corresponding to node n_2 is syntactically directly nested within the synchronized block corresponding to n_1 or the other synchronized statements between n_1 and n_2 are redundant. There is an edge from each method node n_m to the lock node for each outermost synchronized block in m. For each method call within the scope of a synchronized block except calls to Thread.start, there is an edge from the lock node corresponding to the inner most synchronized block that encloses the method call to the method node for the called method.

3. The method node n_m for method m is associated with a set L_m of lock levels. L_m^{init} contains the lock levels of the lock nodes that are children of n_m. Recall that each lock node is associated with a unique lock level in step 1. If n_m has no lock node children, L_m^{init} is empty. Let $called(m)$ denote the set of methods directly called by method m such that the corresponding call sites are not in the scope of a synchronized block in m. The set L_m for each method is computed using a fixed point computation. It is the least fixed point solution to the following set of constraints: for each method m, $L_m = L_m^{init} \cup \bigcup_{m' \in called(m)} L_{m'}$. The right side is monotonic in $L_{m'}$, so the least solution can be computed by a standard fixed-point computation. For each method node n_m, the lock levels in L_m are added to the locks clause of m.

4. For each lock node n with lock level l, and for each lock node n' with lock level l', such that there is an edge from n to n', add the declaration $l > l'$ to the inferred typing. If there is an edge from n to a method node n_m, then for each lock level l' in L_m other than l, add the declaration $l > l'$ to the inferred typing.

5. For each lock node n with lock level l in method m, if n is reachable from a lock node ancestor of n with the same lock level and in a different method, then add to the locks clause of m the final expression denoting the lock acquired by the synchronized statement corresponding to n. (Leaving lock level l in m's locks clause is sometimes unnecessary but harmless).

The complexity of the algorithm is $O((|V_m|^3) + ((|V_m| + |V_l|) \times |E|))$, where $|V_m|$ is the number of method nodes (equal to the number of methods in the program), $|V_l|$ is the number of lock nodes (equal to number of synchronized statements in the program), and $|E|$ is the number of edges in G_L. Our type inference algorithm is correct in the sense that it produces correct typings for all typable programs, as shown in the appendix. For untypable programs, our inference algorithm does not simply halt and report failure; rather, it produces the best type annotations it can for the given program. This is useful for focused run-time checking, as described in Section 6. Our type inference algorithm does not attempt to determine whether the given program is typable; instead, we simply run the type checker on the program with the inferred type annotations.

Figure 4 shows a static lock graph G_L for program in Figure 3. It contains method nodes for deposit, addElement and ensureCapacity. It contains a lock node for each synchronized statement. Each lock node is labeled with the acquired lock and its lock level. For example, n_1 corresponds to the synchronized statement in method deposit, which acquires the lock on this, which has lock level 11 (from step 1 of the type inference algorithm). Each method node n_m is labeled with the set $L(m)$ computed in step 3 of the type inference algorithm.

After computing G_L and L, the type inference algorithm infers the deadlock types for the program in Figure 3 as follows. First the elements of $L(m)$ are used in the locks clause of method m. For example, the locks clause of method deposit contains lock level 11, since $L(\text{deposit}) = \{11\}$. Edges from lock nodes to method nodes introduce lock level orderings. For example, the edge from n_1 to $n_{\text{addElement}}$ introduces the declaration 11 > 12 by step 4. n_2 is an ancestor of n_3 with the same lock level m and in a different method. Therefore, step 5 adds this to the locks clause for method ensureCapacity.

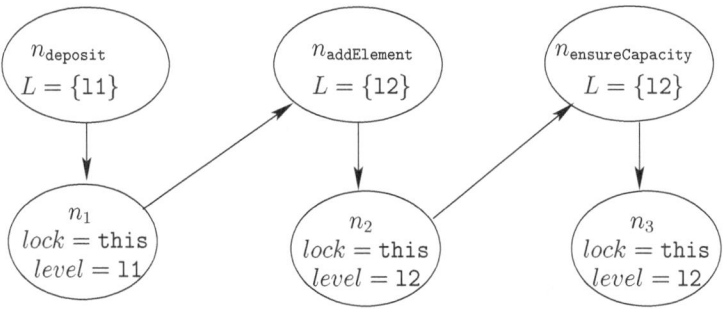

Fig. 4. Static lock graph G_L for program in Figure 3

Theorem 1. *The algorithm above produces a correct typing for a program if the program is typable in the basic deadlock type system of [BLR02].*

The proof for the theorem above is available in the companion technical report [AWS05].

5 Deadlock-Free Extended Parameterized Atomic Java

This section describes how to add basic deadlock types [BLR02], discussed in Section 3, to Extended Parameterized Atomic Java (EPAJ) [SAWS05], our type system that ensures absence of races and atomicity violations. The resulting type system, Deadlock-Free Extended Parameterized Atomic Java (DEPAJ), ensures absence of deadlocks due to locks and hence provides stronger atomicity guarantees. Atomicity types in EPAJ are adopted from Atomic Java [FQ03] and are based on Lipton's theory of reduction [Lip75], which requires that the code to be reduced (*i.e.*, shown to be atomic) cannot be involved in a deadlock. EPAJ and Atomic Java do not consider whether the code can be involved in deadlocks. By adding deadlock types [BLR02] to EPAJ, the resulting type system provides stronger atomicity guarantees. Adding the deadlock types proposed in [BLR02] to EPAJ is straightforward. For brevity, we do not describe atomicity types here, since they are not changed by the addition of deadlock types, although they get a stronger semantics.

EPAJ extends PRFJ to allow each field to have a different guard. Because ownership in EPAJ is per field, not per object (as in PRFJ), PRFJ's notion of rootowner is not well-defined, so we discard it and compensate by allowing formal owner parameters in `accesses` clauses, which are called `requires` clauses in RFJ [FF00] and EPAJ. To make this work, every formal owner parameter f is qualified with a final expression e that indicates the object that f refers to when f is instantiated with `self`. A `guarded_by` clause on a field contains a *lock expression*, which is either a final expression or $f\$e$ where e is a final expression and f is the first formal owner parameter in the type of e. A `guarded_by` clause cannot contain a special owner (explained in Section 3) explicitly, but the formal owner parameter in it may be instantiated with a special owner, providing the same effect. PRFJ allows synchronization only on objects with owner `self`, because only those objects can be roots of ownership trees. In contrast, EPAJ eliminates the concept of root owner and consequently allows synchronization on objects with any owner. Therefore, in DEPAJ, a lock level is associated with each final expression used in a `synchronized` statement. This has the side-effect of allowing different lock levels for different instances of a class in some cases, even in the basic (non-polymorphic) deadlock type system.

The type inference algorithm in Section 4 easily carries over to infer deadlock types in DEPAJ.

6   Focusing Run-Time Checks for Deadlock Detection

Deadlock types enforce a conservative strategy for preventing deadlocks. Therefore, there are deadlock-free programs not typable in this type system. For example, programs which have cycles in the static lock level ordering are untypable even though they may be deadlock-free. An example appears below. The type system also requires all elements of a Collection class to have the same lock level. This may be too restrictive and can lead to untypable programs even though the programs are deadlock-free. For such programs, information gathered from the type system can be used to focus run-time checking, *i.e.*, run-time checking can safely be omitted for parts of the program guaranteed to be deadlock-free by the type system.

To focus the generalized version of the GoodLock algorithm that does not handle gate locks, we find all the cycles of the form $l_1 > l_2... > l_1$ among lock level orderings produced by the deadlock type inference algorithm. We instrument only lock acquires and releases of expressions whose lock level is part of a cycle. Other synchronized expressions do not need to be instrumented. This leads to fewer intercepted events and smaller lock trees that need to be analyzed. It is easy to determine which lock levels are part of cycles. Construct a graph $G = (V, E)$, where each lock level is a node in V and there is an edge from l to l' if the inferred typing declares $l > l'$. A simple depth first search can find all nodes that are part of some cycle.

As an example, we consider a modified version of the elevator program, developed at ETH Zürich and used as a benchmark in [vPG01]. elevator is a simple discrete-event simulation of people going up and down in elevators; we extended it to model the people explicitly as objects. The instances of Person are initially stored in a static Vector field people in the main Elevator class. When some of them make a request to go up or down they are moved from Elevator.people to the upPeople or downPeople vector of the appropriate instance of Floor. An instance of Lift services the request by acquiring the lock on the instance of Floor where the requester(s) are waiting, updating the status flags of the floor, and then moving people from the upPeople or downPeople field of the floor to the appropriate peopleFor vector in the Lift instance based on their destination floor. On reaching the destination floor, the lift moves the people in the corresponding peopleFor vector back to the Elevator.people vector. All the moves between vectors are done using the addAll method of the Vector class. v1.addAll(v2) adds each object in v2 to v1. v1.addAll(v2) acquires the lock on v2 while holding the lock on v1. The modified elevator program is deadlock-free, but not typable with [BLR02]'s basic deadlock types. This is because every Vector class is self-synchronized with some lock level, say l. However, the lock level orderings required by the typing rules as a result of the calls to addAll together create a cycle in the lock ordering. The program is not typable even in the full polymorphic version of the type system, for essentially the same reason. If the vectors in Elevator.people, Lift.peopleFor, and Floor.upPeople are assigned different lock levels l_1, l_2 and l_3 in the polymorphic type system, then the orderings $l_1 > l_2 > l_3 > l_1$ are required. This cycle makes the program untypable. Different instances of Lift are started in a loop, so it is not possible even in the polymorphic type system to assign different lock levels to the peopleFor field of different instances of Lift.

Our type inference algorithm infers orderings of the form $l > l$, where l is the lock level assigned to the self synchronized Vector class. Other locks, including locks on instances of Floor and Lift, are not involved in the cycle. So, we focus run-time checks by intercepting lock acquires and releases only on instances of Vector.

To focus the generalized version of the GoodLock algorithm that handles gate locks, we find all the cycles among lock level orderings produced by the type inference algorithm as discussed above. All lock levels that are comparable to lock levels involved in a cycle in the ordering of lock levels need to be instrumented (not just the lock levels involved in a cycle).

7   Experience

To evaluate our technique, we manually ran the inference algorithm on the five multi-threaded server programs used in [BR01]. The programs are small, ranging from about 100 to 600 lines of code, and totaling approximately 1600 LOC.

Our type inference algorithm successfully inferred complete and correct typings for all five server programs. We compared the inferred typings to Boyapati's manually inserted type annotations. In his code, ChatServer contains 12 deadlock annotations, GameServer contains 8 deadlock annotations, HTTPServer contains 2 deadlock annotations, and QuoteServer and PhoneServer contain none. Thus, approximately 1600 LOC required 22 deadlock annotations; that's approximately 15 annots/KLOC. Our type inference algorithm eliminates the need for the user to write all of those annotations.

Since these programs are guaranteed deadlock free using the types, the need for run-time checking for potential deadlocks is completely eliminated for these programs.

Table 1. The run times of dynamic deadlock detection for modified elevator example

		3 threads	7 threads	15 threads	30 threads	60 threads
Base time		0.23s	0.30s	0.52s	2.60s	6.60s
Full	Size	621	1037	1848	3359	5734
	Unopt	0.76s	0.94s	14.93s	1m23.08s	3m42.9s
	Opt	1.10s	1.66s	17.32s	1m28.05s	4m3.0s
Focused	Size	433	646	1063	1824	2947
	Unopt	0.40s	0.53s	11.71s	34.91s	1m22.66s
	Opt	0.51s	0.72s	12.40s	36.35s	1m28.28s

We implemented the unoptimized and optimized generalized Goodlock algorithms without gate locks described in Section 2 and used them to analyze the modified `elevator` program described in Section 6. The experiments were performed on a Sun Blade 1500 with a 1GHz UltraSPARC III CPU, 2GB RAM, and JDK 5.0. Table 1 shows the running times for the `elevator` program with 3,7,15, 30 and 60 `Lift` threads. The "Base time" row gives the execution time of the original program without any instrumentation. The "Full" and "Focused" rows give the execution results of the program augmented with full and focused run-time checking, respectively. For "Full" and "Focused" rows, sub rows "Size", "Unopt" and "Opt" give the the number of nodes in all runtime lock trees, execution times of the unoptimized algorithm, and optimized algorithm respectively. As discussed in Section 6, focused run-time checking in this example intercepts lock acquires and releases only on instances of `Vector`. The results demonstrate that the focused analysis significantly decreases the run-time overhead of deadlock checking and the size of runtime lock trees. Let Ofull = Full − Base denote the overhead of full checking, and Ofoc = Focused − Base denote the overhead of focused checking. The average speedup (i.e., fractional reduction in overhead) is (Ofull − Ofoc)/Ofull, which is 55.8% for the unoptimized algorithm, and 58.4% for the optimized algorithm. The average size of the runtime lock trees is reduced by 41%. Surprisingly, the optimized algorithm runs slower than the unoptimized algorithm, although its asymptotic worst-case time complexity is better. The main reason is that the optimized algorithm uses more complicated data structures, and for `elevator` example, where the the run-time lock graph is relatively simple, the benefit of caching explored paths falls short of the overhead of data structure maintenance.

8   Related Work

Techniques for run-time detection of deadlocks include the GoodLock algorithm [Hav00], a run-time analysis implemented in Compaq's Visual Threads [Har00] and ConTest [EFG+03]. As discussed in Section 2, the GoodLock algorithm detects only potential deadlocks involving two threads. We generalize it to detect deadlocks involving any number of threads. Bensalem and Havelund independently generalized the GoodLock algorithm to detect such deadlocks [BH05]. They do not consider combining it with static analysis; we do. However, their algorithm eliminates false alarms arising from cycles that contain lock acquires and releases that cannot happen in parallel. Visual Threads can detect potential for deadlocks [Har00], but it is not clear what algorithm is used. ConTest [EFG+03] detects actual deadlocks, not potential deadlocks, and therefore may miss some potential deadlocks; on the other hand, ConTest's scheduling perturbation heuristics make potential deadlocks of all kinds (including deadlocks due to condition synchronization) more likely to manifest themselves as actual deadlocks during testing with ConTest, compared to testing without ConTest. A recent extension to ConTest implements a run-time deadlock checking algorithm that combines information obtained from multiple executions of the program [FNBU]. That technique is compatible with our work, and it would be useful to combine them. ConTest does not use static analysis to optimize run-time checking.

Hatcliff *et al.* [HRD04] verify atomicity specifications using model checking. Their approach is more accurate than type-based analysis of atomicity: it does not produce false alarms, and it fully enforces the condition that deadlock must not occur within an atomic block. Hatcliff et al. point out that previous type systems for atomicity do not enforce this condition. DEPAJ takes a step towards addressing this, by ensuring that lock-induced deadlocks do not occur within atomic blocks. Since their approach is based on model checking, it is computationally expensive and hence practical only for relatively small programs.

Engler *et al.* [EA03], von Praun [vP04], and Williams *et al.* [WTE05] developed inter-procedural static analyses that detect potential deadlocks in programs. These static analyses are also based on checking whether locks are acquired in a consistent order by all threads. These static analyses are more sophisticated and more accurate than basic deadlock types but still produce numerous false alarms (Engler *et al.* and Williams *et al.* partially address this problem by using heuristics to rank or suppress warnings that seem more likely to be false alarms), so it would be useful to use them in conjunction with run-time checking, which generally produces fewer false alarms. Specifically, although these papers do not consider run-time checking, the results of their analyses could be used instead of deadlock types in our technique for focused run-time detection of potential deadlocks.

The idea of using static analysis to optimize run-time checking has been applied to detection of races [vPG01, CLL$^+$02, ASWS05] and atomicity violations [SAWS05, ASWS05] but not to detection of potential deadlocks, to the best of our knowledge.

References

[AS04]   Rahul Agarwal and Scott D. Stoller. Type inference for parameterized race-free Java. In *Proceedings of the Fifth International Conference on Verification, Model Checking and Abstract Interpretation*, volume 2937 of *Lecture Notes in Computer Science*, pages 149–160. Springer-Verlag, January 2004.

[ASS04]   Rahul Agarwal, Amit Sasturkar, and Scott D. Stoller. Type discovery for parameterized race-free Java. Technical Report DAR-04-16, Computer Science Department, SUNY at Stony Brook, September 2004.

[ASWS05]   Rahul Agarwal, Amit Sasturkar, Liqiang Wang, and Scott D. Stoller. Optimized run-time race detection and atomicity checking using partial discovered types. In *Proc. 20th IEEE/ACM International Conference on Automated Software Engineering (ASE)*. ACM Press, November 2005.

[AWS05]   Rahul Agarwal, Liqiang Wang, and Scott D. Stoller. Detecting potential deadlocks with static analysis and runtime monitoring. Technical Report DAR-05-25, Computer Science Department, SUNY at Stony Brook, September 2005. Available at http://www.cs.sunysb.edu/~ragarwal/deadlock/.

[BH05]   Saddek Bensalem and Klaus Havelund. Scalable deadlock analysis of multi-threaded programs. In *Proceedings of the Parallel and Distributed Systems: Testing and Debugging (PADTAD) Track of the 2005 IBM Verification Conference*. Springer-Verlag, November 2005.

[BLR02]   Chandrasekhar Boyapati, Robert Lee, and Martin Rinard. Ownership types for safe programming: Preventing data races and deadlocks. In *Proc. 17th ACM Conference on Object-Oriented Programming, Systems, Languages and Applications (OOPSLA)*, pages 211–230, November 2002.

[BR01]   Chandrasekar Boyapati and Martin C. Rinard. A parameterized type system for race-free Java programs. In *Proc. 16th ACM Conference on Object-Oriented Programming, Systems, Languages and Applications (OOPSLA)*, volume 36(11) of *SIGPLAN Notices*, pages 56–69. ACM Press, 2001.

[CLL$^+$02] Jong-Deok Choi, Keunwoo Lee, Alexey Loginov, Robert O'Callahan, Vivek Sarkar, and Manu Sridharan. Efficient and precise datarace detection for multithreaded object-oriented programs. In *Proc. ACM SIGPLAN Conference on Programming Language Design and Implementation (PLDI)*, pages 258–269. ACM Press, 2002.

[EA03]   Dawson R. Engler and Ken Ashcraft. RacerX: Effective, static detection of race conditions and deadlocks. In *Proc. 24th ACM Symposium on Operating System Principles*, pages 237–252. ACM Press, October 2003.

[EFG$^+$03] Orit Edelstein, Eitan Farchi, Evgeny Goldin, Yarden Nir, Gil Ratsaby, and Shmuel Ur. Framework for testing multi-threaded Java programs. *Concurrency and Computation: Practice and Experience*, 15(3-5):485–499, 2003.

[FF00]   Cormac Flanagan and Stephen Freund. Type-based race detection for Java. In *Proc. ACM SIGPLAN Conference on Programming Language Design and Implementation (PLDI)*, pages 219–232. ACM Press, 2000.

[FF04]   Cormac Flanagan and Stephen Freund. Type inference against races. In *Proc. 11th International Static Analysis Symposium (SAS)*, volume 3148 of *Lecture Notes in Computer Science*. Springer-Verlag, August 2004.

[FNBU]   Eitan Farchi, Yarden Nir-Buchbinder, and Shmuel Ur. Cross-run lock discipline checker for java. Tool proposal for IBM Verification Conference 2005.

[FQ03]   Cormac Flanagan and Shaz Qadeer. A type and effect system for atomicity. In *Proc. ACM SIGPLAN Conference on Programming Language Design and Implementation (PLDI)*, pages 338–349. ACM Press, 2003.

[Har00]   Jerry J. Harrow. Runtime checking of multithreaded applications with Visual Threads. In *Proc. 7th Int'l. SPIN Workshop on Model Checking of Software*, volume 1885 of *Lecture Notes in Computer Science*, pages 331–342. Springer-Verlag, August 2000.

[Hav00]   Klaus Havelund. Using runtime analysis to guide model checking of java programs. In *Proc. 7th Int'l. SPIN Workshop on Model Checking of Software*, volume 1885 of *Lecture Notes in Computer Science*, pages 245–264. Springer-Verlag, August 2000.

[HRD04]   John Hatcliff, Robby, and Matthew B. Dwyer. Verifying atomicity specifications for concurrent object-oriented software using model checking. In *Proceedings of the Fifth International Conference on Verification, Model Checking and Abstract Interpretation (VMCAI 2004)*, Lecture Notes in Computer Science. Springer-Verlag, 2004.

[Lip75]   Richard J. Lipton. Reduction: A method of proving properties of parallel programs. *Communications of the ACM*, 18(12):717–721, 1975.

[SAWS05]   Amit Sasturkar, Rahul Agarwal, Liqiang Wang, and Scott D. Stoller. Automated type-based analysis of data races and atomicity. In *Proc. ACM SIGPLAN 2005 Symposium on Principles and Practice of Parallel Programming (PPoPP)*. ACM Press, June 2005.

[vP04]   Christoph von Praun. *Detecting Synchronization Defects in Multi-Threaded Object-Oriented Programs*. PhD thesis, ETH Zürich, 2004.

[vPG01]   Christoph von Praun and Thomas R. Gross. Object race detection. In *Proc. 16th ACM Conference on Object-Oriented Programming, Systems, Languages and Applications (OOPSLA)*, volume 36(11) of *SIGPLAN Notices*, pages 70–82. ACM Press, October 2001.

[WTE05]   Amy Williams, William Thies, and Michael D. Ernst. Static deadlock detection for Java libraries. In *Proc. 2005 European Conference on Object-Oriented Programming (ECOOP)*, Lecture Notes in Computer Science. Springer-Verlag, July 2005.

Dynamic Deadlock Analysis of Multi-threaded Programs

Saddek Bensalem[1] and Klaus Havelund[2]

[1] Université Joseph Fourier, Verimag, Grenoble, France
[2] Kestrel Technology, Palo Alto, California, USA

Abstract. This paper presents a dynamic program analysis algorithm that can detect deadlock potentials in a multi-threaded program by examining a single execution trace, obtained by running an instrumented version of the program. The algorithm is interesting because it can identify deadlock potentials even though no deadlocks occur in the examined execution, and therefore it scales very well in contrast to more formal approaches to deadlock detection. It is an improvement of an existing algorithm in that it reduces the number of false positives (false warnings). The paper describes an implementation and an application to three case studies.

1 Introduction

The Java programming language [2] explicitly supports concurrent programming through a selection of concurrency language concepts, such as threads and monitors. Threads execute in parallel, and communicate via shared objects that can be locked using synchronized access to achieve mutual exclusion. However, with concurrent programming comes a new set of problems that can hamper the quality of the software. Deadlocks form such a problem category. That deadlocks pose a common problem is emphasized by the following statement in [23]: "*Among the most central and subtle liveness failures is deadlock. Without care, just about any design using synchronization on multiple cooperating objects can contain the possibility of deadlock*".

In this paper we present a dynamic program analysis algorithm that can detect the potential for deadlocks in a program by analyzing a trace (log file) generated from a successful deadlock free execution of the program. The algorithm is interesting because it catches existing deadlock potentials with very high probability even when no actual deadlocks occur during test runs. A basic version of this algorithm has previously been implemented in the commercial tool Visual Threads [16]. This basic algorithm, however, can give false positives (as well as false negatives), putting a burden on the user to refute such. Our goal is to reduce the amount of false positives reported by the algorithm, and for that purpose we have modified it as reported in this paper. Detection of errors in concurrent programs by analysis of successful runs was first suggested for low-level data races in [26]. Other forms of data races have later shown to be

S. Ur, E. Bin, and Y. Wolfsthal (Eds.): Haifa Verification Conf. 2005, LNCS 3875, pp. 208–223, 2006.

detectable using related forms of analysis, such as high-level data races [4] and atomicity violations [5].

Two types of deadlocks have been discussed in the literature [27] [22]: *resource deadlocks* and *communication deadlocks*. In resource deadlocks, a process which requests resources must wait until it acquires all the requested resources before it can proceed with its computation. A set of processes is resource deadlocked if each process in the set requests a resource held by another process in the set, forming a *cycle* of lock requests. In communication deadlocks, messages are the resources for which processes wait. In this paper we focus only on *resource deadlocks*. In Java, threads can communicate via shared objects by for example calling methods on those objects. In order to avoid data races in these situations (where several threads access a shared object simultaneously), objects can be locked using the `synchronized` statement, or by declaring methods on the shared objects `synchronized`, which is equivalent. For example, a thread t can obtain a lock on an object A and then execute a statement S while having that lock by executing the following statement: `synchronized(A){S}`. During the execution of S, no other thread can obtain a lock on A. The lock is released when the scope of the `synchronized` statement is left; that is, when execution passes the curly bracket: '}'. Java also provides the `wait` and `notify` primitives in support for user controlled interleaving between threads. While the `synchronized` primitive is the main source for resource deadlocks in Java, the `wait` and `notify` primitives are the main source for communication deadlocks. Since this paper focuses on resource deadlocks, we shall in the following focus on Java's capability of creating and executing threads and on the `synchronized` statement.

The difficulty in detecting deadlocks comes from the fact that concurrent programs typically are non-deterministic: several executions of the same program on the same input may yield different behaviors due to slight differences in the way threads are scheduled. Various technologies have been developed by the formal methods community to circumvent this problem, such as static analysis and model checking. Static analysis, such as performed by tools like JLint [3], PolySpace [25] and ESC [11], analyze the source code without executing it. These techniques are very efficient, but they often yield many false positives (false warnings) and additionally cannot well analyze programs where the object structure is very dynamic. Model checking has been applied directly to software (in contrast to only designs), for example in the Java PathFinder system (JPF) developed by NASA [18, 29], and in similar systems [14, 21, 10, 6, 28, 24, 12]. A model checker explores all possible execution paths of the program, and will therefore theoretically eventually expose a potential deadlock. This process is, however, quite resource demanding, in memory consumption as well in execution time, especially for large realistic programs.

Static analysis and model checking are both typically complete (no false negatives), and model checking in addition is typically sound (no false positives). The algorithm presented in this paper is neither sound nor complete, but it scales and it is very effective: it finds bugs with high probability and it yields few false positives. The technique is based on trace analysis: a program is instrumented to log synchronization events when executed. The algorithm then examines the log file, building a lock graph, which reveals deadlock potentials by containing *cycles*. The algorithm has been implemented in the Java PathExplorer tool [20], which in addition analyzes Java programs for various forms of data races [26, 4] and conformance with temporal logic properties [7]. Although the implementation is Java specific, the principles and theory presented are universal and apply in full to multi-threaded programs written in languages like C and C++ as well. In fact, two of the case studies involve C++ programs. In this case the programs must currently be manually instrumented. The algorithm was first described in [8].

Some additional closely related work specifically requires mentioning. In earlier work we presented the GoodLock algorithm [17] which attempts to improve the basic lock graph algorithm presented in [16] by reducing false positives in the presence of gate locks (a common lock taken first by involved threads). This algorithm was based on building acyclic lock trees (rather than cyclic lock graphs as in [16]) but it could only detect deadlocks between pairs of threads. The algorithm presented in this paper also tries to reduce false positives in presence of gate locks, but can detect deadlocks between any number of threads, and builds directly on the cyclic graph model in [16]. In addition, the algorithm also reduces false positives arising from code segments that cannot possibly execute concurrently. In parallel work [1], included in these proceedings, an algorithm extending the GoodLock algorithm is suggested for reducing false positives, also in presence of gate locks. This algorithm uses a combination of acyclic lock trees and cyclic lock graphs to represent locking patterns in a program run. In that work a framework is furthermore suggested for using static analysis in combination with dynamic analysis to detect deadlocks. In these pieces of work a single execution trace is used as a basis for the dynamic analysis. Also presented at PADTAD'05 was work by IBM [13] where several execution traces, generated from several runs of the program being tested, are used to create a single locking model that is then analyzed. This approach is useful to reduce false negatives (missed errors).

The paper is organized as follows. Section 2 introduces preliminary concepts and notation used throughout the rest of the paper. Section 3 introduces an example that illustrates the notion of deadlock and the different forms of false positives that are possible. Section 4 presents the basic algorithm suggested in [16] (the algorithm is only explained in few words in [16]). The subsequent two sections 5 and 6 suggest modifications, each reducing false positives. Section 7 shortly describes the implementation of the algorithm and presents the results of three case studies. Finally, Section 8 concludes the paper.

2   Preliminaries

A directed graph is a pair $G = (S, R)$ of sets satisfying $R \subseteq S \times S$. The set R is called the edge set of G, and its elements are called edges. A path p is a non-empty graph $G = (S, R)$ of the form $S = \{x_1, x_2, \ldots, x_k\}$ and

$R = \{(x_1, x_2), (x_2, x_3), \ldots, (x_{k-1}, x_k)\}$, where the x_i are all distinct, except that x_k may be equal to x_1, in which case the path is a cycle. The nodes x_0 and x_k are linked by p; we often refer to a path by the natural sequence of its nodes, writing, say, $p = x_1, x_2, \ldots, x_k$ and calling p a path from x_1 to x_k. In case where the edges are labeled with elements in W, G is triplet (S, W, R) and called a labeled graph with $R \subseteq S \times W \times S$. A labeled path, respectively cycle, is a labeled graph with the obvious meaning. Given a sequence $\sigma = x_1, x_2, \ldots, x_n$, we refer to an element at position i in σ by $\sigma[i]$ and the length of σ by $|\sigma|$. We let $<>$ denote the empty sequence, and the concatenation of two sequences σ_1 and σ_2 is denoted by $\sigma_1 \frown \sigma_2$. We denote by σ^i the prefix $x_1, \ldots, x_i$. Let $M : [A \xrightarrow{m} B]$ be a finite domain mapping from elements in A to elements in B (the $\xrightarrow{m}$ operator generates the set of finite domain mappings from A to B, hence partial functions on A). We let $M \dagger [a \mapsto b]$ denote the mapping M overridden with a mapping to b. That is, the $\dagger$ operator represents map overriding, and $[a \mapsto b]$ represents a map that maps a to b. Looking up the value mapped to by a in M is denoted by $M[a]$. We denote the empty mapping by $[\,]$.

3   An Example

We shall with an example illustrate the three categories of false positives that the basic algorithm reports, but which the improved algorithm will not report. The first category, *single threaded cycles*, refers to cycles that are created by one single thread. *Guarded cycles* refer to cycles that are guarded by a gate lock "taken higher" up by all involved threads. Finally, *thread segmented cycles* refer to cycles between thread segments that cannot possibly execute concurrently. The program in Figure 1 illustrates these three situations, and a true positive. The real deadlock potential exists between threads T_2 and T_3, corresponding

```
Main :
01: new T1().start();
02: new T2().start();
```

T_1 :
```
03: synchronized(G){
04:   synchronized(L1){
05:     synchronized(L2){}
06:   }
07: };
08: t3 = new T3();
09: t3.start();
10: t3.join();
11: synchronized(L2){
12:   synchronized(L1){}
13: }
```

T_2 :
```
14: synchronized(G){
15:   synchronized(L2){
16:     synchronized(L1){}
17:   }
18: }
```

T_3 :
```
19: synchronized(L1){
20:   synchronized(L2){}
21: }
```

Fig. 1. Example containing four lock cycles

to a cycle on L_1 and L_2. The single threaded cycle within T_1 clearly does not represent a deadlock. The guarded cycle between T_1 and T_2 does not represent a deadlock since both threads must acquire the gate lock G first. Finally, the thread segmented cycle between T_1 and T_3 does not represent a deadlock since T_3 will execute before T_1 executes its last two synchronization statements.

When analyzing such a program for deadlock potentials, we are interested in observing all lock acquisitions and releases, and all thread starts and joins. The program can be instrumented to emit such events. A lock trace $\sigma = e_1, e_2, \ldots, e_n$ is a finite sequence of lock and unlock events and start and join events. Let E_σ denote the set of events occurring in σ. Let T_σ denote the set of threads occurring in E_σ, and let L_σ denote the set of locks occurring in E_σ. We assume for convenience that the trace is *reentrant free* in the sense that an already acquired lock is never re-acquired by the same thread (or any other thread of course) before being released. Note that Java supports reentrant locks by allowing a lock to be re-taken by a thread that already has the lock. However, the instrumentation can generate reentrant free traces if it is recorded how many times a lock has been acquired nested by a thread. Normally a counter is introduced that is incremented for each lock operation and decremented for each unlock operation. A lock operation is now only reported if the counter is zero (it is free before being taken), and an unlock operation is only reported if the counter is 0 again after the unlock (it becomes free again).

For illustration purposes we shall assume a non-deadlocking execution trace σ for this program. It doesn't matter which one since all non-deadlocking traces will reveal all four cycles in the program using the basic algorithm. We shall assume the following trace of line numbered events (the line number is the first argument), which first, after having started T_1 and T_2 from the *Main* thread, executes T_1 until the join statement, then executes T_2 to the end, then T_3 to the end, and then continues with T_1 after it has joined on T_3's termination. The line numbers are given for illustration purposes, and are actually recorded in the implementation in order to provide the user with useful error messages. In addition to the lock and unlock events $l(lno, t, o)$ and $u(lno, t, o)$ for line numbers lno, threads t and locks o, the trace also contains events for thread start, $s(lno, t_1, t_2)$ and thread join, $j(lno, t_1, t_2)$, meaning respectively that t_1 starts or joins t_2 in line number lno.

$\sigma =$
$s(1, Main, T_1),\quad s(2, Main, T_2),$
$l(3, T_1, G),\quad l(4, T_1, L_1),\quad l(5, T_1, L_2),\quad u(5, T_1, L_2),\quad u(6, T_1, L_1),\quad u(7, T_1, G),\quad s(9, T_1, T_3),$
$l(14, T_2, G),\quad l(15, T_2, L_2),\quad l(16, T_2, L_1),\quad u(16, T_2, L_1),\quad u(17, T_2, L_2),\quad u(18, T_2, G),$
$l(19, T_3, L_1),\quad l(20, T_3, L_2),\quad u(20, T_3, L_2),\quad u(21, T_3, L_1),$
$j(10, T_1, T_3),\quad l(11, T_1, L_2),\quad l(12, T_1, L_1),\quad u(12, T_1, L_1),\quad u(13, T_1, L_2)$

Occasionally line numbers will be left out when referring to events.

4   Basic Cycle Detection Algorithm

In essence, the detection algorithm consists of finding cycles in a *lock graph*. In the context of multi-threaded programs, the basic algorithm sketched in [16] works as follows. The multi-threaded program under observation is executed, while just *lock* and *unlock* events are observed. A graph of locks is built, with edges between locks symbolizing locking orders. Any cycle in the graph signifies a potential for a deadlock. Hence, we shall initially restrict ourselves to traces including only lock and unlock events (no start or join events). In order to define the lock graph, we introduce a notion that we call a *lock context* of a trace σ in position i, denoted by $\mathcal{C}_L(\sigma, i)$. It is a mapping from each thread to the set of locks owned by that thread at that position. Formally, for a thread $t \in T_\sigma$ we have the following: $\mathcal{C}_L(\sigma, i)(t) = \{o \mid \exists j : j \leq i \wedge \sigma[j] = l(t, o) \wedge \neg \exists k : j < k \leq i \wedge \sigma[k] = u(t, o)\}$. Bellow we give a definition that allows to build the lock graph G_L with respect to an execution trace σ. An edge in G_L between two locks l_1 and l_2 means that there exists a thread t which owns the object l_1 while taking the object l_2.

Definition 1 (Lock graph). *Given an execution trace $\sigma = e_1, e_2, \ldots, e_n$. We say that the lock graph of σ is the minimal directed graph $G_L = (L, R)$ such that : L is the set of locks L_σ, and $R \subseteq L \times L$ is defined by $(l_1, l_2) \in R$ if there exists a thread $t \in T_\sigma$ and a position $i \geq 2$ in σ s. t. $\sigma[i] = l(t, l_2)$ and $l_1 \in \mathcal{C}_L(\sigma, i-1)(t)$.*

In Figure 2 we give an algorithm for constructing the lock graph from a lock trace. In this algorithm, we also use the context $\mathcal{C}_L$ which is exactly the same as in the definition 1. The only difference is that we don't need to explicitly use the two parameters σ and i. The set of cycles in the graph G_L, denoted by $cylces(G_L)$, represents the potential deadlock situations in the program. The lock graph for the example in Figure 1 is also shown in Figure 2. The numbers indicate line numbers where source respectively target locks are taken.

Input: An execution trace σ
G_L is a graph;
$\mathcal{C}_L : [T_\sigma \rightarrow 2^{L_\sigma}]$ is a lock context;
$\mathbf{for}(i = 1 \mathbin{..} |\sigma|)$ do
    $\mathbf{case}\ \sigma[i]\ \mathbf{of}$
        $l(t, o) \rightarrow$
            $G_L := G_L \bigcup \{(o', o) \mid o' \in \mathcal{C}_L(t)\};$
            $\mathcal{C}_L := \mathcal{C}_L \dagger [t \mapsto \mathcal{C}_L(t) \bigcup \{o\}];$
        $u(t, o) \rightarrow$
            $\mathcal{C}_L := \mathcal{C}_L \dagger [t \mapsto \mathcal{C}_L(t) \backslash \{o\}]$
end;
for each c in cycles(G_L) **do**
    print ("deadlock potential:",c);

Fig. 2. The basic algorithm and the lock graph

5   Eliminating Single Threaded and Guarded Cycles

In this section we present a solution that removes false positives stemming from *single threaded cycles* and *guarded cycles*. In [17] we suggested a solution, the

GoodLock algorithm, based on building synchronization trees. However, this solution could only detect deadlocks between pairs of threads. The algorithm to be presented here is not limited in this sense. The solution is to extend the lock graph by labeling each edge between locks with information about which thread causes the addition of the edge and what gate locks were held by that thread when the target lock was taken. A definition of *valid cycles* will then include this information to filter out false positives. First, we define the extended lock graph.

Definition 2 (Guarded lock graph). *Given a trace* $\sigma = e_1, e_2, \ldots, e_n$. *We say that the guarded lock graph of* σ *is the minimal directed labeled graph* $G_L = (L, W, R)$ *such that:* L *is the set of locks* L_σ, $W \subseteq T_\sigma \times 2^L$ *is the set of labels, each containing a thread id and a lock set, and* $R \subseteq L \times W \times L$ *is defined by* $(l_1, (t, g), l_2) \in R$ *if there exists a thread* $t \in T_\sigma$ *and a position* $i \geq 2$ *in* σ *s.t.* $\sigma[i] = l(t, l_2)$ *and* $l_1 \in \mathcal{C}_L(\sigma, i - 1)(t)$ *and* $g = \mathcal{C}_L(\sigma, i - 1)(t)$.

Each edge $(l_1, (t, g), l_2)$ in R is labeled with the thread t that took the locks l_1 and l_2, and a lock set g, indicating what locks t owned when taking l_2. In order for a cycle to be valid, and hence regarded as a true positive, the threads and guard sets occurring in labels of the cycle must be valid in the following sense.

Definition 3 (Valid threads and guards). *Let* G_L *be a guarded lock graph of some execution trace and* $c = (L, W, R)$ *a cycle in* $cycles(G_L)$, *we say that:*

- *threads of* c *are valid if: forall* $e, e' \in R\ e \neq e' \Rightarrow thread(e) \neq thread(e')$
- *guards of* c *are valid if: forall* $e, e' \in R\ e \neq e' \Rightarrow guards(e) \cap guards(e') = \emptyset$

where, for an edge $e \in R$, $thread(e)$, *resp.* $guards(e)$, *gives the first, resp. second, component of the label* (t, g) *of* $e = (l_1, (t, g), l_2)$.

For a cycle to be valid, the threads involved must differ. This eliminates single threaded cycles. Furthermore, the lock sets on the edges in the cycle must not overlap. This eliminates cycles that are guarded by the same lock taken "higher up" by at least two of the threads involved in the cycle. Assume namely that such a gate lock exists, then it will belong to the lock sets of several edges in the cycle, and hence they will overlap at least on this lock. This corresponds to the fact that a deadlock cannot happen in this situation. Valid cycles are now defined as follows:

Definition 4 (Unguarded cycles). *Let* σ *be an execution trace and* G_L *its guarded lock graph. We say that a cycle* $c \in cycles(G_L)$ *is an unguarded cycle if the guards of* c *are valid and threads of* c *are also valid. We denote by* $cycles_g(G_L)$ *the set of unguarded cycles in* $cycles(G_L)$.

We shall in this section not present an explicit algorithm for constructing this graph, since its concerns a relatively simple modification to the basic algorithm – the statement that updates the lock graph becomes:

$$G_L := G_L \bigcup \{(o', (t, \mathcal{C}_L(t)), o) \mid o' \in \mathcal{C}_L(t)\}$$

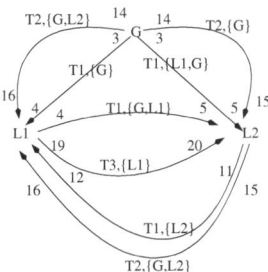

Fig. 3. Guarded lock graph

adding the labels $(t, \mathcal{C}_L(t))$ to the edges. Furthermore, cycles to be reported should be drawn from: $cycles_g(G_L)$.

Let us illustrate the algorithm with an example. We consider again the execution trace σ from Section 3. The guarded graph for this trace is shown in Figure 3. The graph contains the same number of edges as the basic graph in Figure 2. However, now edges are labeled with a thread and a guard set. In particular, we notice that the gate lock G occurs in the guard set of edges $(4, 5)$ and $(15, 16)$. This prevents this guarded cycle from being included in the set of valid cycles since it is not guard valid: the guard sets overlap in G. Also the single threaded cycle $(4, 5) \leftrightarrow (11, 12)$ is eliminated because it is not thread valid: the same thread T_1 occurs on both edges.

6   Eliminating Segmented Cycles

In the previous section we saw the specification of an algorithm that removes false positives stemming from single threaded cycles and guarded cycles. In this section we present the full algorithm that in addition removes false positives stemming from *segmented cycles*. We assume that traces now also contain start and join events. Recall the example in Figure 1 and that the basic algorithm reports a cycle between threads T_1 (line 11-12) and T_3 (line 19-20) on locks L_1 and L_2. However, a deadlock is impossible since thread T_3 is joined on by T_1 in line 10. Hence, the two *code segments*: line 11-12 and line 19-20 can never run in parallel. The algorithm to be presented will prevent such cycles from being reported by formally introducing such a notion of *segments* that cannot execute in parallel. A new directed segmentation graph will record which segments execute before others. The lock graph is then extended with extra label information, that specifies what segments locks are acquired in, and the validity of a cycle now incorporates a check that the lock acquisitions are really occurring in parallel executing segments. The idea of using segmentation in runtime analysis was initially suggested in [16] to reduce the amount of false positives in data race analysis using the Eraser algorithm [26]. We use it in a similar manner here to reduce false positives in deadlock detection.

More specifically, the solution is during execution to associate segment identifiers (natural numbers, starting from 0) to segments of the code that are separated by statements that *start* or *join* other threads. For example, if a thread t_1 currently is in segment s and starts another thread t_2, and the next free segment is $n+1$, then t_1 will continue in segment $n+1$ and t_2 will start in segment $n+2$. From then on the next free segment will be $n+3$. It is furthermore recorded in the segmentation graph that segment s executes before $n+1$ as well as before $n+2$. In a similar way, if a thread t_1 currently is in segment s_1 and joins another thread t_2 that is in segment s_2, and the next free segment is $n+1$, then t_1 will continue in segment $n+1$, t_2 will be terminated, and from then on the next free segment will be $n+2$. It is recorded that s_1 as well as s_2 execute before $n+1$. Figure 5 illustrates the segmentation graph for the program example in Figure 1. In order to give a formal definition of the segmentation we need to define two functions. The first one, $C_S(\sigma)$, *segmentation context* of the trace σ, gives for each position i of the execution trace σ, the current segment of each thread t at that position. Formally, $C_S(\sigma)$ is the mapping with type: $[\mathcal{N} \mapsto [T_\sigma \mapsto \mathcal{N}]]$, associated to trace σ, that maps each position into another mapping that maps each thread id to its current segment in that position. It is defined as follows. Let $C_S^{init} = [0 \mapsto [main \mapsto 0]]$, mapping position 0 to the mapping that maps the main thread to segment 0. Then $C_S(\sigma)$ is defined by the use of the auxiliary function $f_0 : Trace \times Context \times Position \times Current\_Segment \rightarrow Context$:

$C_S(\sigma) = f_0(\sigma, C_S^{init}, 1, 0)$, where the function f_0 is defined by left-to-right recursion over the trace σ as follows:

$$f_0(e \frown \sigma, C_S, i, n) = \begin{cases} f_0(\sigma, C_S, i+1, n) \\ \quad \text{if } e \in \{l(t,o), u(t,o)\}, \\ \\ f_0(\sigma, C_S\dagger[i \mapsto C_S[i-1]\dagger \begin{bmatrix} t_1 \mapsto n+1 \\ t_2 \mapsto n+2 \end{bmatrix}, i+1, n+2) \\ \quad \text{if } e = s(t_1, t_2), \\ \\ f_0(\sigma, C_S\dagger[i \mapsto C_S[i-1]\dagger[t_1 \mapsto n+1], i+1, n+1) \\ \quad \text{if } e = j(t_1, t_2). \end{cases}$$

$f_0(<>, C_S, i, n) = C_S$

The second function needed, $\#_{alloc}$, gives the number of segments allocated in position i of σ. This function is used to calculate what is the next segment to be assigned to a new execution block, and is dependent on the number of start events $s(t_1, t_2)$ and join events $j(t_1, t_2)$ that occur in the trace up and until position i, recalling that each start event causes two new segments to be allocated. Formally we define it as follows : $\#_{alloc}(\sigma, i) = |\sigma^i \downarrow_s| * 2 + |\sigma^i \downarrow_j|$.

We can now define the notion of a directed *segmentation graph*, which defines an ordering between segments. Informally, assume that in trace position i a thread t_1, being in segment $s_1 = C_S(\sigma)(i-1)(t_1)$, executes a start of a thread t_2. Then t_1 continues in segment $n = \#_{alloc}(\sigma, i-1) + 1$ and t_2 continues in segment $n+1$. Consequently, (s_1, n) as well as $(s_1, n+1)$ belongs to the graph, meaning that s_1 executes before n as well as before $n+1$. Similarly, assume that a thread t_1 in position i, being in segment $s_1 = C_S(\sigma)(i-1)(t_1)$, executes a join of a thread t_2, being in segment $s_2 = \mathcal{C}_S(\sigma)(i-1)(t_2)$. Then t_1 continues in segment $n = \#_{alloc}(\sigma, i-1) + 1$ while t_2 terminates. Consequently (s_1, n) as well as (s_2, n) belongs to the graph, meaning that s_1 as well as s_2 executes before n. The formal definition of the segmentation graph is as follows.

Definition 5 (Segmentation graph). *Given an execution trace* $\sigma = e_1, \ldots, e_n$. *We say that a segmentation graph of* σ *is the directed graph* $G_S = (\mathcal{N}, R)$ *where:* $\mathcal{N} = \{n \mid 0 \leq n \leq \#_{alloc}(\sigma, |\sigma|)\}$ *is the set of segments, and* $R \subseteq \mathcal{N} \times \mathcal{N}$ *is the relation given by* $(s_1, s_2) \in R$ *if there exists a position* $i \geq 1$ *s.t.* $\sigma[i] = s(t_1, t_2) \wedge s_1 = \mathcal{C}_S(\sigma)(i-1)(t_1) \wedge (s_2 = \#_{alloc}(\sigma, i-1)+1 \vee s_2 = \#_{alloc}(\sigma, i-1)+2)$, *or* $\sigma[i] = j(t_1, t_2) \wedge (s_1 = \mathcal{C}_S(\sigma)(i-1)(t_1) \vee s_1 = \mathcal{C}_S(\sigma)(i-1)(t_2)) \wedge s_2 = \#_{alloc}(\sigma, i-1)+1$.

The following relation *happens-before* reflects how the segments are related in time during execution.

Definition 6 (Happens-Before relation). *Let* $G_S = (\mathcal{N}, R)$ *be a segmentation graph, and* $G_S^* = (\mathcal{N}, R^*)$ *its transitive closure. Then given two segments* s_1 *and* s_2, *we say that* s_1 *happens before* s_2, *denoted by* $s_1 \triangleright s_2$, *if* $(s_1, s_2) \in R^*$.

Note that for two given segments s_1 and s_2, if neither $s_1 \triangleright s_2$ nor $s_2 \triangleright s_1$, then we say that s_1 *happens in parallel* with s_2. Before we can finally define what is a lock graph with segment information, we need to redefine the notion of lock context, $\mathcal{C}_L(\sigma, i)$, of a trace σ and a position i, that was defined on page 213. In the previous definition it was a mapping from each thread to the set of locks owned by that thread at that position. Now we add information about what segment each lock was taken in. Formally, for a thread $t \in T_\sigma$ we have the following :

$$\mathcal{C}_L(\sigma, i)(t) = \{(o, s) \mid \exists j : j \leq i \wedge \sigma[j] = l(t, o) \wedge \neg(\exists k : j < k \leq i \wedge \sigma[k] = u(t, o)) \wedge \mathcal{C}_S(\sigma)(j)(t) = s\}$$

An edge in G_L between two locks l_1 and l_2 means, as before, that there exists a thread t which owns an object l_1 while taking the object l_2. The edge is as before labeled with t as well as the set of (gate) locks. In addition, the edge is now further labeled with the segments s_1 and s_2 in which the locks l_1 and l_2 were taken by t.

Definition 7 (Segmented and guarded lock graph). *Given an execution trace* $\sigma = e_1, e_2, \ldots, e_n$. *We say that the segmented and guarded lock graph of* σ *is the minimal directed graph* $G_L = (L_\sigma, W, R)$ *such that:*

- $W \subseteq \mathcal{N} \times (T_\sigma \times 2^{L_\sigma}) \times \mathcal{N}$ *is the set of labels* $(s_1, (t, g), s_2)$, *each containing the segment* s_1 *that the source lock was taken in, a thread id* t, *a lock set* g *(these two being the labels of the guarded lock graph in the previous section), and the segment* s_2 *that the target lock was taken in,*
- $R \subseteq L_\sigma \times W \times L_\sigma$ *is defined by* $(l_1, (s_1, (t, g), s_2), l_2) \in R$ *if there exists a thread* $t \in T_\sigma$ *and a position* $i \geq 2$ *in* σ *such that:* $\sigma[i] = l(t, l_2)$ *and* $(l_1, s_1) \in \mathcal{C}_L(\sigma)(i-1)(t)$ *and* $g = \{l' \mid (l', s) \in \mathcal{C}_L(\sigma)(i-1)(t)\}$ *and* $s_2 = \mathcal{C}_S(\sigma)(i-1)(t)$

Each edge $(l_1, (s_1, (t, g), s_2), l_2)$ in R is labeled with the thread t that took the locks l_1 and l_2, and a lock set g, indicating what locks t owned when taking l_2. The segments s_1 and s_2 indicate in which segments respectively l_1 and l_2 were taken.

In order for a cycle to be valid, and hence regarded as a true positive, the threads and guard sets occurring in labels of the cycle must be valid as before. In addition, the segments in which locks are taken must now allow for a deadlock to actually happen. Consider for example a cycle between two threads t_1 and t_2 on two locks l_1 and l_2. Assume further that t_1 takes l_1 in segment x_1 and then l_2 in segment x_2 while t_2 takes them in opposite order, in segments y_1 and y_2 respectively. Then it must be possible for t_1 and t_2 to each take their first lock in order for a deadlock to occur. In other words, x_2 must not happen before y_1 and y_2 must not happen before x_1. This is expressed in the following definition, which repeats the definitions from Definition 3.

Definition 8 (Valid threads, guards and segments). *Let G_L be a segmented and guarded lock graph of some execution trace and $c = (L, W, R)$ a cycle in $cycles(G_L)$, we say that:*

- *threads of c are valid if: forall $e, e' \in R$, $e \neq e' \Rightarrow thread(e) \neq thread(e')$*
- *guards of c are valid if: forall $e, e' \in R$, $e \neq e' \Rightarrow guards(e) \cap guards(e') = \emptyset$*
- *segments of c are valid if: forall $e, e' \in R$, $e \neq e' \Rightarrow \neg (seg_2(e_1) \rhd seg_1(e_2))$*

where, for an edge $e = (l_1, (s_1, (t, g), s_2), l_2) \in R$, $thread(e) = t$, $guards(e) = g$, $seg_1(e) = s_1$ and $seg_2(e) = s_2$.

Valid cycles are now defined as follows.

Definition 9 (Unsegmented and unguarded cycles). *Let σ be an execution trace and G_L its segmented and guarded lock graph. We say that a cycle $c \in cycles(G_L)$ is an unsegmented and unguarded cycle if the guards of c are valid, the threads of c are valid, and the segments of c are valid. We denote by $cycles_s (G_L)$ the set of unsegmented and unguarded cycles in $cycles(G_L)$.*

Figure 4 presents an algorithm for constructing the segmentation graph and lock graph from an execution trace. The set of cycles in the graph G_L, denoted by $cylces_s(G_L)$, see Definition 9, represents the potential deadlock situations in the program. The segmentation graph (G_S) and lock graph (G_L) have the structure as outlined in Definition 5 and Definition 7 respectively. The lock context (C_L) maps each thread to the set of locks owned by that thread at any point in time. Associated with each such lock is the segment in which it was acquired. The segment context (C_S) maps each thread to the segment in which it is currently executing. The algorithm should after this explanation and the previously given abstract definitions be self explanatory. Consider again the trace σ from Section 3. The segmented and guarded lock graph and the segmentation graph for this trace are both shown in Figure 5. The segmentation graph is for illustrative purposes augmented with the statements that caused the graph to be updated. We see in particular that segment 6 of thread T_3 executes before

Input: An execution trace σ
G_L is a lock graph;
G_S is a segmentation graph;
$C_L : [T_\sigma \to 2^{L_\sigma \times \mathbf{nat}}]$ is a lock context;
$C_S : [T_\sigma \to \mathbf{nat}]$ is a segment context;
$n : \mathbf{nat} = 1$ next available segment;
for$(i = 1 \mathinner{..} |\sigma|)$ **do**
    case $\sigma[i]$ **of**
        $l(t,o) \to$
            $G_L := G_L \bigcup \{(o', (s_1, (t, g), s_2), o) \mid$
                    $(o', s_1) \in C_L(t) \wedge$
                    $g = \{o'' \mid (o'', s) \in C_L(t)\} \wedge$
                    $s_2 = C_S(t)\};$
          $C_L := C_L \dagger [t \mapsto C_L(t) \bigcup \{(o, C_S(t))\}];$
        $u(t,o) \to$
            $C_L := C_L \dagger [t \mapsto C_L(t) \backslash \{(o, *)\}];$
        $s(t_1, t_2) \to$
            $G_S := G_S \bigcup \{(C_S(t_1), n), (C_S(t_1), n+1)\};$
            $C_S := C_S \dagger [t_1 \mapsto n, t_2 \mapsto n+1];$
            $n := n + 2;$
        $j(t_1, t_2) \to$
            $G_S := G_S \bigcup \{(C_S(t_1), n), (C_S(t_2), n)\};$
            $C_S := C_S \dagger [t_1 \mapsto n];$
            $n := n + 1;$
end;
for each c in $cycles_s(G_L)$ **do**
    print ("deadlock potential:",c);

Fig. 4. The final algorithm

segment 7 of thread T_1, written as $6 \triangleright 7$. Segment 6 is the one in which T_3 executes lines 19 and 20, while segment 7 is the one in which T_1 executes lines 11 and 12. The lock graph contains the same number of edges as the guarded graph in Figure 3, and the same *(thread,guard set)* labels. However, now edges are additionally labeled with the segments in which locks are taken. This makes the cycle $(19, 20) \leftrightarrow (11, 12)$ segment invalid since the target segment of the first edge (6) executes before the source segment of the second edge (7).

7 Implementation and Experimentation

The algorithm presented in the previous section has been implemented in the Java PathExplorer (JPaX) tool [20]. JPaX consists of two main modules, an *instrumentation module* and an *observer module*. The instrumentation module automatically instruments the bytecode class files of a compiled program by adding new instructions that when executed generate the execution trace consisting of the events needed for the analysis. The observer module reads the event

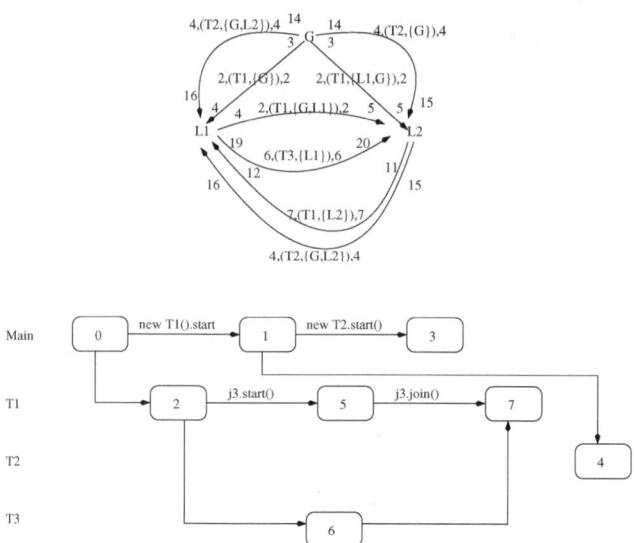

Fig. 5. Segmented lock graph (above) and segmentation graph (below)

stream and dispatches this to a set of observer rules, each rule performing a particular analysis that has been requested, such as deadlock analysis, data race analysis and temporal logic analysis. This modular rule based design allows a user to easily implement new runtime verification algorithms. The Java bytecode instrumentation is performed using the jSpy instrumentation package [15] that is part of Java PathExplorer. jSpy's input is an instrumentation specification, which consists of a collection of rules, where a rule is a predicate/action pair. The predicate is a conjunction of syntactic constraints on a Java statement, and the action is a description of logging information to be inserted in the bytecode corresponding to the statement. As already mentioned, this form of analysis is not complete and hence may yield false negatives by missing to report synchronization problems. A synchronization problem can most obviously be missed if one or more of the synchronization statements involved in the problem do not get executed. To avoid being entirely in the dark in these situations, we added a coverage module to the system that records what lock-related instructions are instrumented and which of these that are actually executed.

JPaX's deadlock analyzer has been applied to three NASA case studies: a planetary rover controller for a rover named K9 programmed in 35 thousand lines of C++; a 7,5 thousand line Java version of the K9 rover controller used as part of an attempt to evaluate Java verification tools; and a planner named Europa programmed in approximately 5-10 thousand lines of C++. In the C++ scenarios ASCII log files were generated which were then read and analyzed by the tool. The Java version of the K9 controller was in particular created to evaluate a range of program verification technologies, among them JPaX, as described in [9]; the other technologies included static analysis and model checking. In this

Java experiment JPaX generally came out well in the comparison with the other tools, as being fast and effective. Errors were seeded by a control team and the study groups had as task to detect the errors with the different tools. The deadlock analysis tool found the seeded deadlocks usually during the first run of the tool. In the C++ version of the K9 rover controller the tool found a real deadlock potential that was unknown to the programmer. Also the first time the tool was run. This experiment was performed by hand instrumenting lock and unlock operations in the program, whereas the observer module could be used unmodified. In the planner Europa, the tool similarly found a real deadlock potential that was unknown to the programming team, also in the first run of the tool. This result caused the team to request an integration of the observer part into their development suite for future use.

The presented algorithm can miss (and hence not report) true positives in some rare cases, where for example a particular execution path causes a gate lock to prevent an error message from being issued, but where another path might contain a deadlock based on the same cycle. Furthermore, it might be argued that cyclic lock patterns represent bad programming style. For this reason, the best application of the algorithm might be to augment warnings given by the basic algorithm with additional information.

8   Conclusions and Future Work

An algorithm has been presented for detecting deadlock potentials in concurrent programs by analyzing execution traces. The algorithm extends an existing algorithm by reducing the amount of false positives reported, and has been implemented in the JPaX tool. Although JPaX analyzes Java programs, it can be applied to applications written in other languages by replacing the instrumentation module. The advantage of trace analysis is that it scales well, in contrast to more formal methods, and in addition can detect errors that for example static analysis cannot properly detect. In current work, we further approach the problem of false positives by developing a framework for generating test cases from warnings issued by this tool. Such test cases will then directly expose the possible deadlocks. An experiment in this direction is described in [17] where a model checker is used to "investigate" deadlock and data race warnings identified using dynamic analysis. Additional current work attempts to extend the capabilities of JPaX with new algorithms for detecting other kinds of concurrency errors, such as various forms of data races and communication deadlocks.

Acknowledgments. We would like to thank the following people for their contribution to the case studies: Chuck Fry (NASA Ames Research Center/QSS Group Inc.) instrumented and detected the deadlock in the Europa planner. Rich Washington (NASA Ames Reserach Center/RIACS, now Google) instrumented and detected the deadlock in the C++ version of the K9 rover controller. Summer intern Flavio Lerda (Carnegie Mellon University) and Masoud Mansouri-Samani

(NASA Ames Research Center/CSC) applied the tool to the Java version of the K9 rover controller during the controlled experiment described in [9].

References

1. R. Agarwal, L. Wang, and S. D. Stoller. Detecting Potential Deadlocks with Static Analysis and Run-Time Monitoring. In *Proceedings of the Parallel and Distributed Systems: Testing and Debugging (PADTAD) track of the 2005 IBM Verification Conference, Haifa, Israel.* Springer-Verlag, November 2005. These proceedings.
2. K. Arnold and J. Gosling. *The Java Programming Language.* Addison-Wesley, 1996.
3. C. Artho and A. Biere. Applying Static Analysis to Large-Scale, Multi-threaded Java Programs. In D. Grant, editor, *13th Australien Software Engineering Conference*, pages 68–75. IEEE Computer Society, August 2001.
4. C. Artho, K. Havelund, and A. Biere. High-level Data Races. *Software Testing, Verification and Reliability (STVR)*, 13(4), December 2003.
5. C. Artho, K. Havelund, and A. Biere. Using Block-Local Atomicity to Detect Stale-Value Concurrency Errors. In *2nd International Symposium on Automated Technology for Verification and Analysis, Taiwan*, October–November 2004.
6. T. Ball, A. Podelski, and S. Rajamani. Boolean and Cartesian Abstractions for Model Checking C Programs. In *Proceedings of TACAS'01: Tools and Algorithms for the Construction and Analysis of Systems*, LNCS, Genova, Italy, April 2001.
7. H. Barringer, A. Goldberg, K. Havelund, and K. Sen. Rule-Based Runtime Verification. In *Proceedings of Fifth International VMCAI conference: Verification, Model Checking and Abstract Interpretation*, volume 2937 of *LNCS*. Springer, January 2004.
8. S. Bensalem and K. Havelund. Reducing False Positives in Runtime Analysis of Deadlocks. Internal report, NASA Ames Research Center, October 2002.
9. G. Brat, D. Drusinsky, D. Giannakopoulou, A. Goldberg, K. Havelund, M. Lowry, C. Pasareanu, W. Visser, and R. Washington. Experimental Evaluation of Verification and Validation Tools on Martian Rover Software. *Formal Methods in System Design*, 25(2), 2004.
10. J. Corbett, M. B. Dwyer, J. Hatcliff, C. S. Pasareanu, Robby, S. Laubach, and H. Zheng. Bandera : Extracting Finite-state Models from Java Source Code. In *Proceedings of the 22nd International Conference on Software Engineering*, Limerich, Ireland, June 2000. ACM Press.
11. D. L. Detlefs, K. Rustan M. Leino, G. Nelson, and J. B. Saxe. Extended Static Checking. Technical Report 159, Compaq Systems Research Center, Palo Alto, California, USA, 1998.
12. O. Edelstein, E. Farchi, Y. Nir, G. Ratsaby, and S. Ur. Multithreaded Java Program Test Generation. *Software Testing and Verification*, 41(1), 2002.
13. E. Farchi, Y. Nir-Buchbinder, and S. Ur. A Cross-Run Lock Discipline Checker for Java. Tool presented at the Parallel and Distributed Systems: Testing and Debugging (PADTAD) track of the 2005 IBM Verification Conference, Haifa, Israel. Tool is available at `http://alphaworks.ibm.com/tech/contest`, November 2005.
14. P. Godefroid. Model Checking for Programming Languages using VeriSoft. In *Proceedings of the 24th ACM Symposium on Principles of Programming Languages*, pages 174–186, Paris, France, January 1997.

15. A. Goldberg and K. Havelund. Instrumentation of Java Bytecode for Runtime Analysis. In *Proc. Formal Techniques for Java-like Programs*, volume 408 of *Technical Reports from ETH Zurich*, Switzerland, 2003. ETH Zurich.
16. J. Harrow. Runtime Checking of Multithreaded Applications with Visual Threads. In *SPIN Model Checking and Software Verification*, volume 1885 of *LNCS*, pages 331–342. Springer, 2000.
17. K. Havelund. Using Runtime Analysis to Guide Model Checking of Java Programs. In *SPIN Model Checking and Software Verification*, volume 1885 of *LNCS*, pages 245–264. Springer, 2000.
18. K. Havelund and T. Pressburger. Model Checking Java Programs using Java PathFinder. *International Journal on Software Tools for Technology Transfer*, 2(4):366–381, April 2000. Special issue of STTT containing selected submissions to the 4th SPIN workshop, Paris, France, 1998.
19. K. Havelund and G. Roşu. Monitoring Java Programs with Java PathExplorer. In *Proceedings of the First International Workshop on Runtime Verification (RV'01)*, volume 55 of *Electronic Notes in Theoretical Computer Science*, pages 97–114, Paris, France, July 2001. Elsevier Science.
20. K. Havelund and G. Roşu. An Overview of the Runtime Verification Tool Java PathExplorer. *Formal Methods in System Design*, 24(2), March 2004. Extended version of [19].
21. G. J. Holzmann and M. H. Smith. A Practical Method for Verifying Event-Driven Software. In *Proceedings of ICSE'99, International Conference on Software Engineering*, Los Angeles, California, USA, May 1999. IEEE/ACM.
22. E. Knapp. Deadlock Detection in Distributed Database Systems. *ACM Computing Surveys*, pages 303–328, Dec. 1987.
23. D. Lea. *Concurrent Programming in Java, Design Principles and Patterns*. Addison-Wesley, 1997.
24. D. Park, U. Stern, J. Skakkebaek, and D. Dill. Java Model Checking. In *Proceedings of the 15th IEEE International Conference on Automated Software Engineering*, pages 253–256, September 2000.
25. PolySpace. An Automatic Run-Time Error Detection Tool. http://www.polyspace.com.
26. S. Savage, M. Burrows, G. Nelson, P. Sobalvarro, and T. Anderson. Eraser: A Dynamic Data Race Detector for Multithreaded Programs. *ACM Transactions on Computer Systems*, 15(4):391–411, November 1997.
27. M. Singhal. Deadlock Detection in Distributed Systems. *IEEE Computer*, pages 37–48, Nov. 1989.
28. S. D. Stoller. Model-Checking Multi-threaded Distributed Java Programs. In *SPIN Model Checking and Software Verification*, volume 1885 of *LNCS*, pages 224–244. Springer, 2000.
29. W. Visser, K. Havelund, G. Brat, S. Park, and F. Lerda. Model Checking Programs. *Automated Software Engineering*, 10(2), April 2003.

Verification of the Java Causality Requirements

Sergey Polyakov and Assaf Schuster

Department of Computer Science,
Technion - Israel Institute of Technology,
Technion City, Haifa

Abstract. The Java Memory Model (JMM) formalizes the behavior of shared memory accesses in a multithreaded Java program. Dependencies between memory accesses are acyclic, as defined by the JMM causality requirements. We study the problem of post-mortem verification of these requirements and prove that the task is NP-complete. We then argue that in some cases the task may be simplified either by considering a slightly stronger memory model or by tracing the actual execution order of Read actions in each thread. Our verification algorithm has two versions: a polynomial version, to be used when one of the aforementioned simplifications is possible, and a non-polynomial version – for short test sequences only – to be used in all other cases. Finally, we argue that the JMM causality requirements could benefit from some fine-tuning. Our examination of causality test case 6 (presented in the public discussion of the JMM) clearly shows that some useful compiler optimizations – which one would expect to be permissible – are in fact prohibited by the formal model.

Keywords: Concurrency, Complexity, Java, Memory Model, Multi-threading, Verification, Shared Memory.

1 Introduction

In the past, memory models where formalized mostly for hardware implementations or abstract hardware/software systems [16, 13, 3, 11]. In recent years there has been an effort to provide memory models for popular programming languages [4, 9, 6, 7]. One reason is that some difficulties arise when implementing multithreading only by means of libraries [8]. The revised version of the Java Memory Model (JMM) [2, 15] is the result of long-term efforts and discussions[17]. It is now expected to be the prototype for the multithreaded C++ memory model [6].

This article studies the problem of post-mortem verification of the JMM causality requirements. Post-mortem methods check that memory behavior is correct for a given execution of the multithreaded program. They analyze *traces* – test prints produced either by running an instrumented code or by running some special debug version of the JVM.

The Java system consists of the compiler, which converts a source code to a *bytecode*, and the Java Virtual Machine (JVM), which executes the bytecode. The Java Memory Model considers the system as a whole: what it describes

S. Ur, E. Bin, and Y. Wolfsthal (Eds.): Haifa Verification Conf. 2005, LNCS 3875, pp. 224–246, 2006.

is the behavior of the *virtual bytecode* – an imaginary bytecode that may be provided by a non-optimizing compiler. Such bytecode is a straight mapping of the Java source code and as such it expresses the programmer's comprehension of the program. This certainly simplifies matters when it comes to analyzing multithreaded programs, but complicates the memory model: The JMM must regulate optimizations performed by the real Java compiler and run-time system (the JVM). It is important that these optimizations not produce *causal loops* – a behavior by which some action "causes itself."

A causal loop might occur if some speculatively executed action affects another action upon which it depends. This phenomenon may adversely affect the behavior even of a correctly synchronized program. Causal acyclicity is therefore implied not only for Java but for most weak memory models. In release consistency [13], for example, causal loops are prevented informally by "respecting the local dependencies within the same processor" [12]. An example of a formal definition of causal acyclicity is given in [5, 12], where it is expressed through the "reach condition."

In weak memory models, causal acyclicity is necessary to guarantee that programs which are data-race free on sequentially consistent platforms will maintain sequentially consistent behavior. In Java, causal acyclicity also provides important safety guarantees for badly synchronized programs: out-of-thin-air values [2] are prohibited, for example. Because the JMM requires that only minimal restrictions be imposed on implementations, it gives a relaxed formalization of causal acyclicity that makes allowances for most compiler optimizations.

The relaxed definition of the JMM is not for free, however. The model does not give any guidelines for the implementor – it is a specification of user demands. While there are relatively simple rules for implementing synchronization [10], no such rules are provided for causality requirements. Therefore, when an implementation uses aggressive optimization methods, it is hard to decide if it meets the specification. Thus, the JMM needs well-defined verification methods.

The JMM guarantees causal acyclicity by requiring that actions be *committed* in some partial order: an action cannot depend on an action that succeeds it. Thus, the problem of verifying the causality requirements may be thought of as a search for some valid commitment order. We found that the problem is NP-complete because there may be a non-polynomial number of partial orders that can be considered as candidates for the commitment order. To make the problem solvable, we propose two approaches. The first one consists in using a stronger memory model, with an additional demand that Read actions on the same variable must be committed in program order. The second approach consists in tracing the actual execution order of Read actions in each thread. The last method can only be applied, however, for JVM implementations in which it is possible to trace the actual execution order of the Read actions, which must also be known to comply with all admissible commitment orders. If neither approach can be used, the problem is still NP-complete with respect to execution length, and only short traces can be verified.

The formal JMM causality requirements are illustrated by the causality test cases [1]. Our manual verification of these test cases have shown that JMM is more restrictive than expected: Test case 6 could not be verified although it must be allowed by the JMM. In our minds, this illustrates a discrepancy between the formal model and its intuitive implications. One of the contributions of this paper is a proposed fix for this problem.

The rest of the article is organized as follows. Section 2 describes our assumptions about test sequences. In section 3 we prove that verification of the JMM causality requirements is NP-hard in the general case and suggest a method for solving the problem. We then provide a verification algorithm (section 4). Finally, in section 5 we discuss some problems and present our conclusions.

2   The Problem of Testing the JMM Causality Requirements

We briefly recount here how a Java program is executed. Readers interested in the complete JMM definition are referred to Appendix A, as well as to [2, 15].

The execution of a Java program is described by a tuple $\langle P, A, \xrightarrow{po}, \xrightarrow{so}, W, V, \xrightarrow{sw}, \xrightarrow{hb} \rangle$, where

- P is the program,
- A is the set of actions,
- $\xrightarrow{po}$ is the program order,
- $\xrightarrow{so}$ is the total order on synchronization actions,
- W is a write-seen function,
- V is a function that assigns a value to each Write,
- $\xrightarrow{sw}$ is a partial order that arranges pairs of actions with release-acquire semantics,
- $\xrightarrow{hb}$ is the happens-before order.

The values of shared variables that appear in the execution must be consistent with various requirements – synchronization, causality, and the like.

We investigate the simplest case of testing the JMM causality requirements. Consider an execution E that has the following properties. It is finite. There are no synchronization actions, actions on final fields, or external actions. There are no happens-before relations that result from executing finalizers. We assume that the threads are spawned at the very beginning of the program; therefore $\xrightarrow{so}$ and $\xrightarrow{sw}$ are irrelevant and $\xrightarrow{hb}$ complies with $\xrightarrow{po}$. Note, however, that initialization Writes may be considered to be performed by a special thread and precede all other actions.

The rest of the execution tuple $(P, A, \xrightarrow{po}, W, V)$ is represented by the program and traces; the program is represented by the virtual bytecode. For simplicity, we use pseudocode in our examples and proofs. We have a separate trace for each thread, and a trace of a thread is called a sequence. Each sequence consists of [GUID:](Read/Write)(x_i, t_i, g_i) actions, where GUID is a unique identifier for the action, x_i is some variable, and t_i is the value which is read / written by the operation. For Read, g_i is the GUID of the Write operation that provided the value that is read. (We assume that the trace provides the write-seen function.) For Write, g_i is its own GUID.

We provide a simplified version of the causality requirements that matches our special case. Execution $E = \langle P, A, \xrightarrow{po}, W, V \rangle$ is *well-formed* if it has the following properties:

1. Each Read of a variable x sees a Write to x.
2. Each thread sequence is identical to that which would be generated by the correct uniprocessor.

A well-formed execution E is consistent with the causality requirements if all actions in A can be iteratively *committed*. Each step i of the commitment process must correspond to the set of committed actions C_i ($C_0 = \emptyset, C_i \subset C_{i+1}$). For each C_i (starting from C_1) there must be a well-formed execution E_i that contains C_i. Series $C_0, \dots, C_i, \dots$ and $E_1, \dots, E_i, \dots$ must have the following properties (less complex than in the general case):

1. $C_i \subseteq A_i$;
2. committed Writes produce final values (the same values as in E);
3. for each E_i , all Reads in C_i see Writes that have just been committed in C_{i-1};
4. for each E_i , all Reads in C_{i-1} see final values;
5. for each E_i ($i > 1$), all Reads that are not in C_{i-1} see default values or values written previously by the same thread. ($C_0 = \emptyset$, C_1 has no Reads and contains all default Writes.)

When we say that Read action r *sees* Write action w in E_i, we imply that the write-seen function of E_i maps r to w. Please note that committing a Write action makes it visible for another thread (two steps later), but previously committed Writes of the same variable remain visible as well (excluding the case when the new committed Write, previously committed Write and the observing Read belong to the same thread). In addition, note that committing a Read action does not abolish values of the same variable that were seen earlier by the thread (the values are permitted to appear anew).

We define the problem of testing of the JMM causality requirements as follows.

INSTANCE: Input in the form of the program and its corresponding thread sequences. The complexity parameter is the length (number of characters) of the input.

QUESTION: Do the given tuple of the program and the thread sequences correspond to some valid Java execution? We prove that the problem is NP-complete even under hard restrictions.

3   Complexity of the Problem

3.1   General Case

Theorem 1. *The problem of verifying JMM causality requirements, restricted to instances in which there are at most two threads and at most one shared variable, is NP-complete.*

PROOF. We use the reduction from 3-Satisfiability (3SAT). Consider a 3SAT instance $\mathcal{F}$ with n variables, $v_1, \ldots, v_n$, and m clauses, $C_1, \ldots, C_m$. The execution E is represented by the program P and the trace sequences T. The program P is shown in Table 1. Unique identifiers for the actions are given in square brackets. Only memory access actions have identifiers. For clarity, we denote shared variables as alphabetical characters other than r (a, b, ...) and the values on the JVM stack as the letter "r" with index (r_1, r_2, ...). Initially, $a = 0$ and all values on the JVM stack are 0.

Table 1. The program (pseudocode)

Thread 1	Thread 2
[0:] $a = 1;$	[1:] $r_{a1} = a;$
	[2:] $r_{b1} = a;$
	$r_1 = r_{a1} - r_{b1}$;
	...
	[2*n-1:] $r_{an} = a;$
	[2*n:] $r_{bn} = a;$
	$r_n = r_{an} - r_{bn};$
	$<$ **comment:** *in the following **if** statements*
	r_{cji} *may be any of* $r_1 \ldots r_n;$
	r_{cji} *simulates an appearance of a variable in place i of clause j;*
	s_{ji} *is **1** if the variable appears without negation, and **-1** otherwise* $>$
	if (!($s_{11} == r_{c11}$ || $s_{12} == r_{c12}$ || $s_{13} == r_{c13}$))
	$r_0 = 1;$
	if (!($s_{21} == r_{c21}$ || $s_{22} == r_{c22}$ || $s_{23} == r_{c23}$))
	$r_0 = 1;$
	...
	if (!($s_{m1} == r_{cm1}$ || $s_{m2} == r_{cm2}$ || $s_{m3} == r_{cm3}$))
	$r_0 = 1;$
[2*n+3:] $r_{n+2} = a;$	[2*n+1:] $r_{n+1} = a;$
if ($r_{n+2} = 2$)	if ($r_0 == 0$ || $r_{n+1} == 3$)
[2*n+4:] $a = 3;$	[2*n+2:] $a = 2;$

The trace sequences are shown in Table 2. Components g_i of the operations are not presented in the table because each `Write` operation provides a unique value and there is no ambiguity.

Table 2. The trace sequences

Thread 1	Thread 2
[0:] Write (a, 1)	[1:] Read (a, 1)
	[2:] Read (a, 1)
	[3:] Read (a, 1)
	[4:] Read (a, 1)
	. . .
	[2*n-1:] Read (a, 1)
	[2*n:] Read (a, 1)
[2*n+3:] Read (a, 2)	[2*n+1:] Read (a, 3)
[2*n+4:] Write (a, 3)	[2*n+2:] Write (a, 2)

Note that correspondence between the operations in the program and the actions in the trace is provided here by matching of the unique identifiers. There is a one-to-one correspondence because there are no loops. If loops are present, indexes may be added to GUIDs (here there is no need for this).

An assignment to v_i is simulated by the following components:

1. operation $a = 1$ by Thread 1 and two operations $r_{ai} = a$ and $r_{bi} = a$ by Thread 2;
2. corresponding actions *[0:] Write (a, 1)* by Thread 1 and *[2*i-1:] Read (a, 1)*, *[2*i:] Read (a, 1)* by Thread 2;
3. commitment order between actions *[2*i-1:] Read (a, 1)* and *[2*i:] Read (a,1)* that belong to Thread 2.

Stack variable r_i may receive the values 1, -1 or 0. An assignment of the value *true* to v_i is simulated by assigning the value 1 to r_i, and an assignment of the value *false* to v_i is simulated by assigning the value -1 to r_i. If r_i receives the value 0, this means that v_i cannot be *true* or *false* in any clause. r_i may receive the value 1 or -1 only if the two actions Read (a, 1) by Thread 2 belong to different commitment sets. Please note: the JMM does not require that commitment order of the accesses to the same variable a comply with the program order.

An OR for the clause j is simulated by the program operations *if (!(s_{j1} == r_{cj1} || s_{j2} == r_{cj2} || s_{j3} == r_{cj3})) and r_0= 1;* by Thread 2;

AND is simulated by the operations *if (r_0== 0 || r_{n+1}== 3) and $a = 2$;* as well as by action *[2*n+2] Write (a, 2)* by Thread 2.

Lemma 1. *Let $\mathcal{F}$ be the instance of a 3SAT problem and let E be the execution constructed as described above. Then E complies with the JMM causality requirements iff $\mathcal{F}$ is satisfiable.*

PROOF.

$\Leftarrow$ Suppose $\mathcal{F}$ is satisfiable. Then, there exists a satisfying assignment Å for $\mathcal{F}$. We construct the commitment order $\mathbb{C}$ for E in which for each i, $r_{ai} = a$ directly precedes $r_{bi} = a$ if v_i is *true* in Å and, conversely, $r_{bi} = a$ directly precedes $r_{ai} = a$ if v_i is *false* in Å. The process of commitment is described in Table 3.

Table 3. Commitment order

Action	Final Value	First Committed in	First Sees Final Value In	Comment
a = 1	1	C_1	E_1	
...	...	...	...	
$r_{ai} = a$	1	C_2	E_3	simulates an assignment
$r_{bi} = a$	1	C_4	E_5	of the value *true* to v_i
...	...	...	...	
$r_{aj} = a$	1	C_4	E_5	simulates an assignment
$r_{bj} = a$	1	C_2	E_3	of the value *false* to v_j
...	...	...	...	
a = 2	2	C_3	E_3	commits before "a = 3" because $r_0 == 0$
$r_{n+2} = a$	2	C_3	E_4	
a = 3	3	C_4	E_4	
$r_{n+1} = a$	3	C_4	E_5	

$\Rightarrow$ Conversely, suppose E is a positive instance. Assume $\mathbb{C}$ is a valid commitment order for E. If $r_{ai} = a$ precedes $r_{bi} = a$ in $\mathbb{C}$, then the satisfying assignment for v_i is assumed *true*, and if $r_{bi} = a$ precedes $r_{ai} = a$ in $\mathbb{C}$, then the satisfying assignment for v_i is assumed *false*. If r_{ai} and r_{bi} commit in the same step, then the assignment for v_i is neither *true* nor *false*.

Suppose that some clause j is unsatisfied. Then, by our construction, r_0 is assigned value 1. In this case, *[2*n + 2:] Write (a, 2)* cannot be committed before *[2*n + 4:] Write (a, 3)*. Because *[2*n + 4:] Write (a, 3)* can commit only after *[2*n + 2:] Write (a, 2)*, we may conclude that E is not a positive instance – a contradiction to our assumption. □

Lemma 2. *The problem of verifying JMM causality requirements is in NP.*

PROOF. Given a series of commitment sets $C_1, \ldots, C_k$ and validating executions $E_1, \ldots, E_k$, the JMM causality requirements may be checked in polynomial time by simulating each execution E_i (when shared memory Reads are replaced by reading constants) and checking properties 1-5 on page 227. □

By lemmas 1, 2 and because the construction may be done in polynomial time, theorem 1 holds. □

3.2 Preset Restrictions on the Commitment Order

The previous section gives rise to the intuition that the problem is NP-hard due to uncertainty of commitment order. Let us search for some additional restrictions on this order for the purpose of making the problem tractable. We will use a memory model that is stronger than the JMM but, possibly, close enough to it to be applicable for some implementations.

The first possibility is to restrict the commitment order of Read actions so that it must comply with the program order. We reject this restriction, however,

Initially, x = y = z = 0	
Thread 1	Thread2
r3 = z;	r3 = y;
r2 = x;	z = r3;
y = r2;	x = 1
r1 == r2 == r3 == 1 is a legal behavior	

Initially, = 0	
Thread 1	Thread2
r1 = x	z2 = x
x = 1	x = 2
r1 == 2, r2 == 1 is a legal behavior	

Fig. 1. Causality test cases 7(left), 16 (right)

because it prohibits reasonable Java executions (see, for example, causality test case 7 in [1], depicted in Figure 1).

The second possibility is to restrict the commitment order of actions on the same variable so that it must comply with the program order. We reject this restriction as well, because it prohibits the execution that is allowed by causality test case 16 [1], depicted in Figure 1.

The third possibility is to restrict the commitment order of Read actions on the same variable so that it must comply with the program order. This restriction does not help in its general case, because the problem is still NP complete (see Appendix B). However, if the number of shared variables and threads is bounded, the restriction makes the task polynomial. It may be solved using the frontier graph approach, described in section 3.3, with the following modification: The sequence of accesses to the same variable in each thread must be treated as a separate thread when building frontiers.

3.3   Tracing the Commitment Order of Actions for Each Thread

Let us try another approach. Actions in the trace may be augmented with their prospective numbers in the thread's commitment order. These numbers can be determined as follows: Each thread must hold a local counter. The counter is initialized to zero. When the value of some memory access a is *decided*, the current value of the counter is assigned to a as its number and the counter is incremented. Action numbers may not comply with the program order because of compiler and run-time optimizations. The method is applicable if, in the particular architecture, the moment when the value is decided can be fixed.

Having the modified trace, we reformulate the problem: Do the given tuple of the program and the thread sequences correspond to some valid Java execution whose commitment order complies with the numbering of actions in the trace?

When the number of threads is bounded, the problem may be solved by the polynomial algorithm using the *frontier graph* approach. Frontier graphs were used by Gibbons and Korach [14] for verifying sequential consistency. The problem of verifying sequential consistency is NP-complete in the general case, but some private cases (for example, when the number of threads is bounded and the write-seen function provided) may be solved in polynomial time using frontier graphs. Here we explain briefly the main concepts. Readers interested in a more in-depth description should refer to [14].

For the set of k thread sequences with n memory access operations, the *frontier* defines some set of k actions, each action belonging to a different thread.

Each action in the frontier is considered the last action performed by the thread at some moment in the execution. The last action may be null if the thread has not performed any action yet. If the order of actions in any thread sequence is fixed, the frontier defines some prefix. (For sequential consistency, the program order is considered, and for the JMM, the commitment order is considered.) There are $O(n^k/k^k)$ frontiers.

The *frontier graph* is defined as follows. Each possible frontier is mapped as a vertex in the graph, and it is marked "legal" or "illegal," depending on some problem-dependent validation rules. Then, also depending on the problem, some frontier F_j may be a legal extension of another frontier F_i; in this case the frontier graph must contain the edge (F_i, F_j). The graph has a polynomial number of vertexes and edges and can be traversed in polynomial time. The execution is considered consistent with the memory model if there is a directed path (of legal vertexes only) from the starting frontier (having k null operations) to the terminating frontier (consisting of the last operation in each thread sequence).

We can use this approach to provide a polynomial algorithm for the JMM. This algorithm would work for the case where commitment order for each thread is fixed and the number of threads is bounded. (Note that the write-seen function is provided by the execution trace as well.)

3.4 Tracing the Commitment Order of Read Actions for Each Thread

The method in section 3.3 may be further improved. The idea is that the commitment order of Writes does not matter. For each execution E_i having a set Ψ of committed Reads, there exists the minimum set Ω of Writes that must be committed in E_{i-1}. Ω includes the set of Writes that Reads from Ψ see in the final execution E. In addition, Ω includes each Write that is the last preceding (in $\xrightarrow{po}$) Write of the same variable for some Read from Ψ. There is no reason to commit additional Writes in E_{i-1}, because this would only restrict the set of possible values that Reads in E_i may see. Therefore, the set of committed Writes is unambiguously defined in E_{i-1} with respect to E_i.

Let us evaluate the complexity of the method. Consider an execution that has k threads and n shared memory accesses, among them n_r Reads. There are $O((\frac{n_r}{k})^k)$ frontiers. For two frontiers, the complexity of deciding if the first is a legal extension of the second is $O(n^2)$. See Algorithm 1 in section 4. Since the number of frontier pairs is $O((\frac{n_r}{k})^{2k})$, the total complexity does not exceed $O(n^2 * (\frac{n_r}{k})^{2k})$.

In closing, we add that an augmented trace may be easily produced when running the test on a simulator. Note that if the simulator can be programmed to generate the global commitment order, the verification problem is trivial.

3.5 Short Test Sequences

If the commitment order of Reads cannot be traced, the verification task is NP-complete with respect to the execution length. Therefore, only short test sequences may be used. Consider an execution that has n shared memory accesses, n_r Reads and n_w Writes. The execution may be verified by checking all possible commitment orders that have at most n steps (the complexity is $O(n^2 * (\frac{n+n-1}{n}))$. In section 4 we suggest a method that is of complexity $O(n^2 * 2^{2*n_r})$, and is preferable when n_r is not too close to n.

4   The Verification Algorithm

The verification algorithm does not differ much for the two cases we consider: 1) that in which either the commitment order of Read actions on the same variable is known to comply with the program order and the number of shared variables and threads is bounded, or the commitment order of Reads is traced and the number of threads is bounded; 2) in the general case. The algorithm we present is therefore unified. We do remark, in our explanations, on some of the differences between the two cases. In case (1) the algorithm is polynomial; if it is not polynomial, only short tests may be used.

Consider an execution tuple $E = <P, A, \overset{po}{\rightarrow}, W, V>$. In order to validate E, we perform the following:

1. We construct a group Γ of candidate execution *skeletons* that may be used for building a validating sequence. Each skeleton is characterized by the set f of Read actions that are in the second step of commitment. These are Reads that *must* see their final values (the values that they see in E) in any execution that has the skeleton. Only Reads in f may see non-default Writes belonging to other threads. Other Reads may see only the default Writes or Writes that precede them in the program order (these Writes may or may not write final values). For convenience, we divide the group of skeletons into subgroups $F(E, i)$, where $i = |f|$, $1 \leq i \leq n_r$. Γ differs in the two variations of the algorithm (non-polynomial and polynomial). In the first case, it includes skeletons corresponding to all possible f. In the second case, Reads in each f must form a prefix defined by some frontier.
2. We associate a *simulated* execution $S(E, f)$ with each skeleton defined by f. $S(E, f)$ is obtained by running each thread separately in program order (all values are initialized to the default) while some actions are replaced as follows. Each Read action rd in f is replaced by the action of reading the constant value. This constant value is equal to the final value returned by rd in E. For example, an action $r_3 = b$ with a final value of 5 is replaced by the action $r_3 = 5$. All other Reads are left unchanged. This simulation gives us the set of Writes that may be committed (these are Writes that write final values in $S(E, f)$). $S(E, f)$ is valid if each Read in f can see a Write that may be committed.
3. We check if it is possible to build, from the simulated executions, a chain that complies with the commitment rules. This chain must have the following properties:

(a) each simulated execution in the chain must be valid;

(b) `Writes` seen by `Reads` in f_{i+1} may be committed in all S_j in the chain where $i \leq j$.

In the pseudocode in Algorithm 1 we use **R** (the reaching set) as a set of valid simulated executions $S(E, f_i)$ (found in the previous step) that have the following property: Execution E_i that is simulated by $S(E, f_i)$ may "reach" E by committing its actions step-by-step in a series of executions E_i, E_{i+1}, ..., E. **N** (next) is a set of simulated executions (found in the current step) that will replace **R** in the next step of the algorithm. **P** (passed) is a set of simulated executions that were in **R** in previous steps of the algorithm. **I** is a set that contains an initial execution: one where no `Read` is required to see its final value. We start from the *final* execution S_f where all `Reads` see their final values and then attempt to reach an *initial* execution S_i.

We further prove that the algorithm works. The intuition of the proof is as follows. `Reads` are committed in two steps, and they can see non-default values written in other threads beginning with the second step after committing the corresponding `Write`. In our chain of simulated executions, however, `Reads` can see all `Writes` in the previous step. Therefore this chain is more "rare" than the chain of executions used to validate E. The proof provides a mapping between the two chains.

$\Leftarrow$ First, we prove that if any execution is valid with respect to the causality requirement, the algorithm confirms this. Suppose that for some execution E there exists a chain of sets of committed actions C_0, C_1, C_2, ... and a chain V of corresponding executions E_1, E_2, ... that are used for validating E. We build a sub-chain S of V that starts from E_1. For each execution E_i in S, the next execution in S is the first execution after E_i in V that has more `Reads` returning values written by other threads than E_i. S has the following properties: (1) all executions in S are well-formed; (2) if any `Read` sees the final non-default value in some execution E_i ($i > 1$) in S, then (a) the corresponding `Write` writes the final value in E_{i-1} and (b) the last preceding (by program order) `Write` to the same variable (if one exists) writes the final value in E_{i-1} and E_i.

Therefore we can replace each execution in S with the simulated one and obtain a valid path from the initial to the final simulated execution. (A `Read` that returns a value written by another thread is considered to be a skeleton.) Our algorithm would find this path because it checks all possible simulated executions.

$\Rightarrow$ Conversely, if the algorithm finds that some execution is valid with respect to the causality requirement, this execution is actually valid. Suppose we have a set of our simulated executions such that there exists a path from the initial to the final execution through valid executions only. For each pair $(S(E, f_i), S(E, f_{i+1}))$ of subsequent simulated executions in the path, there exist commitment sets C_i, C_i', C_{i+1} and corresponding executions E_i, E_i, E_{i+1} (where E_i is simulated by $S(E, f_i)$ and E_{i+1} is simulated by $S(E, f_{i+1})$), such that the following conditions hold:

Algorithm 1.   Verification of the Java causality requirements

---

$\text{JCV}(E)$

$\triangleright$ initialization
1 $\mathbf{N} \leftarrow \phi$
2 $\mathbf{P} \leftarrow \phi$
3 $\mathbf{R} \leftarrow \{S(E, f') \mid f' \in F(E, n_r)\}$ $\triangleright$ $\mathbf{R}$ contains the final execution
4 $\mathbf{I} \leftarrow \{S(E, \phi)\}$ $\triangleright$ $\mathbf{I}$ contains the initial execution

$\triangleright$ iterations
5 **while** $\mathbf{R} \neq \phi$
6   **for each** $r = S(E, f) \in \mathbf{R}$
7   **for each** $e \in \{F(E,i) \mid i < |f|\} \setminus (\mathbf{N} \cup \mathbf{P} \cup \mathbf{R})$
8   **do if** $\text{EXTENDS}(e, r)$
9   **then do**
10   **if** $e \in \mathbf{I}$
11   **return** *true*
12   **else**
13             $\mathbf{N} \leftarrow \mathbf{N} \cup \{e\}$
14   $\mathbf{P} \leftarrow \mathbf{P} \cup \mathbf{R}$
15   $\mathbf{R} \leftarrow \mathbf{N}$
16   $\mathbf{N} \leftarrow \phi$
17 **return** *false*

$\text{EXTENDS}(v, u)$ $\triangleright$ v is extended by u

$\triangleright$ check that: (1) v is valid; (2) `Reads` committed in v stay committed in u;
$\triangleright$ (3) `Writes` seen by other threads in u may be committed in v
$\triangleright$ (4) for each `Read` r that sees a final value in u, the last preceding `Write`
$\triangleright$ of the same variable by the same thread may be committed in v and u
1 **for each** `Read` r in v
2   **if** r sees the last preceding `Write` of the same variable by the same thread
3   **do** nothing
4   **else if** r sees a default `Write` **and** there is no preceding `Write`
               of the same variable by the same thread
5   **do** nothing
6   **else if** r sees in v and u a non-default `Write` of a final value by some other thread
               that may be committed in v and in u **and**
               the last `Write` to the same variable that precedes r in program order,
               if it exists, writes final value in v and in u
7   **do** nothing
8   **else return** *false*

---

9 **return** *true*

---

1. $C_i \subseteq C_i' \subseteq C_{i+1}$ and $C_i \subseteq A_i$, $C_i' \subseteq A_i$, $C_{i+1} \subseteq A_{i+1}$;
2. all `Reads` that see in E_{i+1} final non-default values written by other threads are committed in C_i'.

We can consider $\ldots, E_i, E_i, E_{i+1}, \ldots$ to be a chain of executions that may be used for validating E and $\ldots, C_i, C_i', C_{i+1}, \ldots$ as a chain of commitment sets.   □

4.1   Examples

In the following examples we illustrate the non-polynomial algorithm for short test sequences.

Consider the code in Figure 2 (taken from Figure 16 in [2]) as an example.

```
Initially, x = y = z = 0
Thread 1          Thread2
------------------------------
1: r3 = x;        5: r2 = y;
   if (r3 == 0)   6: x = r2;
2:      x = 42;
3: r1 = x;
4: y = r1;
       r1 == r2 == r3 == 42 is a legal behavior
```

Fig. 2. An example of a valid execution

In Figure 3 we depict simulated executions, one for each possible set of Read actions that are required to see final values. Each execution is described by a table. The table contains information about the Read and Write accesses, represented by their GUIDs. A Read access is marked 'f' if it is required to return the final value produced by the same Write as in the final execution, and '*' otherwise. A Write access w is marked 'c' if it produces its final value and therefore *may* be committed; in addition, it is marked 'n' if it is needed to validate any Read that is required to see w in the execution. Any access is marked '-' if it is absent in the execution. We mark an execution *initial* if all its Reads are marked either '*' or '-'.

If any execution contains a Write that is marked n but not marked c, this execution is invalid. We assume that execution E_j extends execution E_i ($E_i \rightarrow E_j$) if the following hold:

1. all Reads that are marked 'f' in E_i are marked 'f' in E_j;
2. any Write that is marked 'n' in E_j is marked 'c' in E_i;
3. any Write that is marked 'n' in E_i is marked 'n' in E_j.

In our example, there is a path from the initial to the final execution through valid executions only: $E_1 \rightarrow E_2 \rightarrow E_8$. Therefore, the example is valid.

Now consider the example in Figure 4 (taken from Figure 17 in [2]).

The behavior is not allowed because either the final execution is invalid (in the case where the write-seen function for (1:) returns (7:)), or there is no valid path from the initial to the final execution. Figure 5 illustrates this example.

E_8	Reads	Writes
valid	1 : f	2 : -
final	3 : f	4 : c n
	5 : f	6 : c n

E_5	Reads	Writes	E_6	Reads	Writes	E_7	Reads	Writes
valid	1 : ⋆	2 : c	invalid	1 : f	2 : -	invalid	1 : f	2 : -
	3 : f	4 : c n		3 : ⋆	4 : n		3 : f	4 : c
	5 : f	6 : c n		5 : f	6 : c n		5 : ⋆	6 : n

E_2	Reads	Writes	E_3	Reads	Writes	E_4	Reads	Writes
valid	1 : ⋆	2 : c	invalid	1 : ⋆	2 : c	invalid	1 : f	2 : -
	3 : ⋆	4 : c n		3 : f	4 : c		3 : ⋆	4 :
	5 : f	6 : c		5 : ⋆	6 : n		5 : ⋆	6 : n

E_1	Reads	Writes
valid	1 : ⋆	2 : c
initial	3 : ⋆	4 : c
	5 : ⋆	6 :

Fig. 3. Running the example in Figure 2

Initially, x = y = z = 0

Thread 1	Thread2	Thread 3	Thread 4
1: r1 = x;	3: r2 = y;	5: z = 42;	6: r0 = z;
2: y = r1;	4: x = r2;		7: x = r0;

r0 == 0, r1 == r2 == 42 is not a legal behavior

Fig. 4. An example of a prohibited execution

E_8	Reads	Writes
invalid	1 : f	2 : c n
final	3 : f	4 : c
	6 : f	5 : c
		7 : n

Fig. 5. Invalid final execution

Initially, x = y = 0

Thread 1	Thread2
1: r1 = x;	3: r2 = y;
2: y = r1;	4: x = r2;

r1 == r2 == 1 is not a legal behavior

Fig. 6. Another example of a prohibited execution

```
E₄   Reads│Writes
valid 1 : f │2 : c n
final 3 : f │4 : c n
```

```
E₂      Reads│Writes    E₃      Reads│Writes
invalid 1 : f │2 : c     invalid 1 : ⋆ │2 : n
        3 : ⋆ │4 : n             3 : f │4 : c
```

```
E₁     Reads│Writes
valid   1 : ⋆ │2 :
initial 3 : ⋆ │4 :
```

Fig. 7. Running the example in Figure 6

Finally, consider the example in Figure 6 (causality test case 4 from [1]). The behavior is prohibited because a valid path from execution E_1 to execution E_4 does not exist (see Figure 7).

5   Discussion and Conclusions

The JMM specification reduces the complex task of validating an execution for causal acyclicity to a series of relatively simple commitment steps. Each such step deals with some well-formed intermediate execution. Well-formed executions must obey intra-thread consistency: each thread must execute as if it runs in program order in isolation, while the memory model determines the values of shared variables. Furthermore, because the set of values in the intermediate execution is predefined in the previous commitment steps, it cannot be influenced by optimizations. If we take all of this into account, we may disregard optimizations and assume that each thread in the intermediate execution runs as if there is no reordering. This assumption makes it possible to check the causality requirement by a series of relatively simple simulations.

It is the simplicity of the simulations that make the model practical. Without it, it would be hard to tell which compiler optimizations are allowed for intermediate executions. Note that we may reason in terms of compiler optimizations (as in [1]) in order to explain a behavior of some execution but not to prove that the behavior is allowed. This does not mean that the JMM prohibits compiler optimizations (and Theorem 1 in [4] proves the converse). What it does mean is that, when the values transferred by the memory model between the different threads are predefined and the intra-thread semantics respected, the simulated execution is indistinguishable from the one in which each thread executes in program order. We conclude that the executions should be simulated in program order; otherwise, the JMM must be defined in terms of compiler transformations. This conclusion gives rise to some problems with causality test case 6 [1], which is shown below in Figure 8:

Initially, A = B = 0

Thread 1	Thread2
1: r1 = A;	3: r2 = B;
if (r1 -- 1)	if (r2 == 1)
2: B = 1;	4: A = 1;
	if (r2 == 0)
	5: A = 1;

r1 == r2 == 1 is a legal behavior

Fig. 8. Causality test case 6

The test is allowed ([1]) even though our algorithm shows that it fails (see Figure 9).

E_4	Reads	Writes
valid	1 : f	2 : c n
final	3 : f	4 : c n
		5 : -

E_2	Reads	Writes
invalid	1 : f	2 : c
	3 : ⋆	4 : n -
		5 : c

E_3	Reads	Writes
invalid	1 : ⋆	2 : n
	3 : f	4 : c
		5 : -

E_1	Reads	Writes
valid	1 : ⋆	2 :
initial	3 : ⋆	4 : -
		5 : c

Fig. 9. Running causality test case 6. There is no path from E_1 to E_4.

Two problems preclude us from adopting $E_1 \rightarrow E_2 \rightarrow E_4$ as a valid path. First, Write action (5:) does not appear explicitly in the final execution. But we must commit it for the purpose of producing a value for Read (1:) in E_2. However, by definition, the final execution E contains all the committed actions:

$$A = \bigcup (C_0,\ C_1,\ C_2,\ \ldots\ ).$$

The implication follows that E might include some "virtual" actions that don't really appear. If so, this should be stated explicitly in the JMM specification.

Furthermore, in order for causality test case 6 to succeed, the write-seen function for (1:) must be allowed to return (4:) in E_4, and it must be allowed to return (5:) in E_2. But the write-seen function has to be the same in the final execution E and in any execution E_i (that is used to validate E) for all Reads in the second commitment step. (Recall that these Reads are in C_{i-1}.) This step, however, is the only one in which Reads are allowed to see non-default Writes by other threads. Therefore, for executions E_4 and E_2, (4:) and (5:) must be allowed to appear in this step. This in turn rules out the aforementioned requirement – that the write-seen function be the same – for the case of "virtual" Writes. This case is not addressed in the JMM specification.

Perhaps there is another solution to this problem, one which does not require weakening the constraints on the commitment process. Let us see what happens if we consider different actions to be one and the same. In our example there are two actions, (4:) and (5:), which are Writes of the same constant value, only one of which may appear in any execution. In this particular case, we would have good reason to merge the actions. Doing so, however, might not help in more general cases. Thus, we suggest a more complicated formalization: Different Writes of the same variable by the same thread that are not separated by synchronization actions may be considered to be the same action from that moment in the commitment process when their values are determined to be the same.

In this paper we have proven that verification of the Java causality requirements is NP-complete in the general case. We suggest a verification algorithm that may be used in its two versions for the following two cases: the common, non-polynomial version for short test sequences, and the special, polynomial version for some particular systems. We assume that no synchronization actions or final variables are used. We further add that their use makes testing more complicated because each Read might see a number of hb-consistent Writes. We conclude that the JMM specification requires some additional clarifications and adjustments, which may in turn lead to changes in our algorithm.

References

1. Causality Test Cases.
   http://www.cs.umd.edu/~pugh/java/memoryModel/CausalityTestCases.html.
2. JSR-133: Java Memory Model and Thread Specification.
   http://www.cs.umd.edu/~pugh/java/memoryModel/jsr133.pdf, August 2004.
3. S. V. Adve and M. D. Hill. A unified formalization of four shared-memory models. *IEEE Trans. on Parallel and Distributed Systems,* 4(6):613–624, 1993.
4. Sarita V. Adve, Jeremy Manson, and William Pugh. The Java Memory Model. In *POPL'05,* 2005.
5. S.V. Adve, K. Gharachorloo, A. Gupta, J.L. Hennessy, and M.D. Hill. Sufficient System Requirements for Supporting the PLpc Memory Model. Technical Report #1200, University of Wisconsin-Madison, 1993.
6. Andrei Alexandrescu, Hans Boehm, Kelvin Henney, Doug Lea, and Bill Pugh. Memory model for multithreaded C++.
   http://www.open-std.org/jtc1/sc22/wg21/docs/papers/2004/n1680.pdf, 2004.
7. Hans Boehm, Doug Lea, and Bill Pugh. Implications of C++ Memory Model Discussions on the C Language.
   http://www.open-std.org/jtc1/sc22/wg14/www/docs/n1131.pdf, 2005.
8. Hans-J. Boehm. Threads cannot be implemented as a library. *SIGPLAN Not.,* 40(6):261–268, 2005.

9. William Kuchera Charles. The UPC Memory Model: Problems and Prospects. In *Proceedings of the International Parallel and Distributed Processing Symphosium*. IEEE, 2004.

10. Lea Doug. The JSR-133 Cookbook for Compiler Writers. *http://gee.cs.oswego.edu/dl/jmm/cookbook.html*.

11. K. Gharachorloo, S. V. Adve, A. Gupta, J. L. Hennessy, and M. D. Hill. Programming for different memory consistency models. *Journal of Parallel and Distributed Computing*, 15(4):399–407, 1992.

12. K. Gharachorloo, S.V. Adve, A. Gupta, J.L. Hennessy, and M.D. Hill. Specifying System Requirements for Memory Consistency Models. Technical Report #CSL-TR-93-594, Stanford University, 1993.

13. K. Gharachorloo, D. Lenoski, J. Laudon, P. Gibbons, A. Gupta, and J. Hennessy. Memory consistency and event ordering in scalable shared-memory multiprocessors. *Proceedings of the 17th Intl. Symp. on Computer Architecture*, pages 15–26, May 1990.

14. Phillip B. Gibbons and Ephraim Korach. New results on the complexity of sequential consistency. Technical report, AT&T Bell Laboratories, Murray Hill NJ, September 1993.

15. J. Gosling, B. Joy, G. Steele, and G. Bracha. *The Java Language Specification*. Addison-Wesley, third edition, 2005.

16. Prince Kohli, Gil Neiger, and Mustaque Ahamad. A Characterization of Scalable Shared Memories. Technical Report GIT-CC-93/04, College of Computing, Georgia Institute of Technology, Atlanta, Georgia 30332-0280, January 1993.

17. The Java Memory Model mailing list archive. *http://www.cs.umd.edu/~pugh/java/memoryModel/archive*.

A   The JMM Definition

Java threads communicate through shared memory (heap) by the following *inter-thread actions* (here and below, actions and virtual bytecode commands will be printed in `typewriter` font):

- `Read` marks the moment that a value is accepted by a thread. It corresponds to `getfield`, `getstatic` or array load (`aaload`, `faload`, `iaload`, etc.) bytecode;
- `Write` marks the moment that a value is issued by a thread. It corresponds to `putfield`, `putstatic` or array store (`aastore`, `fastore`, `iastore`, etc.) bytecode;
- `Lock` corresponds to `monitorenter` bytecode or to acquiring a lock when leaving the wait state;
- `Unlock` corresponds to `monitorexit` bytecode or to releasing a lock when entering the wait state;

`Read` and `Write` actions may access normal, *volatile* or *final* variables (a variable is a field or an array element). *Synchronization* actions operate in the same way as release or acquire [13, 3] actions. They include `Locks` and `Unlocks`, `Reads` and `Writes` of volatile variables, as well as actions that start a thread or detect that one has stopped. Each execution has a total order over all its synchronization actions, called *synchronization order*. Final fields don't change after initialization except in special cases of reflection and deserialization.

The JMM contains the following assumptions about memory access granularity. References to objects and all variable types except for *double* and *long* are treated as atomic memory locations. Accesses to these locations do not interfere with one another. A variable of type double or long is treated as two distinct 32-bit locations. Each access for reading or writing a double or long variable is treated as two Read or Write actions, respectively.

An execution E of a Java program P is described by a tuple $\langle P, A, \xrightarrow{po}, \xrightarrow{so}, W, V, \xrightarrow{sw}, \xrightarrow{hb} \rangle$, where:

- P is the program (virtual bytecode);
- A is the set of actions, each one represented by the thread to which it belongs, its type, the variable or monitor it operates, and an arbitrary unique identifier;
- $\xrightarrow{po}$ is the program order (the order of actions in each thread as defined by the program)
- $\xrightarrow{so}$ is the synchronization order;
- W is a function that, for each Read action r, gives the Write action seen by r (it is called a write-seen function);
- V is a function that assigns a value to each Write;
- $\xrightarrow{sw}$ is a synchronizes-with partial order. The synchronizes-with order arranges pairs of actions with release-acquire semantics (the order in each such pair complies with $\xrightarrow{so}$). The following actions are paired as release with acquire: an Unlock with each subsequent Lock of the same monitor; a Write to a volatile variable with each subsequent Read of the same variable; the write of the default value with the beginning of each thread; the initialization of a thread with the start of that thread; the interruption of a thread with its "realization" that it has been interrupted; the termination of a thread with the detection of this by another thread;
- $\xrightarrow{hb}$ is the happens-before order. For non-final variables it is similar to the happens-before-1 relation in the Data-Race-Free-1 [3] memory model. There is a happens-before edge from the end of the constructor of any object o to the beginning of the finalizer for o. In addition, for any two actions a and b, $a \xrightarrow{hb} b$ if one of the following holds: $a \xrightarrow{po} b$, $a \xrightarrow{so} b$, or there exists another action c such that $a \xrightarrow{hb} c$ and $c \xrightarrow{hb} b$.

The JMM requires that the execution E be *well-formed*. Each thread must execute as a correct uniprocessor; each Read of some variable must see a Write of this variable; Lock and Unlock actions must be correctly nested. Furthermore, $\xrightarrow{hb}$ must be a valid partial order. In addition, $\xrightarrow{hb}$ and $\xrightarrow{so}$ must be consistent with the write-seen function W: in $\xrightarrow{hb}$ or $\xrightarrow{so}$, no Read r may precede its corresponding Write w and no Write of the same variable may intervene between w and r.

JMM also requires that the execution E meet *causality* requirements in order to prevent causal loops. E must be iteratively built by *committing* all its actions.

Each step i of the commitment process must correspond to the set of committed actions C_i ($C_0 = \emptyset$, $C_i \subset C_{i+1}$). For each C_i (starting from C_1) there must be a well-formed execution E_i that contains C_i. Series $C_0, \ldots, C_i, \ldots$ and $E_1, \ldots, E_i, \ldots$ must have the following properties:

- for each E_i, $\xrightarrow{hb}$ and $\xrightarrow{so}$ between committed actions must be the same as in E;
- for each E_i, committed Writes must be the same as in E, and the same holds true for Reads in the "second step" of the commitment process (those belonging to C_{i-1}); all other Reads must see Writes that happen-before them; a Read may be committed starting with the step following one in which its corresponding Write (in E) commits; Reads in the "first step" of commitment (they are in C_i - C_{i-1}) must see previously committed (those belonging to C_{i-1}) Writes;
- for each E_i, if there is a release-acquire pair from the *sufficient set of synchronization edges* that happens-before some new committed action, this release-acquire pair must be present in all subsequent E_j. (A sufficient set of synchronization edges is a unique minimal set of $\xrightarrow{so}$ edges that produces $\xrightarrow{hb}$ when transitively closed with $\xrightarrow{po}$);
- if some action is committed in C_i, all external actions that happen-before it must be committed in C_i. (An external action is one that may be observed outside of an execution: printing and so forth.)

The JMM considers two additional issues that we don't deal with here. These include value transfer in programs that don't terminate in a finite period of time and *final field semantics* (which may weaken or strengthen the happens-before relation for final field accesses). In addition, [2] and chapter 17 of [15] cover some other aspects of multithreading: waits, notification, interruption and finalization.

B   A Special Case of the Verification Problem

In the following theorem, we prove that restricting only the commitment order for Read actions on the same variable does not simplify the problem.

Theorem 2. *The problem of verifying JMM causality requirements, restricted to instances in which there are at most two threads, and the commitment order complies with the program order for Read actions on the same variable, is NP-complete.*

PROOF. We use the reduction from 3-Satisfiability (3SAT). Consider a 3SAT instance $\mathcal{F}$ with n variables, $v_1, \ldots, v_n$, and m clauses, $C_1, \ldots, C_m$. The execution E is represented by the program P and the trace sequences T. The program P is shown in Table 4. Notation is the same as in the proof of theorem 1. Initially, $a_i = b_i = c = 0$ and all values on the JVM stack are 0.

The trace sequences are shown in Table 5.

The operations in the program and the actions in the trace are made to correspond by matching their unique identifiers.

Table 4. The program with $2 * n + 1$ variables (pseudocode)

Thread 1	Thread 2
[0:] $\quad$ $a_1 = 1$; [1:] $\quad$ $b_1 = 1$;	[2:] $\quad$ $r_{a1} = a_1$; [3:] $\quad$ $r_{b1} = b_1$; $\quad$ $r_1 = r_{a1}\text{-}r_{b1}$;
. . .	. . .
[4*n-4:] $\quad$ $a_n = 1$; [4*n-3:] $\quad$ $b_n = 1$;	[4*n-2:] $r_{an} = a_n$; [4*n-1:] $r_{bn} = b_n$; $\quad$ $r_n = r_{an}\text{-}r_{bn}$;
	$< $ ***comment:*** *in the following* ***if*** *statements* r_{cji} *may be any of* $r_1 \ldots r_n$; r_{cji} *simulates an appearance of a variable in place* ***i*** *of clause* ***j***; s_{ji} *is* ***1*** *if the variable appears without negation, and* ***-1*** *otherwise* $>$

	if (!($s_{11} == r_{c11}$ || $s_{12} == r_{c12}$ || $s_{13} == r_{c13}$))
	 $\quad$ $r_0 = 1$;
	 if (!($s_{21} == r_{c21}$ || $s_{22} == r_{c22}$ || $s_{23} == r_{c23}$))
	 $\quad$ $r_0 = 1$;
	 . . .
	 if (!($s_{m1} == r_{cm1}$ || $s_{m2} == r_{cm2}$ || $s_{m3} == r_{cm3}$))
	 $\quad$ $r_0 = 1$;

[4*n+2:] $r_{n+2} = c$; $\quad$ if ($r_{n+2} = 2$) [4*n+3:] $\quad$ $c = 3$;	[4*n:] $r_{n+1} = c$; $\quad$ if ($r_0 == 0$ || $r_{n+1} == 3$) [4*n+1:] $\quad$ $c = 2$;

Table 5. The trace sequences of the program with $2 * n + 1$ variables

Thread 1	Thread 2
[0:] Write (a_1, 1)	[2:] Read (a_1, 1)
[1:] Write (b_1, 1)	[3:] Read (b_1, 1)
[4:] Write (a_2, 1)	[6:] Read (a_2, 1)
[5:] Write (b_2, 1)	[7:] Read (b_2, 1)
. . .	. . .
[4*n-4:] Write (a_n, 1)	[4*n-2:]Read (a_n, 1)
[4*n-3:] Write (b_n, 1)	[4*n-1:] Read (b_n, 1)
[4*n+2:] Read (c, 2)	[4*n:] Read (c, 3)
[4*n+3:] Write (c, 3)	[4*n+1:] Write (c, 2)

An assignment to v_i is simulated by the following components:

1. operations $a_i = 1$ and $b_i = 1$ by Thread 1 and two operations $r_{ai} = a_i$ and $r_{bi} = b_i$ by Thread 2;
2. corresponding actions *[4 * i − 4:] Write (a_i, 1)*, *[4 * i − 3:] Write (b_i, 1)* by Thread 1 and
 *[4 * i − 2:] Read (a_i, 1)*, *[4 * i − 1:] Read (b_i, 1)* by Thread 2;
3. commitment order between actions *[4 * i − 2:] Read (a_i, 1)* and *[4 * i − 1:] Read (b_i, 1)* that belong to Thread 2;

Stack variable r_i may receive the values 1, -1 or 0. An assignment of the value *true* to v_i is simulated by assigning the value 1 to r_i, and an assignment of the value *false* to v_i is simulated by assigning the value -1 to r_i. If r_i receives the value 0, this means that v_i cannot be used as *true* or *false* in any clause. r_i may receive the value 1 or -1 only if the two actions Read (a_j, 1) and Read (b_j, 1) by Thread 2 belong to different commitment sets. Please note: the JMM does not require that commitment order of the memory accesses comply with the program order.

An OR for the clause j is simulated by the program operations

if (!(s_{j1} == r_{cj1} || s_{j2} == r_{cj2} || s_{j3} == r_{cj3})) and r_0 = 1; by Thread 2;

AND is simulated by the operations *if (r_0 == 0 || r_{n+1} == 3) and c = 2;* as well as by action *[4*n+1] Write (c, 2)* by Thread 2.

Lemma 3. *Let $\mathcal{F}$ be the instance of a 3SAT problem and let E be the execution constructed as described above. Then E complies with the JMM causality requirements iff $\mathcal{F}$ is satisfiable.*

PROOF.

$\Leftarrow$ Suppose $\mathcal{F}$ is satisfiable. Then, there exists a satisfying assignment Å for $\mathcal{F}$. We construct the commitment order $\mathbb{C}$ for E in which for each i, $r_{ai} = a_i$ directly precedes $r_{bi} = b_i$ if v_i is *true* in Å and, conversely, $r_{bi} = b_i$ directly precedes $r_{ai} = a_i$ if v_i is *false* in Å. The process of commitment is described in Table 6.

Table 6. Commitment order

Action	Final Value	First Committed in	First Sees Final Value In	Comment
...	...	...	...	
$a_j = 1$	1	C_1	E_1	
$b_j = 1$	1	C_1	E_1	
...	...	...	...	
$a_j = 1$	1	C_1	E_1	
$b_j = 1$	1	C_1	E_1	
...	...	...	...	
$r_{ai} = a_j$	1	C_2	E_3	simulates an assignment
$r_{bi} = b_j$	1	C_4	E_5	of the value *true* to v_i
...	...	...	...	
$r_{aj} = a_j$	1	C_4	E_5	simulates an assignment
$r_{bj} = b_j$	1	C_2	E_3	of the value *false* to v_j
...	...	...	...	
c = 2	2	C_3	E_3	commits before "c = 3" because r_0 == 0
$r_{n+2} = c$	2	C_3	E_4	
c = 3	3	C_4	E_4	
$r_{n+1} = c$	3	C_4	E_5	

⇒ Conversely, suppose E is a positive instance. Assume $\mathbb{C}$ is a valid commitment order for E. If $r_{ai} = a_i$ precedes $r_{bi} = a_i$ in $\mathbb{C}$, then the satisfying assignment for v_i is assumed *true*, and if $r_{bi} = b_i$ precedes $r_{ai} = a_i$ in $\mathbb{C}$, then the satisfying assignment for v_i is assumed *false*. If $r_{ai} = a_i$ and $r_{bi} = b_i$ commit in the same step, then the assignment for v_i is neither *true* nor *false*.

Suppose that some clause j is unsatisfied. Then, by our construction, r_0 is assigned value 1. In this case, *[4*n + 1:] Write (c, 2)* cannot be committed before *[4*n + 3:] Write (c, 3)*. Because *[4*n + 3:] Write (c, 3)* can commit only after *[4*n + 1:] Write (c, 2)*, we may conclude that E is not a positive instance – a contradiction to our assumption. ☐

By lemmas 3, 2 and because the construction may be done in polynomial time, theorem 2 holds. ☐

Choosing Among Alternative Futures

Steve MacDonald[1], Jun Chen[1], and Diego Novillo[2]

[1] School of Computer Science, University of Waterloo, Waterloo, Ontario, Canada
{stevem, j2chen}@uwaterloo.ca
http://plg.uwaterloo.ca/~stevem
http://www.cs.uwaterloo.ca/~j2chen
[2] Red Hat Inc.
dnovillo@acm.org
http://people.redhat.com/dnovillo

Abstract. Non-determinism is a serious impediment to testing and debugging concurrent programs. Such programs do not execute the same way each time they are run, which can hide the presence of errors. Existing techniques use a variety of mechanisms that attempt to increase the probability of uncovering error conditions by altering the execution sequence of a concurrent program, but do not test for specific errors. This paper presents some preliminary work in deterministically executing a multithreaded program using a combination of an intermediate compiler form that identifies the set of writes of a shared variable by other threads are visible at a given read of that variable and aspect-oriented programming to control program execution. Specifically, the aspects allow a read of a shared variable to return any of the reaching definitions, where the desired definition can be selected before the program is run. As a result, we can deterministically run test cases. This work is preliminary and many issues have yet to be resolved, but we believe this idea shows some promise.

1 Introduction

Testing concurrent programs is difficult because of non-determinism; each execution of the program is different because of changes in thread interleavings. Of the large possible number of interleavings, only a few may cause errors. In particular, we are concerned about *race conditions*, when the interleaving violates assumptions about the order of certain events in the program. For this paper, we define a race condition as two operations that must execute in a specific order for correctness but where there is insufficient synchronization to guarantee this order. An important characteristic of this definition is that mutual exclusion is not necessarily a solution; two updates of a variable in different threads, even if protected by locks, can still cause a race under this definition. This is a more general definition that considers timing-dependent errors.

This paper presents preliminary work on deterministically executing a concurrent program using a combination of two technologies. The first technology is CSSAME [19,20,21,22], an intermediate compiler form for explicitly-parallel shared-memory programs that identifies the set of writes to a shared variable by other threads are visible at a given read of that variable. These writes are called *concurrent reaching definitions*

S. Ur, E. Bin, and Y. Wolfsthal (Eds.): Haifa Verification Conf. 2005, LNCS 3875, pp. 247–264, 2006.

for that particular use of the shared variable. The second is aspect-oriented programming, which allows us to intercept field accesses and method calls in object-oriented programs and inject code at these points [14,13]. Using these technologies, we have created a technique for deterministic execution with three desirable characteristics. First, for a given race condition we can deterministically execute each order, allowing all paths to be tested. In contrast, many other tools are dependent on the timing characteristics of a specific execution of the program which means some orders may not be properly tested. Other tools introduce code into the program and can subtly change its timing, preventing the race condition from appearing during execution. Our deterministic approach does not rely on such timing, and does not suffer from this problem. Second, our method requires no existing execution trace of the program since we are deterministically executing the program, not replaying it. Third, the method works even if accesses to shared variables are not protected by locks.

This deterministic execution can be used in several ways. First, not all race conditions represent errors; in some cases, each order may be acceptable. However, in testing it is important that each order be enumerated to verify correctness. This technique can be used for this enumeration, ensuring each case is handled. Second, other testing methods enumerate over a large number of test cases by generating different interleavings, but cannot guarantee that specific, vital tests are run. This work could be used as an initial sanity test, to explicitly check basic functionality. Third, deterministic execution could be used to support incremental debugging. A user can construct schedules where the order differs in a small number of ways. If one schedule is faulty and another correct, then the difference may be used to locate the error.

This paper is organized as follows. Section 2 presents related research. Section 3 describes the CSSAME form and the compiler that produces it. A brief description of aspect-oriented programming is given in Section 4. In Section 5 we describe how we combine CSSAME and aspects to deterministically execute a concurrent program. Examples of the use of this technique are given in Section 7. Outstanding issues and future work are presented in Section 8, and the paper concludes with Section 9.

2   Related Work

One common testing technique is to introduce noise into multithreaded programs to force different event orders, usually adding conditional sleeps or yields on synchronization or shared variable accesses [25,9]. One research effort added noisemaking using aspect-oriented programming [8]. These techniques rely on properly seeding the program, determining which delays to execute and, in the case of sleep, how long to delay. This technique is not intended to test a specific problem, but rather increases the probability that race conditions will appear if the program is run many times.

Another interesting technique that has been used is based on ordering the execution of the atomic blocks in a multithreaded program [4]. This effort assumes that all accesses to shared data are protected by locking. Under these conditions, the behaviour of a concurrent program can be captured by enumerating over all possible orders of the atomic blocks in the program. This captures the more general idea of a race condition in this work. However, it does not address unsynchronized accesses to data. Also, this work uses a customized JVM that includes checkpointing and replay facilities.

An improvement on these schemes is *value substitution* [1]. Rather than perturb the thread schedule with noise, value substitution tracks reads and writes to shared data. When a read operation is executed, the value that is returned can be taken from any already-executed write operation that is visible to the read operation, simulating different thread schedules. To ensure that the substituted values are consistent, a visibility graph is produced to maintain event ordering. There are two main weaknesses of this algorithm. First, the amount of data needed for the visibility graph is prohibitively large. Second, the substituted values must be from writes that have already executed, which can limit test coverage. The second weakness was addressed through *fidgeting* [2], where the choice of substituted value is delayed until the value is used in a program statement that cannot be re-executed (such as output or conditionals). This delay increases the possible substitutions for a read and allows more schedules to be tested.

Another common strategy is program replay, in systems like DejaVu [5] among others. These systems capture the state of a program execution and use it to reproduce the execution. If the captured execution is an erroneous one, then it can be rerun multiple times to debug the problem. However, it may take many executions of the program to produce an erroneous schedule. Further, capturing the state can perturb the execution, reducing the probability of the error (or even removing it altogether).

A variation of replay called *alternative replay* uses a visibility graph to produce alternate schedules from a saved program execution [2]. This work is similar is style to *reachability testing* [11]. Alternative replay takes a partial execution state of a program and runs it up to some event e. Before e is run, the replay algorithm can look ahead in the visibility graph and may reorder events to produce an interleaving that is different from the recorded one. For example, if e is a read operation, alternative replay may trace the visibility graph to find a write operation w that executed later in the program but could have run before e, making w a potential reaching definition for the read. In baseline value substitution, because w happened later, its value cannot be substituted in the read. Alternative replay rewrites the visibility graph so w happens first. From this point, the program is run normally. Reachability testing systematically enumerates over all possible thread interleavings by tracing every execution and generating alternative schedules from those as well. Both of these systems rely on being able to establish a total order of the events in a program (at least for those that are replayed at the start of the program execution). In contrast, the properties of the CSSAME form permits us to achieve the same effects without imposing this total order, by having a history of each write to a variable. The necessary properties are discussed in Section 5. Further, we can control the execution of the complete program or any portion, not just starting from the beginning of the execution.

Other systems detect race conditions at runtime, such as Eraser [24]. These tools check that accesses to shared data are performed while a lock is held. However, these tools can only detect problems for a specific execution of the program. Since data races may not happen on each execution, these tools may miss race conditions. Some of these tools may also produce false alarms. Furthermore, these tools only check that shared data accesses are protected by locks, and do not address the more general notion of a race condition that is used in this paper.

The idea of incremental debugging by schedule comparison has been explored using *Delta Debugging* [6]. The technique starts with a failing schedule and then tries to construct a correct execution using a combination of inserting noise and splicing scheduling information from successful runs. The correct and incorrect schedules are compared to determine their differences and locate the source of the error. The process is repeated until the most likely source of the error is located.

Another approach to locating errors in concurrent code is model checking, implemented by tools like Java Pathfinder [] and many others. Most model checkers do not execute the application, but rather create an internal representation (usually a finite state machine) and analyze the representation to verify properties of the original program. This internal representation may be in a modeling language rather than the original source code, and it generally elides much of the detail in the program to reduce the size of the representation and make analysis tractable. JPF-2, the second generation of JPF, implements its own Java virtual machine to run the application to locate errors. JPF-2 uses the Eraser algorithm described above to locate simple data races. Since they implement their own JVM, they also control the thread scheduler and use this to enumerate over the different set of thread interleavings, using backtracking to save re-executing a program from the beginning for each test. The enumeration is similar to that used in reachability testing. Our aspect-oriented approach cannot implement backtracking. However, our approach is considerably simpler and does not require a new JVM.

3   CSSAME – Concurrent Single Static Assignment with Mutual Exclusion

CSSAME (Concurrent Static Single Assignment with Mutual Exclusion, pronounced *sesame*) [19,20,21,22] is an intermediate compiler form used to analyze explicitly parallel shared-memory programs (programs where parallelism is explicitly expressed by introducing library calls or language constructs into the application code). CSSAME is a variant of the Concurrent Static Single Assignment form [17] that includes additional analysis based on the synchronization structure of the program. Both forms extend the Static Single Assignment form.

The SSA form has the property that each variable is assigned only once in the program, and that every use of a variable is reached by one definition. However, control flow operators can result in multiple reaching definitions. To resolve this problem, SSA includes *merge operators* or ϕ *functions*. Figure 1 shows an example of the SSA form.

CSSAME extends SSA to include π functions that merge concurrent reaching definitions from other threads in an explicitly parallel program. A concurrent reaching definition is a write to a shared variable by one thread that may be read by a particular use of that variable. The value that is read will be one of these concurrent reaching definitions. One important contribution of the CSSAME form is that it prunes the set of reaching definitions based on the synchronization structure of the program. This pruning is based on two observations regarding the behaviour of shared variable accesses inside critical sections [22]. First, a definition can only be observed by other threads if it reaches the exit of a critical section. Second, if a definition and use occur within the same critical section, then concurrent definitions from other threads cannot be

$$a = 0$$
$$\text{if (condition)}$$
$$a = 1$$
$$\text{print}(a)$$

(a) Original source code

$$a_1 = 0$$
$$\text{if (condition)}$$
$$a_2 = 1$$
$$a_3 = \phi(a_1, a_2)$$
$$\text{print}(a_3)$$

(b) SSA form

Fig. 1. An example of ϕ functions in SSA

observed. This pruning process reduces the number of dependencies and provides more opportunities for compiler optimization.

An example of the CSSAME form is shown in Figure 2. In T_3, the use of the variable a can only be reached by one definition, which requires all concurrent reaching definitions be merged using a π function. When generating the arguments for the π function, note that the definition a_1 in T_1 cannot reach this use because of synchronization; a_1 does not reach the exit of the critical section because of definition a_2.

The set of reaching definitions for unprotected accesses to shared memory are dictated by the underlying memory model. Odyssey, the compiler that implemented the CSSAME form, assumes a sequentially consistent memory store. However, different memory models can be accommodated by changing the placement of and arguments to the π functions [19].

The Odyssey compiler works with a superset of C that includes explicit parallelism in the form of cobegin/coend constructs and parallel loops. The cobegin/coend construct is used in this paper. In this construct, the body of each thread is indicated using a switch-statement-like syntax, as shown in Figure 2. Each thread must be finished at the coend statement. In addition, Odyssey supports locks for mutual exclusion. Event variables and barriers are also available, but Odyssey does not take this synchronization into account when adding terms to π functions.

```
cobegin
    T₁:  Lock(L)
         a = 1
         a = 2
         Unlock(L)
    T₂:  Lock(L)
         a = 3
         Unlock(L)
    T₃:  Lock(L)
         print(a)
         Unlock(L)
coend
```

(a) Original Odyssey source code

```
cobegin
    T₁:  Lock(L)
         a₁ = 1
         a₂ = 2
         Unlock(L)
    T₂:  Lock(L)
         a₃ = 3
         Unlock(L)
    T₃:  Lock(L)
         a₄ = π(a₂, a₃)
         print(a₄)
         Unlock(L)
coend
```

(b) CSSAME form

Fig. 2. An example of π functions in CSSAME

4   Aspect-Oriented Programming

Aspect-oriented programming was created to deal with *cross-cutting concerns* in source code [14]. A cross-cutting concern is functionality in a system that cannot be encapsulated in a procedure or method. Instead, this functionality must be distributed across the code in a system, resulting in tangled code that is difficult to maintain.

A common example of such a concern is method logging, where every public method logs its entry and exit times. Such code must be placed at the start and end of every public method; there is no way to write this code once and have it applied to every method. Worse, every developer of the application must be aware of this requirement when adding new methods or changing protection modifiers on existing ones, and changes to the logging interface may require many changes throughout the code.

Aspect-oriented programming was created to address these cross-cutting concerns. Such tangled code is encapsulated into an *aspect*, which describes both the functionality of the concern and the *join points*, which indicates where the aspect code should appear in the original source code. The aspect code and application code are merged in a process called *aspect weaving* at compile time to produce a complete program.

```
aspect LoggingAspect {
    Logger logger = new Logger();

    pointcut publicMethods(): call(public * *(..));

    before() : publicMethods() {
        logger.entry(thisJoinPoint.getSignature().toString());
    }
    after() : publicMethods() {
        logger.exit(thisJoinPoint.getSignature().toString());
    }
}
```

Fig. 3. Logging aspect code in AspectJ

Figure 3 shows the aspect code for the logging example using AspectJ [13], which supports aspect-oriented programming in Java. Like a class, an aspect can have instance variables, in this case an instance of the logging class. The pointcut represents a set of join points. In this example, the pointcut represents any call to any public method with any return type through the use of wildcards. Following that are two pieces of *advice*, which is the aspect code to be woven into the application code. The first piece of advice represents code that should be run before the call is made, which logs the entry. The argument is a string of the signature of the method, obtained through the join point in the aspect. The second piece of advice is run after the method returns, logging its exit.

There is a wide variety of point cuts other than method calls that can be intercepted, including constructor executions, class initialization, and object initialization. For this paper, the most important of these is the ability to define a pointcut that intercepts accesses to fields, both reads and writes to instance and static variables.

5   Controlling the Future: Controlling the Execution of the CSSAME Form Using Aspects

The key observation is to note that the π functions in the CSSAME form shows, for a given use of a variable, all possible concurrent reaching definitions from other threads contained within the code that the compiler analyzes. Concurrent reaching definitions hidden from the compiler (*i.e.*, contained in library code) cannot be analyzed and limit our ability to properly detect race conditions. Assuming that no accesses are hidden, the CSSAME form identifies all potential race conditions in the program. Further, the terms in the π function show all possible values that can reach a given variable use. If one of these values leads to an error, then it forms a race condition. Thus, a necessary condition for a race condition is a term in a π function that, if selected for a particular use, causes an error. Finding race conditions requires these terms be identified, and removing race conditions requires these terms be removed. In some cases, simply detecting an extra term in the π function may be enough to find a race condition if the race is the error [19].

However, there are legitimate reasons for multiple definitions to appear in a π function that do not represent concurrency errors. If the race is not an error, then the terms in the π function show the possible execution interleavings for a given variable use. In such cases, it is important to be able to test multiple paths through the code, including paths caused by concurrent reaching definitions. Unfortunately, the π functions also represent the non-deterministic parts of a concurrent program; the exact reaching definition (or, rather, the term whose value is returned in the π function) is determined at runtime based on the order of events during the execution of the program.

For testing purposes, it is desirable to be able to deterministically execute a concurrent program. For a given π function with n terms, we would like to run the program exactly n times to check that each reaching definition results in a correct execution. This determinism has three main benefits. First, it reduces the number of times the program must be run to test it. If non-determinism is still present, the program may need to be run many times to ensure all orderings are covered. Worse, some orderings may not occur during testing because of the deterministic nature of some thread schedulers, instead appearing only after the application has been deployed on a system with a different scheduler [1]. Second, determinism provides better support for debugging. Once a bug is detected, it must usually be replicated several times before the source of the error can be located and fixed. Once a fix is applied, the program must be retested to ensure the fix is correct. If the error appears infrequently, this debugging process is difficult. Third, determinism can eliminate the possibility of hiding the bug when trying to locate it. Adding code to monitor the execution can alter the order of events and mask the problem, only for it to reappear once the monitoring code is removed.

The first part of our approach to removing the non-determinism in the execution of a concurrent program is to execute the code from the CSSAME form. That is, for the code in Figure 2(a) we generate the code in Figure 2(b), where each variable is assigned once and has only one reaching definition. Also included in the transformed code are the π functions for concurrent reaching definitions and ϕ functions for control flow.

To generate different test cases for Figure 2(b), we need to control the value returned by the π function. Again, the arguments to the π function are the set of potentially

S. MacDonald, J. Chen, and D. Novillo

reaching definitions at this point in the execution of the program. The return value must be one of the arguments, a_2 or a_3, which is assigned to a_4 and printed in the next line.

This leads to the second part of our approach, which is to control the π functions using aspects. To produce different test cases, the user selects the desired reaching definition from the terms in the π function. Different π functions are distinguished by adding the source line number to the function name. The call to this function is intercepted by an aspect, which can override the return value to return the desired definition.

The benefit of using an aspect to control the execution of the CSSAME form is the same benefit as aspects in general: separation of concerns. Using this approach, we can generate the CSSAME form once, independently of any specific test case. Each test case is written as a separate aspect. Keeping the test cases separate from the code makes the development of each simpler.

However, we need to ensure that the reaching definition has been written before the read can take place. This task is accomplished by having our test aspect intercept all writes to a field. This is sufficient in Java because only fields can be shared between threads. Here, we benefit from the single assignment nature of the CSSAME form. Each instance variable is represented by a set of individual subscripted variables. We can save the values for all writes to a field since they are now separate variables. Furthermore, we treat each variable as a latch. This latch is easily implemented as a class that maintains a boolean variable indicating if the value has been set. A read of the field must block until the latch has been set. A write sets the latch value and unblocks readers attempting to obtain the value for the π function.

As an example, consider the code in Figure 2(b). The print statement in T_3 has two possible outcomes: 2 (from a_2) or 3 (from a_3). The aspect code for the test case where T_3 reads the value from a_2, with additional support code, is shown in Figure 4. In addition, the source code includes implementations for the π functions which are subsumed by the aspect. To get the π function to return the value for a_3, we need only change the value of the choice variable in the advice for the π function to latch_a_3.

An important characteristic of this approach is that it is not sensitive to any timing in the program execution. If the desired term in a π function has not yet been written, the reading thread blocks. We also exploit the single assignment property of CSSAME. Each write is to a separate instance of a variable, meaning we have a complete history of each write. Thus, we can return any already-written term in a π function, even if the same variable has been updated many times. Further, the read value is assigned to a specific instance of a variable, meaning that it cannot be subsequently overwritten by another thread, which removes race conditions on variables involved in test cases. This facilitates testing and debugging of concurrent programs.

The CSSAME form also includes ϕ functions to merge multiple definitions resulting from control flow operators. The ϕ functions cannot be removed as they may be used as terms in a π function. Thus, they must also be executed. Again, we use our aspect. The aspect captures all writes to a variable and includes advice for the ϕ functions. For this case, it is important to note that a ϕ function only includes terms from the current thread and not concurrent definitions. Thus, the aspect maintains a single thread-local variable that represents the last update of the variable by the current thread. The advice for a ϕ function returns this value.

```
public aspect TestCase1 {
    // The args() clause lets advice use values of the arguments.
    // Pointcut for the π function.
    pointcut piFunc(int a, int b) :
        (call private int π(int, int) && args(a, b));

    // Pointcuts for setting the field a₂ and a₃.
    // Since a₁ is not visible in a π function, don't capture it.
    pointcut set_a₂(int n) :
        set(protected int Example.a₂) && args(n);
    pointcut set_a₃(int n) :
        set(protected int Example.a₃) && args(n);

    // Advice that wraps around execution of π function.
    // The body of that function is replaced by this code.
    // Return the selected argument after it has been written.
    int around(int a₂, int a₃) : piFunc(int, int) && args(a₂, a₃) {
            ItemLatch choice = latch_a₂;
            synchronized(choice) {
                while(!choice.ready) {
                    choice.wait();
                }
            }
            return(choice.value);
    }

    // Trip the latch when the field is set.
    after(int n) : set_a₂(n) {
        synchronized(latch_a₂) {
            latch_a₂.ready = true;
            latch_a₂.value = n;
            latch_a₂.notifyAll();
        }
    }
    after(int n) : set_a₃(n) {
        synchronized(latch_a₃) {
            latch_a₃.ready = true;
            latch_a₃.value = n;
            latch_a₃.notifyAll();
        }
    }
    ItemLatch latch_a₂ = new ItemLatch();
    ItemLatch latch_a₃ = new ItemLatch();
}
```

```
public class ItemLatch {
    public ItemLatch() {
        ready = false;
    }
    public boolean ready;
    public int value;
}
```

(a) Aspect that selects a_2 for π function   (b) ItemLatch helper class

Fig. 4. Aspect and helper code for a test case in Figure 2(b)

6   Current Prototype

Our goal is to produce a Java version of this technique, including tool support for producing and running test cases selected by the user. However, to demonstrate the idea, we are currently prototyping the idea by manually translating the explicitly-parallel superset of C supported by Odyssey into Java, and using AspectJ to write aspects.

Currently, we start with an Odyssey program using the cobegin/coend parallel construct. The compiler produces the CSSAME form for the program, which looks like the code in Figure 2(b). Once we have this code, it is translated into its Java equivalent.

Translating Odyssey programs to Java is straightforward. First, we create a class for the complete Odyssey program. For the cobegin/coend construct, each thread body in the construct is translated to a different inner class implementing the `Runnable` interface, making it a valid thread body. Finally, all shared variables in the construct are converted to instance variables, since local variables are not shared in Java threads. By using inner classes for the thread bodies, the threads are able to share the instance variables without the need for accessor methods or public instance variables. Locks become synchronized blocks. Finally, we add dummy methods for the π and ϕ functions.

This translation is currently done by hand. Later, we hope to have an implementation of the CSSAME form for Java, possibly using the SSA support in Shimple [27], part of the Soot framework for Java [28].

Once we have translated Odyssey code into Java, we construct the aspect to control the execution of the program. At this time, this aspect is also written manually. However, the aspect code is straightforward to write. Individual test cases are selected as explained in Section 5, by manually changing the aspect. In the future, we expect to build tool support to generate the aspects for selected test cases.

7   Examples of Alternative Futures

In this section, we examine several of the example programs based on examples from [1]. The objective is to show how to use the π functions to highlight errors in concurrent code. In these cases, these errors can be located by examining the terms in the π function and noting race conditions. Using our aspect-oriented approach, we can also execute the applications to produce the different answers.

7.1   Example 1

Our first example, in Figure 5, is based on Figure 7 from [1]. In the original, a short thread sets a connection variable to null. A second thread starts by sleeping then processing a long method call before setting the connection variable to some non-null value. A third thread waits for the second to complete and then uses connection, relying on a non-null value. We simulate this example using integer arithmetic, where the first thread sets a shared integer to zero, a second does some work before setting the variable to some non-zero value, and the third performs division using the shared variable. The waiting is performed with a barrier that only synchronizes the last two threads.

When a Java version of this code is run, the value of n seen by T_3 is the write by T_2, so the output is 7. The barrier, combined with the long execution time of T_2 compared

```
                                        int n_0 = 1
                                        cobegin
      int n = 1                            T_1:  Lock(L)
      cobegin                                   n_1 = 0
         T_1:  Lock(L)                          Unlock(L)
               n = 0                      T_2:  Lock(L)
               Unlock(L)                        sleep(10)
         T_2:  Lock(L)                          n_2 = 3
               sleep(10)                        Unlock(L)
               n = 1                            barrier(2)
               Unlock(L)                  T_3:  int result
               barrier(2)                       barrier(2)
         T_3:  int result                       Lock(L)
               barrier(2)                       n_3 = π(n_1, n_2)
               Lock(L)                          result = 21/n_3
               result = 21/n                    print(result)
               print(result)                    Unlock(L)
               Unlock(L)                  coend
      coend
```

(a) Odyssey code   (b) CSSAME form

Fig. 5. An example based on Figure 7 from [1]

to T_1, hides the race condition at runtime. However, the π function in T_3 exposes this condition, noting that the value of n in the division can be the zero value set in T_1.

In [1], the race condition is considered to be the error. In that work, the authors can select either n_1 or n_2 in T_3 with a 0.50 probability, assuming both writes have run. In most cases, the error will present itself in two executions. However, this probability decreases as the number of concurrent writes increases; for w writes in different threads, the probability decreases to $\frac{1}{w}$. This will likely require more than w runs. In contrast, we need exactly w executions to capture all possible orders for such a race.

Furthermore, the assumption here is that n_1 is not a legitimate reaching definition for n_3 in T_3. Another possibility is that this race condition is not an error, but assigning the value of zero (which causes the division by zero error) is the mistake. In this example, the consequences of this mistake are obvious and happen immediately. However, more subtle mistakes may require the user to rerun this particular error case many times to locate and correct the problem. Our aspect can deterministically return n_1 in the π function, allowing us to rerun this test as many times as needed to fix the error.

However, a small change in this example makes it difficult for value substitution to produce an incorrect execution. Value substitution allows a read operation to take a value from any already-executed write operation that is visible [1]. In this example, all writes to n generally happen before its use in T_3. If we instead move the sleep from T_2 to T_1, the race condition is still unlikely to occur yet value substitution is also unlikely to produce a read of n with a value of zero. Alternative replay is one way to solve this problem [2]. In our approach, the CSSAME form may be enough to spot the problem, by noting that n_1 is a term in the π function. If this term is expected, then we can use an aspect to test the cases for both n_1 and n_2 deterministically rather than probabilistically.

258   S. MacDonald, J. Chen, and D. Novillo

It is worth noting that the current version of Odyssey does not correctly handle barrier synchronization. In this example, it is clear that T_3 should not be able to see the initial value of n set before the cobegin statement, as it must execute after T_2. However, Odyssey adds n_0 as another term in the π function. When we implement the CSSAME form for Java, we will augment its set of supported synchronization mechanisms to include barriers and event synchronization.

7.2 Example 2

The second example is shown in Figure 6, which is based on Figure 9 from [1]. In the original, an integer array of length 100 is created, with all elements initialized to 0. The main thread launches 100 other threads, where thread i increments the value at index i. After launching the threads, the main thread sleeps for some time and then sums up the elements in the array, without joining with the threads. Our example simplifies this example by using four separate variables and uses a fifth thread to perform the summation. We cannot use the main thread here because of the semantics of the cobegin/coend construct; when the coend statement is reached, all of the threads have completed execution, which removes the concurrency error that we wish to introduce.

The intended result of this code is a sum of 4. With a large enough sleep value, this is the obtained result because threads T_1 through T_4 finish before T_5 reads the values. In fact, any non-zero sleep suffices. If we remove the sleep, then the order in which threads are launched plays the largest role in determining the outcome. Because the threads are so short, they generally run to completion once launched. Without the sleep, T_5 usually sees the updates from all threads launched before it. If launched in program order, T_5 almost always sees the updates from all earlier threads. In a simple test running the program 1000 times, the program reported a sum of 3 only 5 times.

However, from the π functions in Figure 6(b), the race condition in the code becomes clear. Each read of the four summed variables in T_5 may obtain either the incremented value from one of the other threads or the initial value of 0. From this, we can see that

$$\text{sum} = n0 = n1 = n2 = n3 = 0$$
cobegin
    T_1: $n0 \mathrel{+}= 1$
    T_2: $n1 \mathrel{+}= 1$
    T_3: $n2 \mathrel{+}= 1$
    T_4: $n3 \mathrel{+}= 1$
    T_5: sleep(10)
          $\text{sum} = n0 + n1 + n2 + n3$
          print(sum);
coend

(a) Odyssey code

$$\text{sum} = n0_1 = n1_1 = n2_1 = n3_1 = 0$$
cobegin
    T_1: $n0_2 \mathrel{+}= 1$
    T_2: $n1_2 \mathrel{+}= 1$
    T_3: $n2_2 \mathrel{+}= 1$
    T_4: $n3_2 \mathrel{+}= 1$
    T_5: sleep(10)
          $n0_3 = \pi_0(n0_1, n0_2)$
          $n1_3 = \pi_1(n1_1, n1_2)$
          $n2_3 = \pi_2(n2_1, n2_2)$
          $n3_3 = \pi_3(n3_1, n3_2)$
          $\text{sum} = n0_3 + n1_3 + n2_3 + n3_3$
          print(sum);
coend

(b) CSSAME form

Fig. 6. An example based on Figure 9 from [1]

the sum may be any value between 0 and 4. That the sum returns 4 is a function of the thread scheduler and not because the program has correct synchronization.

Note that this race condition is not solved by adding mutual exclusion to the program to protect the accesses to the four variables. The race condition results because T_5 can run at any time with respect to the other threads; the solution is to add an event variable or a barrier to control T_5 and ensure it runs after T_1 through T_4.

The above argument assumes that the intended result is 4, in which case static analysis and user observation is sufficient to detect the problem. However, it may be the case that the intent of the code is that the sum must be between 0 and 4, where any value is acceptable. In this case, one potential mistake is that a variable is decremented in one of threads T_1 through T_4. This error can be detected by controlled execution of the program. We can create an aspect that returns the initial value for each of the summed variables except one, where we use the value set by the updating thread. If any thread mistakenly decrements the variable, the sum will be negative. We can test this by running the program only four times, once for each of the setting threads, by returning the appropriate value for each π function in our aspects.

In the most general case, we may need to know what the set of possible values for sum to verify that the program works correctly regardless of the computed value. In this case, there are 16 possible combinations of the four summed values, which we can enumerate over with 16 executions by again controlling the returned values for each π function.

8   Issues and Future Research

Inconsistent Futures. Our approach does have a significant drawback in that it is possible to select an inconsistent execution of a program. Figure 7 shows an example of this problem. In the second mutex region in thread T_2, a and b must have the same value, either 1 or 2. As a result, the print statement can only produce the output "11" or "20", based on the control flow statements in the program.

The problem is that the π functions do not capture the dependency between a and b. Instead, the π functions make it appear that the choice between a_1 and a_2 in π_a is independent of the choice between b_1 and b_2 in π_b. A user who is trying to control the execution of this program could select a_1 in the former case and b_2 in the latter. In this particular example, b_2 is always assigned, so it is possible to construct an aspect that results in the output "12" which is invalid for this program.

Making this case more difficult is that this particular execution is invalid because of the synchronization code. Different synchronization code (say, where the threads obtain different locks to update a and b) or a lack of synchronization code with a sequentially consistent memory could legitimately produce this result.

This case is difficult to detect because both b_1 and b_2 are assigned in this case. The problem is easier to detect when the user-selected return value for a π function is not assigned because of other control flow statements. For example, if b_2 was not assigned in T_2, the latch for it would never open and T_2 would block in the aspect, trying to return that value in π_b. Eventually all running threads would be blocked, which would indicate that an invalid execution had been selected.

One possible solution is to construct the visibility graph from [1] as the program executes to verify the integrity of the selected test case at run-time. A second option

```
                                            cobegin
                                                T₁:  Lock(L)
                                                     a₁ = 1
              cobegin                                b₁ = 1
                  T₁:  Lock(L)                       Unlock(L)
                       a = 1                    T₂:  int z₀ = 0
                       b = 1                          Lock(L)
                       Unlock(L)                      a₂ = 2
                  T₂:  int z = 0                      b₂ = 2
                       Lock(L)                        Unlock(L)
                       a = 2                           Lock(L)
                       b = 2                           a₃ = πₐ(a₁, a₂)
                       Unlock(L)                       if (a₃ == 1)
                       Lock(L)                              b₃ = π_b(b₁, b₂)
                       if (a == 1)                          z₁ = b₃
                            z = b                       z₂ = φ(z₀, z₁)
                       print(a, z)                      print(a₃, z₂)
                       Unlock(L)                        Unlock(L)

              coend                             coend


              (a) Original code                (b) CSSAME form
```

Fig. 7. A program where a user can select an inconsistent history

may be to use a *program slice* [26], which can be extended to concurrent programs [18]. A program slice shows the subset of statements in a program that may affect the final value of a particular variable at some specific point in a program. We may be able to use the slice to help detect inconsistent test cases during test case construction.

Other SSA/CSSAME Limitations. In addition to the inconsistent futures problem, the SSA form has other limitations (which are inherited by CSSAME and other derivatives of SSA). Some of these have been considered by other research, and others are still open questions. The three problems we will consider are issues with arrays, handling loops, and synchronization analysis.

The SSA form was created to analyze programs with scalars. A simple approach for dealing with arrays is to consider the array as one object. However, the resulting analyses are too coarse-grained to be useful for many programs. Instead, the Array SSA form allows a more precise element-by-element analysis of arrays, where the ϕ functions perform an element-level merge of its arguments [15]. This merge is based on a timestamp associated with each element. A concurrent version of Array SSA that includes π terms for merging writes by different threads is proposed in [7]. These analyses could be added to our compiler to handle arrays.

Another problem with SSA is loops. The *static* part of static single assignment means that each statement that is an assignment uses a separate variable, but it is possible for that statement to be executed several times in a loop. An example is shown in Figure 8. At the end of the loop in T_1 in the CSSAME form (Figure 8(b)), the values in a_2 and b_2 represent the current values for a and b that should be used after the loop terminates. (For simplicity, we assume the loop executes at least once in this example.)

```
                                      a₁ = 1
                                      b₁ = 1
                                      cobegin
                                          T₁: Lock(L)
      a = 1                                  while(condition) {
      b = 1                                      a₂ = φ(a₁, a₃)
      cobegin                                    b₂ = φ(b₁, b₃)
          T₁: Lock(L)                            b₃ = a₂ + b₂
             while(condition) {                  a₃ = a₂ * 2
                 b = a + b                    }
                 a = a * 2                    Unlock(L)
             }
             Unlock(L)                     T₂: Lock(L)
                                             a₄ = π(a₁, a₂)
          T₂: Lock(L)                        print(a₄)
             print(a)                        b₄ = π(b₁, b₂)
             print(b)                        print(b₄)
             Unlock(L)                       Unlock(L)
      coend                            coend
```

$$a_1 = 1$$
$$b_1 = 1$$
$$\text{cobegin}$$
$$T_1: \text{Lock}(L)$$
$$\text{while(condition)} \{$$
$$a_2 = \phi(a_1, a_3)$$
$$b_2 = \phi(b_1, b_3)$$
$$b_3 = a_2 + b_2$$
$$a_3 = a_2 * 2$$
$$\}$$
$$\text{Unlock}(L)$$
$$T_2: \text{Lock}(L)$$
$$a_4 = \pi(a_1, a_2)$$
$$\text{print}(a_4)$$
$$b_4 = \pi(b_1, b_2)$$
$$\text{print}(b_4)$$
$$\text{Unlock}(L)$$
$$\text{coend}$$

(a) Original code   (b) CSSAME form

Fig. 8. An CSSAME example with a loop, based on an example from [3]

If the same lock protects both the loop in T_1 and all other accesses to a and b, as is the case in this example, then this analysis is sufficient even though the loop body may be executed many times. Only the last write of a_2 and b_2 will reach the exit of the critical section. Any earlier writes cannot reach the uses of the two variables in T_2.

However, if the synchronization in T_1 is removed (or if the two writes to a and b are protected individually rather than the complete loop), the CSSAME form is insufficient. The scheduler may interrupt T_1 at any time while it is executing in the loop and run T_2, which will see values for a_2 and b_2 from the current iteration. It is now possible for the values written by any iteration of the loop to be visible in T_2, not just the final ones. We must instead capture each such value (or *write instance*) to enumerate over the set of values that could be printed in T_2. These instances could be captured in an array rather than a single variable, a technique used in *dynamic single assignment* (DSA) [29]. However, the DSA form is applicable to only a small subset of programs, such as multimedia applications [29], suggesting DSA will be inappropriate for this work.

One added difficulty is that there may be dependencies between the write instances that are visible to a thread because of the memory consistency model. For example, in Figure 8(b), if the value for a_2 is taken from iteration i in the loop in T_1, then the value for b_2 must be taken from iteration $i - 1$ or later (since we have assumed a sequentially consistent memory - other consistency models may differ). Event variables or barrier synchronization may introduce similar dependencies.

The final issue we want to address is limitations in analyzing synchronization. This analysis is used to remove terms in π functions where it can be proven that a definition cannot reach a particular use of a shared variable. Where Odyssey cannot prove that a statement is synchronized, it must assume the access is unprotected. This assumption can lead to extraneous terms in *pi* functions. Since the aspects use these terms to determine the set of interleavings to be tested, it can result in extra test cases being run and may lead to more cases of inconsistent futures.

With the `synchronized` block in Java, such analysis is relatively straightforward. However, with the introduction of `java.util.concurrent` in Java 1.5 (or the use of Lea's `util.concurrent` library [16], on which this new package is built), other synchronization primitives like locks and semaphores have been introduced. With these primitives, it is possible to construct irregular synchronization code that are used in real concurrent program but that can be difficult to analyze. Odyssey includes some novel techniques for identifying these locking patterns based on adding and analyzing reaching definitions for the locks used in the code [21].

Of course, if additional synchronization primitives are added, Odyssey must be extended to recognize the primitives and apply their semantics in its analysis.

Tool Support. The goal of this project is to provide tool support for creating and running test cases for concurrent programs. There are several features that a tool should have.

First, we would like to work with the original source code. Although the CSSAME form is useful for a compiler, the transformation may not be as clear to programmers. As a result, we will need to map the statements in the CSSAME form into original source code statements. This is already done by the Odyssey compiler.

Second, tool support should help the user construct test cases, particularly in the presence of control flow statements. Given a write to a shared variable, it may be useful for the user to view the path that must be taken through the code to execute that statement. This will allow the user to specify input to the program that causes the program to run as desired. This problem may be addressed by the program slices discussed earlier. Furthermore, we could automatically generate the aspect code corresponding to each concurrent test. The idea would be to allow users to select the appropriate concurrent reaching definition for a subset of the reads in the program to create the conditions in which they are interested. Ideally, the user can specify these conditions and the program input, and a complete test case can be constructed from this information.

Third, an important consideration is that we clearly cannot require the user to specify the complete program execution by determining the return value for every π function. It is important that the user be able to specify only those terms needed to produce the desired test case. A typical program may well have thousands of these functions, far more than a user can reasonably manage. As well, only data needed to produce the test case should be maintained by the aspect executing the code, to reduce the memory footprint. The CSSAME form introduces multiple copies of a variable, which may require substantial memory. Clearly, we will need a way to reduce the volume of information that must be maintained. This is the topic of ongoing research.

One option is to construct this project as a plugin in the Eclipse IDE for Java [23], and model it after the JUnit framework [12] for constructing unit tests. This would allow a user to construct a complete test suite and run it easily.

Benchmarks. Another direction for future work is to use this technique on a variety of concurrent programs. One source of such programs is a concurrent benchmark effort currently underway [10]. This effort also includes the creation of a concurrent testing framework in which different static and dynamic techniques are combined into more complete and powerful tools. This research may find a place within this framework.

Correcting Concurrent Bugs. Since a race condition appears as an undesired term in a π function, the CSSAME form can be useful in helping a user determine when the problem has been corrected. After the program has been changed, the terms of the relevant π function can be examined once again to ensure the problematic reaching definition is no longer visible.

9   Conclusions

This paper presented some preliminary work in deterministically executing a concurrent program. The method is based on using aspect-oriented programming to control the execution of the CSSAME intermediate compiler form. The main advantages of this approach are that we can deterministically execute the program covering all potential orders for a given race condition, that we require no execution trace, and that we can execute the program even if accesses to shared variables are not synchronized.

Although this work is preliminary and there are still many issues to be resolved, we believe it holds some promise in the field of testing and debugging concurrent programs.

Acknowledgements

This research was supported by the Natural Science and Engineering Research Council of Canada and the University of Waterloo. We would also like to thank the anonymous referees for their comments and suggestions.

References

1. M. Biberstein, E. Farchi, and S. Ur. Choosing among alternative pasts. In *Proc. 2003 Workshop on Parallel and Distributed Systems: Testing and Debugging*, 2003.
2. M. Biberstein, E. Farchi, and S. Ur. Fidgeting to the point of no return. In *Proc. 2004 Workshop on Parallel and Distributed Systems: Testing and Debugging*, 2004.
3. M. Brandis and H. Mössenböck. Single-pass generation of static single-assignment form for structured languages. *ACM Transactions on Programming Languages and Systems*, 16(6):1684-1698, 1994.
4. D. Bruening. Systematic testing of multithreaded java programs. Master's thesis, Dept. of Electrical Engineering and Computer Science, Massachusetts Institute of Technology, 1999.
5. J.-D. Choi and H. Srinivasan. Deterministic replay of java multithreaded applications. In *Proc. SIGMETRICS Symposium on Parallel and Distributed Tools*, pages 48–59, 1998.
6. J.-D. Choi and A. Zeller. Isolating failure-inducing thread schedules. In *Proc. 2002 International Symposium on Software Testing and Analysis*, pages 210-220, 20023
7. J.-F. Collard. Array SSA for explicitly parallel programs. In *Proc. 5th International Euro-Par Conference*, *LNCS* vol. 1685, pages 383-390. Spring-Verlag, 2005.
8. S. Copty and S. Ur. Multi-threaded testing with AOP is easy, and it finds bugs! In *Proc. 11th International Euro-Par Conference*, *LNCS* vol. 3648, pages 740-749. Springer-Verlag, 2005.

9. O. Edelstein, E. Farchi, Y. Nir, G. Ratsaby, and S. Ur. Multithreaded java program test generation. *IBM Systems Journal*, 41(1):111–125, 2002.

10. Y. Eytani, K. Havelund, S. Stoller, and S. Ur. Toward a benchmark for multi-threaded testing tools. *Concurrency and Computation: Practice and Experience*, 2005. To appear.

11. G.-H. Hwang, K.-C. Tai, and T.-L. Huang. Reachability testing: An approach to testing concurrent software. *International Journal of Software Engineering and Knowledge Engineering*, 5(4):493-510, 1995.

12. JUnit. *JUnit: Testing Resources for Extreme Programming*. http://www.junit.org.

13. G. Kiczales, E. Hilsdale, J. Hugunin, M. Kersten, J. Palm, and W. Griswold. An overview of AspectJ. In *Proc. Fifteenth European Conference on Object–Oriented Programming, LNCS* vol. 2072, pages 327–353. Springer–Verlag, 2001.

14. G. Kiczales, J. Lamping, A. Mendhekar, C. Maeda, C. Videira Lopes, J.-M. Loingtier, and J. Irwin. Aspect-oriented programming. In *Proc. 11th European Conference on Object–Oriented Programming, LNCS* vol. 1241, pages 220–242. Springer-Verlag, 1997.

15. K. Knobe and V. Sarkar. Array SSA form and its use in parallelization. In *Proc. 25th ACM SIGPLAN Symposium on Principles of Programming Languages*, pages 107-120, 1998.

16. D. Lea. *Overview of package util.concurrent Release 1.3.4*, 2004. Available at http://gee.cs.oswego.edu/dl/classes/EDU/oswego/cs/dl/util/concurrent/intro.html.

17. J. Lee, S. Midkiff, and D. Padua. Concurrent static single assignment form and constant propagation for explicitly parallel programs. In *Proc. 10th Workshop on Languages and Compilers for Parallel Computing*, 1997.

18. M. G. Nanda and S. Ramesh. Slicing concurrent programs. In *Proc. 2000 ACM SIGSOFT International Symposium on Software Testing and Analysis*, pages 180–190, 2000.

19. D. Novillo. *Analysis and Optimization of Explicitly Parallel Programs*. PhD thesis, Department of Computing Science, University of Alberta, 2000.

20. D. Novillo, R. Unrau, and J. Schaeffer. Concurrent ssa form in the presence of mutual exclusion. In *Proc. 1998 International Conf. on Parallel Programming*, pages 356-364, 1998.

21. D. Novillo, R. Unrau, and J. Schaeffer. Identifying and validating irregular mutual exclusion synchronization in explicitly parallel programs. In *Proc. 6th International Euro-Par Conference, LNCS* vol. 1900, pages 389-394. Springer-Verlag, 2000.

22. D. Novillo, R. Unrau, and J. Schaeffer. Optimizing mutual exclusion synchronization in explicitly parallel programs. In *Proc. Fifth Workshop on Languages, Compilers, and Runtime Systems for Scalable Computers*, pages 128–142, 2000.

23. Object Technology International, Inc. *Eclipse Platform Technical Overview*, 2003. Available at: http://www.eclipse.org/whitepapers/eclipse-overview.pdf.

24. S. Savage, M. Burrows, G. Nelson, P. Sobalvarro, and T. Anderson. Eraser: A dynamic data race detector for multithreaded programs. *ACM Transactions on Computer Systems*, 15(4):391-411, 1997.

25. S. Stoller. Testing concurrent java programs using randomized scheduling. *Electronic Notes in Theoretical Computer Science*, 70(4), 2002.

26. F. Tip. A survey of program slicing techniques. *Journal of Programming Languages*, 3(3):121-189, 1995.

27. N. Umanee. *A Brief Overview of Shimple*, 2003. http://www.sable.mcgill.ca/soot/tutorial.

28. R. Vallée-Rai, E. Gagnon, L. Hendren, P. Lam, P. Pominville, and V. Sundaresan. Optimizing java bytecode using the soot framework: Is it feasible? In *Proc. 9th International Conference on Compiler Construction*, pages 18–34, 2000.

29. P. Vanbroekhoven, G. Janssens, M. Bruynooghe, H. Corporaal, and F. Catthoor. Advanced copy propagation for arrays. In *Proc. 2003 ACM Conference on Languages, Compilers, and Tools for Embedded Systems*, pages 24-33, 2003.

Author Index

Lecture Notes in Computer Science

For information about Vols. 1–3789

please contact your bookseller or Springer

Vol. 3835: G. Sutcliffe, A. Voronkov (Eds.), Logic for Programming, Artificial Intelligence, and Reasoning. XIV, 744 pages. 2005. (Sublibrary LNAI).

Vol. 3834: D.G. Feitelson, E. Frachtenberg, L. Rudolph, U. Schwiegelshohn (Eds.), Job Scheduling Strategies for Parallel Processing. VIII, 283 pages. 2005.

Vol. 3833: K.-J. Li, C. Vangenot (Eds.), Web and Wireless Geographical Information Systems. XI, 309 pages. 2005.

Vol. 3832: D. Zhang, A.K. Jain (Eds.), Advances in Biometrics. XX, 796 pages. 2005.

Vol. 3831: J. Wiedermann, G. Tel, J. Pokorný, M. Bieliková, J. Štuller (Eds.), SOFSEM 2006: Theory and Practice of Computer Science. XV, 576 pages. 2006.

Vol. 3829: P. Pettersson, W. Yi (Eds.), Formal Modeling and Analysis of Timed Systems. IX, 305 pages. 2005.

Vol. 3828: X. Deng, Y. Ye (Eds.), Internet and Network Economics. XVII, 1106 pages. 2005.

Vol. 3827: X. Deng, D.-Z. Du (Eds.), Algorithms and Computation. XX, 1190 pages. 2005.

Vol. 3826: B. Benatallah, F. Casati, P. Traverso (Eds.), Service-Oriented Computing - ICSOC 2005. XVIII, 597 pages. 2005.

Vol. 3824: L.T. Yang, M. Amamiya, Z. Liu, M. Guo, F.J. Rammig (Eds.), Embedded and Ubiquitous Computing – EUC 2005. XXIII, 1204 pages. 2005.

Vol. 3823: T. Enokido, L. Yan, B. Xiao, D. Kim, Y. Dai, L.T. Yang (Eds.), Embedded and Ubiquitous Computing – EUC 2005 Workshops. XXXII, 1317 pages. 2005.

Vol. 3822: D. Feng, D. Lin, M. Yung (Eds.), Information Security and Cryptology. XII, 420 pages. 2005.

Vol. 3821: R. Ramanujam, S. Sen (Eds.), FSTTCS 2005: Foundations of Software Technology and Theoretical Computer Science. XIV, 566 pages. 2005.

Vol. 3820: L.T. Yang, X.-s. Zhou, W. Zhao, Z. Wu, Y. Zhu, M. Lin (Eds.), Embedded Software and Systems. XXVIII, 779 pages. 2005.

Vol. 3819: P. Van Hentenryck (Ed.), Practical Aspects of Declarative Languages. X, 231 pages. 2005.

Vol. 3818: S. Grumbach, L. Sui, V. Vianu (Eds.), Advances in Computer Science – ASIAN 2005. XIII, 294 pages. 2005.

Vol. 3817: M. Faundez-Zanuy, L. Janer, A. Esposito, A. Satue-Villar, J. Roure, V. Espinosa-Duro (Eds.), Nonlinear Analyses and Algorithms for Speech Processing. XII, 380 pages. 2006. (Sublibrary LNAI).

Vol. 3816: G. Chakraborty (Ed.), Distributed Computing and Internet Technology. XXI, 606 pages. 2005.

Vol. 3815: E.A. Fox, E.J. Neuhold, P. Premsmit, V. Wuwongse (Eds.), Digital Libraries: Implementing Strategies and Sharing Experiences. XVII, 529 pages. 2005.

Vol. 3814: M. Maybury, O. Stock, W. Wahlster (Eds.), Intelligent Technologies for Interactive Entertainment. XV, 342 pages. 2005. (Sublibrary LNAI).

Vol. 3813: R. Molva, G. Tsudik, D. Westhoff (Eds.), Security and Privacy in Ad-hoc and Sensor Networks. VIII, 219 pages. 2005.

Vol. 3812: C. Bussler, A. Haller (Eds.), Business Process Management Workshops. XIII, 520 pages. 2006.

Vol. 3811: C. Bussler, M.-C. Shan (Eds.), Technologies for E-Services. VIII, 127 pages. 2006.

Vol. 3810: Y.G. Desmedt, H. Wang, Y. Mu, Y. Li (Eds.), Cryptology and Network Security. XI, 349 pages. 2005.

Vol. 3809: S. Zhang, R. Jarvis (Eds.), AI 2005: Advances in Artificial Intelligence. XXVII, 1344 pages. 2005. (Sublibrary LNAI).

Vol. 3808: C. Bento, A. Cardoso, G. Dias (Eds.), Progress in Artificial Intelligence. XVIII, 704 pages. 2005. (Sublibrary LNAI).

Vol. 3807: M. Dean, Y. Guo, W. Jun, R. Kaschek, S. Krishnaswamy, Z. Pan, Q.Z. Sheng (Eds.), Web Information Systems Engineering – WISE 2005 Workshops. XV, 275 pages. 2005.

Vol. 3806: A.H. H. Ngu, M. Kitsuregawa, E.J. Neuhold, J.-Y. Chung, Q.Z. Sheng (Eds.), Web Information Systems Engineering – WISE 2005. XXI, 771 pages. 2005.

Vol. 3805: G. Subsol (Ed.), Virtual Storytelling. XII, 289 pages. 2005.

Vol. 3804: G. Bebis, R. Boyle, D. Koracin, B. Parvin (Eds.), Advances in Visual Computing. XX, 755 pages. 2005.

Vol. 3803: S. Jajodia, C. Mazumdar (Eds.), Information Systems Security. XI, 342 pages. 2005.

Vol. 3802: Y. Hao, J. Liu, Y.-P. Wang, Y.-m. Cheung, H. Yin, L. Jiao, J. Ma, Y.-C. Jiao (Eds.), Computational Intelligence and Security, Part II. XLII, 1166 pages. 2005. (Sublibrary LNAI).

Vol. 3801: Y. Hao, J. Liu, Y.-P. Wang, Y.-m. Cheung, H. Yin, L. Jiao, J. Ma, Y.-C. Jiao (Eds.), Computational Intelligence and Security, Part I. XLI, 1122 pages. 2005. (Sublibrary LNAI).

Vol. 3799: M. A. Rodríguez, I.F. Cruz, S. Levashkin, M.J. Egenhofer (Eds.), GeoSpatial Semantics. X, 259 pages. 2005.

Vol. 3798: A. Dearle, S. Eisenbach (Eds.), Component Deployment. X, 197 pages. 2005.

Vol. 3797: S. Maitra, C. E. V. Madhavan, R. Venkatesan (Eds.), Progress in Cryptology - INDOCRYPT 2005. XIV, 417 pages. 2005.

Vol. 3796: N.P. Smart (Ed.), Cryptography and Coding. XI, 461 pages. 2005.

Vol. 3795: H. Zhuge, G.C. Fox (Eds.), Grid and Cooperative Computing - GCC 2005. XXI, 1203 pages. 2005.

Vol. 3794: X. Jia, J. Wu, Y. He (Eds.), Mobile Ad-hoc and Sensor Networks. XX, 1136 pages. 2005.

Vol. 3793: T. Conte, N. Navarro, W.-m.W. Hwu, M. Valero, T. Ungerer (Eds.), High Performance Embedded Architectures and Compilers. XIII, 317 pages. 2005.

Vol. 3792: I. Richardson, P. Abrahamsson, R. Messnarz (Eds.), Software Process Improvement. VIII, 215 pages. 2005.

Vol. 3791: A. Adi, S. Stoutenburg, S. Tabet (Eds.), Rules and Rule Markup Languages for the Semantic Web. X, 225 pages. 2005.

Vol. 3790: G. Alonso (Ed.), Middleware 2005. XIII, 443 pages. 2005.